The past fifty years have witnessed major achievements in ecological physiology, the study of physiological adaptations that improve survival or permit organisms to exploit extreme environments. *New Directions in Ecological Physiology* outlines novel conceptual approaches to the study of physiological adaptation in animals, approaches that will stimulate the continued growth of this field.

Twenty leading ecological physiologists and evolutionary biologists have contributed critical evaluations of developments in their respective areas, highlighting major conceptual advances as well as research questions yet to be answered. The volume is organized into three parts: The first deals with comparisons of different species and populations; the second, with comparisons of individuals within a population; the last, with interacting physiological systems within individual animals.

New Directions in Ecological Physiology, by encouraging critical debate about general issues and directions of growth in this field, is intended to foster the invigoration of ecological physiology in particular and of organismal biology in general.

Martin E. Feder is Associate Professor of Anatomy and in The Committee on Evolutionary Biology, The University of Chicago; Albert F. Bennett is Professor of Developmental and Cell Biology and Professor of Ecology and Evolution, University of California, Irvine; Warren W. Burggren is Professor of Zoology, University of Massachusetts, Amherst; and Raymond B. Huey is Professor of Zoology, University of Washington, Seattle.

NEW DIRECTIONS IN ECOLOGICAL PHYSIOLOGY

NEW DIRECTIONS IN ECOLOGICAL PHYSIOLOGY

Editors

MARTIN E. FEDER
The University of Chicago

ALBERT F. BENNETT
University of California, Irvine

WARREN W. BURGGREN
University of Massachusetts

RAYMOND B. HUEY
University of Washington

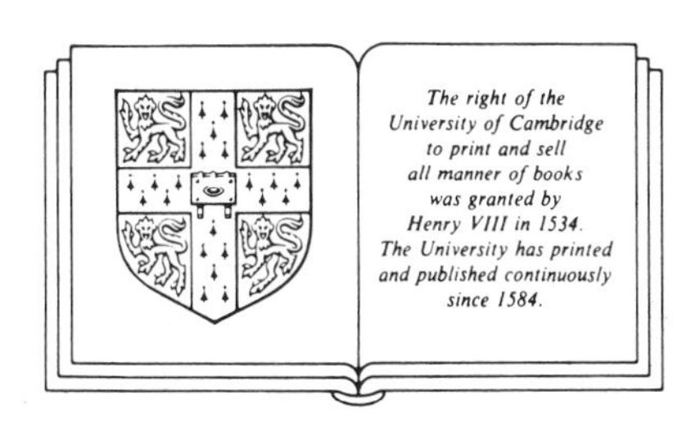

CAMBRIDGE UNIVERSITY PRESS

Cambridge
New York New Rochelle Melbourne Sydney

Published by the Press Syndicate of the University of Cambridge
The Pitt Building, Trumpington Street, Cambridge CB2 1RP
32 East 57th Street, New York, NY 10022, USA
10 Stamford Road, Oakleigh, Melbourne 3166, Australia

First published 1987

Printed in the United States of America

Library of Congress Cataloging-in-Publication Data
New directions in ecological physiology.
Includes index.
1. Adaptation (Physiology). 2. Ecology.
3. Physiology. I. Feder, Martin E.
QH546.N48 1987 574.5 87–24252

British Library Cataloguing in Publication Data
New directions in ecological physiology.
1. Physiology 2. Animal ecology
I. Feder, Martin E.
591.1 QP33.5

ISBN 0 521 34138 8 hard covers
ISBN 0 521 34938 9 paperback

Contents

Contributors and discussants *page* vii
Preface ix

1 The accomplishments of ecological physiology 1
Albert F. Bennett

PART ONE: COMPARISONS OF SPECIES AND POPULATIONS 9

2 Interspecific comparison as a tool for ecological physiologists 11
George A. Bartholomew
Discussion 35

3 The analysis of physiological diversity: the prospects for pattern documentation and general questions in ecological physiology 38
Martin E. Feder
Discussion 71

4 Phylogeny, history, and the comparative method 76
Raymond B. Huey
Discussion 98

5 A multidisciplinary approach to the study of genetic variation within species 102
Dennis A. Powers
Discussion 130

6 Comparisons of species and populations: a discussion 135
William R. Dawson

PART TWO: INTERINDIVIDUAL COMPARISONS 145

7 Interindividual variability: an underutilized resource 147
Albert F. Bennett
Discussion 166

8 The importance of genetics to physiological ecology 170
Richard K. Koehn
Discussion 186

9 Genetic correlation and the evolution of physiology 189
Stevan J. Arnold
Discussion 212

10 The misuse of ratios to scale physiological data that vary allometrically with body size 216
Gary C. Packard and Thomas J. Boardman
Discussion 236

11 Interindividual comparisons: a discussion 240
Douglas J. Futuyma

PART THREE: INTERACTING PHYSIOLOGICAL SYSTEMS 249

12 Invasive and noninvasive methodologies in ecological physiology: a plea for integration 251
Warren W. Burggren
Discussion 272

13 The use of models in physiological studies 275
Peter Scheid
Discussion 285

14 Symmorphosis: the concept of optimal design 289
Stan L. Lindstedt and James H. Jones
Discussion 305

15 Assigning priorities among interacting physiological systems 310
Donald C. Jackson
Discussion 326

16 Physiological changes during ontogeny 328
James Metcalfe and Michael K. Stock
Discussion 340

17 Interacting physiological systems: a discussion 342
David J. Randall

18 New directions in ecological physiology: conclusion 347
Martin E. Feder

Author index 353
Subject index 360

Contributors and discussants

STEVAN J. ARNOLD, Department of Biology, The University of Chicago, 940 East 57th Street, Chicago, Illinois 60637

GEORGE A. BARTHOLOMEW, Department of Biology, University of California, Los Angeles, California 90024

ALBERT F. BENNETT, School of Biological Sciences, University of California, Irvine, California 92717

STEVEN BISHOP, Regulatory Biology Program, National Science Foundation, Washington, D.C. 20550

THOMAS J. BOARDMAN, Department of Statistics, Colorado State University, Fort Collins, Colorado 80523

WARREN W. BURGGREN, Department of Zoology, University of Massachusetts, Amherst, Massachusetts 01003

JAMES P. COLLINS, Population Biology and Physiological Ecology Program, National Science Foundation, Washington, D.C. 20550. [Present address: Department of Zoology, Arizona State University, Tempe, Arizona 85287]

MARK W. COURTNEY, Population Biology and Physiological Ecology Program, National Science Foundation, Washington, D.C. 20550

WILLIAM R. DAWSON, Division of Biological Sciences, University of Michigan, Ann Arbor, Michigan 48109

JAMES L. EDWARDS, Biological Research Resources Program, National Science Foundation, Washington, D.C. 20550

MARTIN E. FEDER, Department of Anatomy and The Committee on Evolutionary Biology, The University of Chicago, 1025 East 57th Street, Chicago, Illinois 60637

GREGORY L. FLORANT, Department of Biology, Swarthmore College, Swarthmore, Pennsylvania 19081

DOUGLAS J. FUTUYMA, Department of Ecology and Evolution, State University of New York, Stony Brook, New York 11794

LEWIS GREENWALD, Regulatory Biology Program, National Science Foundation, Washington, D.C. 20550. [Present address: Department of Zoology, Ohio State University, Columbus, Ohio 43210]

RAYMOND B. HUEY, Department of Zoology NJ-15, University of Washington, Seattle, Washington 98195

DONALD C. JACKSON, Division of Biology and Medicine, Brown University, Providence, Rhode Island 02912

JAMES H. JONES, Department of Physiological Sciences, School of Veterinary Medicine, University of California, Davis, California 95616

RICHARD K. KOEHN, Department of Ecology and Evolution, State University of New York, Stony Brook, New York 11794

STAN L. LINDSTEDT, Department of Zoology and Physiology, The University of Wyoming, University Station, Box 3166, Laramie, Wyoming 82071

JAMES METCALFE, Department of Medicine, Heart Research Laboratory – L464, Oregon Health Sciences University, 3181 S.W. Sam Jackson Park Road, Portland, Oregon 97201

GARY C. PACKARD, Department of Zoology, Colorado State University, Fort Collins, Colorado 80523

DENNIS A. POWERS, Department of Biology, The Johns Hopkins University, 3400 North Charles Street, Baltimore, Maryland 21218 [mailing address]; and Center for Marine Biotechnology, University of Maryland, Baltimore, Maryland

DAVID J. RANDALL, Department of Zoology, University of British Columbia, 6270 University Boulevard, Vancouver, British Columbia V6T 2A9, Canada

PETER SCHEID, Institut für Physiologie, Ruhr-Universität, Universitätstrasse 150, D-4630 Bochum, West Germany

MICHAEL K. STOCK, Department of Medicine, Heart Research Laboratory – L464, Oregon Health Sciences University, 3181 S.W. Sam Jackson Park Road, Portland, Oregon 97201

BRUCE L. UMMINGER, Division of Cellular Biosciences, National Science Foundation, Washington, D.C. 20550

Preface

Modern physiology has two primary objectives. The first is the understanding of the central unifying principles by which physiological function is governed. Such unifying principles transcend particular taxa, environments, developmental stages, and so on. The second is the understanding of physiological diversity in biological molecules, cells, tissues, organs, and organisms. This objective concerns the origin of physiological diversity, the constraints upon it, and its consequences for physiological function. This book is about the attempts of one subdiscipline of physiology, ecological physiology, to attain this second objective. Ecological physiology examines physiological diversity in relation to the environments in which organisms live or have lived. It is also termed "physiological ecology" and "ecologically relevant physiology"; we will use these terms interchangeably. The terms themselves, however, fail to express the breadth of the enterprise, which embraces aspects of behavior, morphology, biochemistry, and evolutionary biology, among other fields.

As summarized by the first two chapters of this volume, ecological physiology has made considerable progress in understanding physiological diversity and in establishing itself as a scientific discipline. Along with the achievements of ecological physiology, however, have come the perceptions that little of real novelty remains to be investigated. There is a sense that "the cream has been skimmed off the top" of the analysis of physiological diversity. Along with enormous increases in our understanding of physiological adaptation has come the perception that "adaptation" may be trite, flawed, or inadequate as a conceptual basis for future progress. As Albert Bennett wrote in the proposal for support of the workshop that preceded this volume:

> I have frequently encountered questions concerning the significance of specific projects and the general goals and directions of the field. In my conversations with many people, I sense a real lack of perceived focus and direction in the field as a whole. People are generally aware of where we have been but not where we are going. There is no sense of movement towards general questions or goals and often a sense that these do not exist.

Whether these perceptions are accurate ones is an important question that is taken up at length in the following chapters.

This volume stemmed from a mutual conviction of the editors that ecological physiology is very much alive and well. We were concerned, however, that misplaced perceptions of lack of directions, foci, and goals were proving increasingly inimical to the vitality of ecological physiology. Furthermore, we felt that these perceptions were often understandable because practitioners of ecological physiology seldom made explicit statements of general goals or likely avenues for future developments. We decided, therefore, that the time had come for a public discussion of goals and future directions in ecological physiology. By virtue of our own research interests, we confined the project to the ecological physiology of animals.

We invited participants to a workshop with the understanding that each would contribute a manuscript based upon the presentation at the workshop. To set an appropriate tone for the workshop, the editors posed some conceptual questions to each participant; some of these questions are included verbatim in the following chapters. The workshop was held in late May, 1986, in Washington, D.C. Each of three morning sessions, corresponding to the three sections in this book, included four to five presentations on assigned topics. A discussion, led by assigned discussants, occupied each entire afternoon. The discussions were recorded in their entirety; the assigned discussants edited the transcripts of these recordings that appear at the end of many chapters of this book. Manuscripts based on the morning presentations were subsequently reviewed both by the editors and by reviewers who were not present at the workshop. The editors are extremely grateful for these lengthy and uniformly constructive external reviews. The workshop was supported by the National Science Foundation under Grant BSR 86–07794. Any opinions, findings, and conclusions or recommendations expressed in this publication are those of the authors and do not necessarily reflect the views of the National Science Foundation.

This book should be considered a collection of suggestions for avenues of research in ecological physiology. It conveys what a small number of ecological physiologists found compelling about their discipline in 1986. It is clearly not a comprehensive collection of suggestions, and we make no claims that the suggestions we have included will necessarily correspond to the best possible avenues for the future development of ecological physiology. As George Bartholomew has said, it is sometimes difficult to anticipate which of our contributions will be significant. We do, however, believe it important that the dialogue represented by this book be continued and expanded. Ecological physiology has indeed suffered from the lack of explicit discussion of its current and future goals. If the "new directions" contained herein are not the best ones, then the field needs to consider which directions *are* best. We hope that this book will promote this kind of consideration, particularly among the graduate students who will create the ecological physiology of the future. Moreover, we hope this book will intrigue practitioners of related disciplines in involving themselves in the enterprise of ecological physiology.

1 The accomplishments of ecological physiology

ALBERT F. BENNETT

Introduction

Strange animals living in unusual situations have always fascinated humans. We are particularly interested in and impressed by animals with functional capacities that exceed our own, such as Arctic mammals and small birds that can exist in bitter cold, desert animals that can survive intense heat or forego drinking water, hibernators that can become torpid for weeks, divers that can stay under water for hours, or animals that can lift and carry many times their own mass. Our natural curiosity leads us to ask how they can do that, or to paraphrase the title of a book by Knut Schmidt-Nielsen (1972), How do these animals work? The primary task and subject of the field known as ecological physiology (or physiological ecology or environmental physiology) is to understand how animals are designed with reference to their natural environments and evolutionary histories.

To place the field in perspective, this chapter reviews very briefly where it originated and where it has been, what its major accomplishments are, and where it finds itself today. This is a short and personal essay on those topics.

Ecological physiology arose from a blending of the traditions of comparative physiology and natural history (Chapter 2). The former traces its traditions to laboratory physiology in the latter half of the nineteenth and early twentieth centuries. It has developed a large and impressive data base on the physiological capabilities of numerous different kinds of animals, often concentrating on those that best exemplify some physiological process (Krogh, 1929). It subscribes to a rigorous protocol of experimental design and analysis. The field of natural history, also already well-developed in the nineteenth century, contributed a knowledge of animals, sometimes very unusual animals, and their habits under natural conditions in the field. The merger of these two areas resulted in experimental scientists undertaking research on nontraditional animals and referring the results of that research to the behavior and functional responses of species under natural conditions.

Animal ecological physiology began to assume a separate existence and prominence in the late 1940's with the seminal contributions of George Bartholomew, Knut Schmidt-Nielsen, and Per Scholander. These and other investigators and their students were responsible for an explosion of knowledge during the past forty years. Ecological physiology is now a firmly established area of biological investigation with hundreds of practitioners. The center of activity of the field has always been the United States, but it is also well represented in Canada, Europe, Australia, and Africa. It forms an important component in numerous scientific societies, including the American Physiological Society, the American Society of Zoologists, the Ecological Society of America, and the Society for Experimental Biology. Ecological physiologists also play prominent roles in scientific societies associated with particular animal taxa, especially those concerned with mammals, birds, reptiles, and amphibians. Papers in this area are published in numerous scientific journals, including the *American Journal of Physiology*, *Comparative Biochemistry and Physiology*, the *Journal of Comparative Physiology*, the *Journal of Experimental Biology*, the *Journal of Experimental Zoology*, the *Journal of Thermal Biology*, *Physiological Zoology*, *Respiration Physiology*, various taxonomic journals (e.g., *Condor*, *Copeia*), ecological journals (e.g., *Ecology*, *Oecologia*), as well as more general scientific publications. Its research is supported by two different programs of the National Science Foundation: Population Biology and Physiological Ecology in the Division of Biological Systems and Resources, and Regulatory Biology in the Division of Cellular Biosciences. The field has grown and matured rapidly and has assumed an important and well-defined place in modern biological studies.

The insights of ecological physiology

I would like to enumerate briefly what I consider to be major insights and accomplishments to emerge from this field. Any such designation is bound to be hazardous and eclectic. However, I have tried to select themes that have been particularly well developed in ecological physiology and have implications far beyond that particular area of biological investigation.

Energy availability and utilization are important constraints on animal function.

The measurement of energy intake and utilization by animals has been a major area of study within ecological physiology. This area can assume such a prominence that it is sometimes regarded as the principal subject and accomplishment of the entire field (e.g., Townsend and Calow, 1981). Through energetics, ecological physiology finds its strongest links with ecology and behavior. In these fields, energy exchange assumes a major role in many hypotheses (e.g., life history strategy and foraging theory). Most often, ecological physiology has provided the methodology and data to test such hypotheses empirically. It consistently asks how much processes or activities cost and proceeds to make the appropriate determinations. The field has been successful in energetic studies primarily because of its insistence on quantification of energy exchange.

Energy availability has often been found to impose constraints on animals. These constraints may arise either because of the relatively large amount of energy required in comparison to that available from the environment, or because of limitations of the physiological processing capacities of the animal. Some physiological processes and behaviors, such as endothermy in the cold and sustained flight, are so expensive that they require nearly all of an animals's energy intake or are beyond that which can be taken from the environment. Other constraints on energy use may reflect internal limitations of physiological rate processes. The ability to undertake sustainable behaviors, for instance, is limited by the capacity for oxygen transport and utilization. In ectothermic animals, maximal oxygen consumption may be low and may be even further diminished at low body temperatures. Ecological physiology often undertakes studies to measure the capacity of functional processes in order to define the limits within which function is permitted.

Body temperature regulation is expensive in time and energy. Its alternative, temperature conformity, entails variability in all physiological processes.

Interest in thermoregulation and its consequences has been an immense area of ecological physiology. A large set of empirical observations on the thermal biology of animals has been assembled. In addition, sophisticated biophysical models have been constructed to predict thermal exchange. This research has led to an appreciation of the considerable complexity of thermal environments and the importance of microclimates rather than macroclimates in determining thermal balance. One of the major lessons of ecological physiology is that the influence of temperature on energy exchange and behavior in natural environments is liable to be large and cannot be neglected.

Physiologists have known for more than 200 years that rate processes are very sensitive to temperature. But it remained for ecological physiologists to elaborate the consequences of temperature conformity and thermoregulation for nearly all aspects of animal life. If rate processes are temperature dependent, then exposure to cold depresses many different time-dependent phenomena, including metabolic, growth, and behavioral capacities. Some organisms accept this dependence; others go to extraordinary lengths with behavioral and physiological adjustments to overcome the effects of low and/or variable body temperatures. Many undergo long-term physiological adaptation (acclimation, acclimatization) to their thermal regime. Others regulate temperature behaviorally, expending a large amount of effort and limiting activity times to certain portions of the day or night. Still others regulate temperature metabolically by varying heat production. For mammals, birds, and some insects, this thermoregulation is extraordinarily expensive. It represents by far the largest component of their energy budgets and diverts energy from other important processes such as growth and reproduction.

Body size affects nearly every biological variable.

Comparative studies on hundreds of different species found their natural comparative base in the examination of the effects of size on biological variation. The original summaries by Kleiber (1932) and Benedict (1938) concerning the effect of size on basal metabolic rate have now been supplemented with observations on nearly all physiological and many ecological variables, including such factors as heart rate, glomerular filtration rate, territory size, and life-span. These have been summarized by Peters (1983), Schmidt-Nielsen (1984), and Calder (1984).

Most functions do not depend on animal size in a simple linear manner. If one animal is twice the mass of another, it does not require twice as much food, water, or space in which to obtain them. Rather, nearly all physiological and ecological factors can be described as exponential functions of body mass. Metabolic rate, for instance, scales with mass to the 0.75 power, a relationship that has profound consequences for energy requirements and thermoregulation. Ecological physiology has generated an enormous series of predictive equations that can be used to estimate the magnitude of almost any function in many animals, given only the taxon and the animal's mass. The scaling of different morphological and physiological variables has proved to be a fruitful area for examination of their functional bases and general design criteria of animals (e.g., Taylor and Weibel, 1981). Further, allometric equations can be used as a basis for examination of the physiological and/or ecological causes of deviations from general relationships. Many groups of desert-adapted mammals and birds, for instance, have a lower metabolic rate than would be anticipated on the basis of their size. Both the ecological consequences and physiological underpinnings of such deviations have been important as documentation of environmental adaptation.

Behavior is an important component of functional adjustment to the environment.

A basic strength of laboratory physiology is its ability to isolate and control experimental variables. However, an important variable that is often lost in laboratory investigations is behavioral response. A principal contribution of ecological physiology has been the appreciation that behavior is often the primary and sometimes the crucial means by which an animal copes with an environmental challenge. This appreciation has reemphasized the necessity of validating conclusions based on laboratory experiments with field observations on animals.

For many animals, behavioral discretion is often the better part of regulatory valor. Animals can exist in a variety of very challenging environments, thermally, energetically, osmotically. Sometimes they cope with them by conforming to the environmental variable or by regulating the variable internally, often at great energetic expense. But often ecological physiologists have found that a third option is used: animals may avoid those environmental challenges, either behaviorally or physiologically, by effectively removing themselves from certain challenging aspects of the environment. This avoidance may lead to a variety of adaptive patterns, even within restricted taxa, to a common environmental challenge (Chapter 2). Animals that hibernate or become torpid avoid acute or chronic energetic or thermal stress. Animals that migrate avoid the problems of continuous residence in seasonally inhospitable environments. Just because an animal lives in a particular environment does not mean that it experiences the full force of that environment, particularly as perceived by a human. Ecological physiologists have found that avoidance is a common response and again one that can be evaluated only by observations on animals in natural environments.

The organism is a compromise. The result of natural selection is adequacy and not perfection.

This view is discussed in detail in Chapter 2. More than many other field of biology, ecological physiology has appreciated the limitations imposed on animals by their environments, their resources, and their phylogenetic history. Not all possibilities are open to animals, and they sometimes have to make the best of bad situations.

Animals **are** *adapted to their environments.*

A considerable amount of effort in the field has gone into the documentation of the correspondence between physiological capacity and natural environment. The issue at question has never been the tautological demonstration that animals can occur where they do. Rather the problem has been how animals are able to occupy different niches and environments, whether they possess adaptive specializations in their form and functional capacities or whether there is a much less specific match between animal and environment. Studies in ecological physiology have provided many textbook examples of adaptation, for example, the kidney of a kangaroo rat, the nasal gland of a sea bird, the pelage of an Arctic fox. As a whole, this body of information, including less spectacular but nevertheless crucial adjustments, provides overwhelming documentation of the fact of adaptation of species to occupy specific environments. We now take this conclusion for granted because of the mass of documentary evidence provided by this and other areas of organismal biology.

The state of the art

Ecological physiology is no longer a new or even a young field of biological science. It has lost the shine of first discovery and has confronted limitations that were not apparent in its initial promise. In short, it has grown from its adolescence to its maturity, a sobering and unexciting experience for everyone. To a large extent, it has been a victim of its own successes.

Ecological physiologists set out to describe how animals function in their natural environments and how they adapted to them evolutionarily. They have been extraordinarily successful in doing precisely that. They have characterized the physical challenges imposed by nearly every environment on earth, from the deep sea to high altitudes, from humid tropical forests to hot deserts to cold polar regions. They have outlined, sometimes in minute detail, the interplay between organism and environment in a number of different animal taxa. The field has been successful because of its enormous background of information on the physiological responses of many different kinds of animals, its technical excellence and precision, its empirical basis, and its appreciation of the importance of observations on animals in natural environments. It has accomplished many of the goals that it originally set out to meet. In this, it has done better than many other areas of biology.

However, up to this time, the field has progressed largely by exploiting adaptations of novel animals to novel environmental situations. These will be progressively harder to come by, and certain areas of the field already seem to be engaged in collecting further examples of phenomena that are already well understood. The surprises, the unusual and unexpected examples, will be slower in coming in the future than they have been in the past. When they do appear, new phenomena are rapidly and thoroughly explored, but they do not form the basis of fundamentally new approaches or ways of looking at things. Witness, for example, the discovery of thermal vent communities in the deep sea. These proved to be populated with animals of previously unknown phyla, with metabolic patterns and physiologies previously undescribed. Yet, in spite of their novelty, their inaccessibility, and their discovery only in the mid-1970's, we already know much about their physiology, about how they make a living and extract energy from a seemingly unpromising environment. Even the discovery of totally new animals in a completely new environment did not produce any qualitative change in the field.

Our future does not lie in the discovery of new animals or new environments. Perhaps these await us in space, and our progeny can renew this exploratory phase of our science. There are certainly still many opportunities for ecological physiologists to investigate new animals on earth, particularly invertebrates and especially insects. But I reiterate my belief that the discovery of new adaptive patterns will be infrequent. Considerable progress will be made, as it always is, on the heels of advances in instrumentation and technique. In particular, improvements in computer technology, telemetry instrumentation, and applied statistics will greatly improve our ability to ask questions and get answers. But these again are refinements of current paradigms.

It is time to begin searching for new directions for studies in ecological physiology. This is not to say that traditional approaches are complete or trivial or should be abandoned. It is not to say that we know all there is to know about physiological adaptation in a majority or even in a significant fraction of the total number of animal species. But it is to say that the broad outlines of physiological adaptation have already been sketched and that past accomplishments do not form a totally adequate agenda for future work. Unless we are to be content with fitting more examples into well-worn analytical paradigms, we must expand our horizons to new questions and new sorts of studies.

If we are content to continue only in traditional approaches, we run the risk of becoming outmoded. Consider, for example, the field of comparative anatomy. It was a vigorous discipline up to thirty years ago. It was a classical area of biology, a staple of every undergraduate biology curriculum, a cornerstone of evidence for evolutionary thought. Yet rather suddenly it became irrelevant to modern biological thought and ceased to be an active area of investigation. Its extinction did not result from their running out of animals to describe: the anatomy of most species of animals is still uninvestigated. Comparative anatomy passed from the scene because it succeeded in its descriptive mission and did not develop new insights, because it was producing new information quantitatively but not qualitatively. Fortunately, anatomical studies have become reoriented and reorganized as the field of functional morphology. The near demise of comparative anatomy is a cautionary tale for ecological physiology.

This book describes some potential new techniques and new directions in ecological physiology. They are not meant to be exclusive or to supplant existing approaches totally. They are suggestions of possible ways in which the field may wish to proceed in the future. I am impressed with the diversity and vitality of these approaches: ecological physiology is not wanting for new ideas. Our future will be at least as interesting as our past.

Acknowledgments

I thank W. Burggren, M. Feder, and R. Huey for helpful comments on the manuscripts. Financial support was provided by NSF Grants BSR 86-00066, BSR 86-07794, and DCB 85-02218.

References

Benedict, F. G. (1938) *Vital Energetics: A Study in Comparative Basal Metabolism.* Washington, D.C.: Carnegie Institute of Washington.

Calder, W. A., III. (1984) *Size, Function, and Life History.* Cambridge, Mass.: Harvard University Press.

Kleiber, M. (1932) Body size and metabolism. *Hilgardia* 6:315-353.

Krogh, A. (1929) Progress of physiology. *Am. J. Physiol.* 90:243-251.

Peters, R. H. (1983) *The Ecological Implications of Body Size.* Cambridge University Press.

Schmidt-Nielsen, K. (1972) *How Animals Work.* Cambridge University Press.

Schmidt-Nielsen, K. (1984) *Scaling: Why Is Animal Size So Important?* Cambridge University Press.

Taylor, C. R., and Weibel, E. R. (1981) Design of the mammalian respiratory system. I. Problem and strategy. *Respir. Physiol.* 44:1-10.

Townsend, C. R., and Calow, P. (1981) *Physiological Ecology: An Evolutionary Approach to Resource Use.* Sunderland, Mass.: Sinauer.

PART ONE

Comparisons of species and populations

Part One explores the utility of physiological comparisons involving species or populations. The species has long been the primary unit of comparison in ecological physiology and in biology in general. Traditionally, particular species were chosen for analysis because of their occupancy of some extreme or otherwise unusual environment, their distribution across an interesting environmental gradient, or their possession of some tractable physiological attribute (the August Krogh Principle). With the rise of the New Synthesis in the 1940's, a few ecological physiologists (e.g., John Moore) began to study physiological patterns *within* as well as among species. These interspecific and interpopulational comparisons remain at the heart of ecological physiology – after all, physiological diversity is most conspicuous at these levels. Nevertheless, a critical evaluation of such comparisons is appropriate.

Part One begins appropriately with a discussion by George A. Bartholomew of the historical and contemporary significance of among-species comparisons (Chapter 2). Bartholomew outlines several formats for interspecific comparisons that have proven productive in the past, and he argues that the careful application of such approaches is still useful and often necessary. In a preface to this discussion, Bartholomew outlines his own historical roots and working philosophy for ecological physiology. Given his role as one of the founders and leaders of the field, these features have academic and historical significance.

Chapter 3 by Martin E. Feder is both a complement and a counterpoint to Chapter 2. Feder argues that the documentation of patterns of physiological adaptations to the environment, the historical "mission" of ecological physiology, is now approaching the point of diminishing returns in providing novel insights and conceptual advances. To remedy this situation, Feder suggests that ecological physiology should also pursue the investigation of numerous general questions of biological importance, which could enliven studies in ecological physiology.

Chapter 4 by Raymond B. Huey builds on part of Feder's commentary by showing how an explicitly evolutionary and historical perspective can invigorate interspecific and intraspecific analyses of patterns of physiological adaptations. Huey stresses that recent developments in comparative methodology offer opportunities for exploring a variety of old as well as new questions. He further shows why a phylogenetic perspective can be useful to comparative physiologists, even if they are *not* specifically investigating evolutionary issues.

Chapter 5 by Dennis A. Powers has two central messages. First, comparisons among populations, especially when evaluated within an historical and phylogenetic perspective, offer an underexploited resource for ecological physiologists. Second, interdisciplinary approaches provide a crucial link for understanding adaptation at all levels, from molecular to evolutionary. Powers's own work is an elegant example of this approach, one that is simultaneously reductionist yet synthetic.

2 Interspecific comparison as a tool for ecological physiologists

GEORGE A. BARTHOLOMEW

A personal view of ecological physiology: origins, assumptions, and themes

Physiological ecology is one skein in the fabric of biology. Like other areas of biological specialization it has a life history with identifiable stages and an indeterminate life expectancy. The editors of this volume asked me to discuss some of the earlier stages in the life history of physiological ecology as well as to evaluate interspecific comparisons as a tool. I have not studied the history of either physiology or ecology. My statements reflect only the views of a participant in these two fields. Like most working scientists, I am unaware of many of the historical factors that have shaped the area of my professional concern. Science is too large and complex for me to see more than a fraction of the inputs that have shaped even the small part of it in which I operate. What I can do is to look at my own background and attempt to identify the factors that led me to physiological ecology. Perhaps the factors that have shaped my view of biology, or factors related to them, have also helped to shape the field of physiological ecology. However, my confidence in this conjecture is limited.

Because of the success of the natural sciences and the technology associated with them, for the past few centuries a significant fraction of our species has lived in a state of rapid cultural change. The rate of change has been particularly high for us scientists. Indeed, we live in a state of continuous revolution. As an elder member of the community of physiological ecologists, I am almost overwhelmingly aware that during this continuing intellectual revolution, seniority is more likely to be correlated with obsolescence than with wisdom.

Despite the rapidity of scientific change, science is an intensely historical activity. Scientists' perceptions of their disciplines and their ideas about which questions are of interest are apt to be strongly affected by the persons under whom they studied and by the points of view that were current during their formative years. A substantial portion of physiological ecology is based on the work of persons who were trained during the late 1930's and the 1940's, and the work of their students. To understand how certain questions became important in physiological ecology, it should be helpful to examine the intellectual milieu of that period. I can best do so by using my own background as an example.

The world of ideas, descriptions, and measurement in which a scientist operates is given focus and coherence by individuals, and sustained by groups with common or overlapping interests. Every generation of scientists has temporal thresholds prior to which the immediacy and relevance of theories and data fall off rapidly. For today's undergraduate biology students, papers published before 1975 are largely of historical interest.

I was an undergraduate during the 1930's; my intellectual world stems from that decade. However, some of the teachers who shaped my view of the biological world had their intellectual roots in the nineteenth century. For example, my genetics teacher was Richard Goldschmidt, who was an active scientist when Mendelian genetics was rediscovered. I learned about invertebrates from Henry Bigelow, who started out as Alexander Agassiz's research assistant. My undergraduate studies of biology were dominated by the views of professors trained during the first third of the present century. My initial perceptions of vertebrate zoology were shaped by the systematic, taxonomic, and distributional orientation held by Joseph Grinnell, a natural historian who founded the Museum of Vertebrate Zoology at the University of California at Berkeley, and by four of his followers who were my mentors (Seth Benson, Jean Linsdale, Raymond Hall, and Alden Miller). My perceptions of paleontology were formed by Joseph Camp and Alfred Romer, who were undergraduates before World War I.

My perceptions of evolutionary biology have somewhat less remote origins. They grow out of a background that is still real, at least to some present-day graduate students. In the 1940's, a series of books on evolution appeared that had a permanent influence on my world view: Huxley's *New Systematics* (1940) and *Evolution, the Modern Synthesis* (1942), Dobzhansky's *Genetics and the Origin of Species* (1937), Mayr's *Systematics and the Origin of Species* (1942), and Simpson's *Tempo and Mode in Evolution* (1944). I remember vividly the cool response I received in "The Vertebrate Review," a graduate seminar run by the staff of the Museum of Vertebrate Zoology at Berkeley, when I presented and espoused Huxley's point of view that students of geographic variation would be well advised to study clines rather than to continue naming subspecies.

After receiving a Master's degree at Berkeley under the supervision of Alden Miller, one of the leading avian systematists of his generation, I moved to Harvard to continue graduate work with the hope of studying the functional biology of organisms. Because of the death of my first advisor and the interruptions caused by World War II, I had four different Ph.D. supervisors while I was at Harvard. In sequence they were Glover Allen (mammalogy and ornithology), A. S. Romer (paleontology), K. S. Lashley (physiological psychology), and George Clarke (marine ecology). Each of these persons strongly affected my biological development, but not one of them was as important as a group of graduate students with whom I associated at Harvard before and after the war and whose points of view shaped my pattern of research and helped me to define my own approach to biology. They were Donald Griffin, Oliver Pearson, Charles Lyman, Peter Morrison, and Edgar Folk. At that time they were studying respectively, echolocation and navigation, reproductive physiology and energy metabolism, seasonal changes in coat color, blood chemistry, and daily rhythms of activity. Griffin identified himself as an "experimental naturalist," a title appealing to me then and also now.

During World War II, I worked in magnetic mines warfare as a physicist in the Bureau of Naval Ordinance, and perforce developed a practical knowledge of electrical instrumentation and measurement. After returning to graduate school in 1946, I vividly remember deciding on a research niche in biology that my interest in natural history and my wartime experience with instrumentation would allow me to occupy – studying the physiological adjustments of organisms to the physical environment. These are my scientific roots. They include substantial exposure to natural history, systematics, zoogeography, paleontology, behavior, and physiology, and an empirical familiarity with electrical instrumentation. Together they have nourished a point of view that makes it comfortable for me to work at the intersection of physiology, ecology, and behavior, to deal with organisms, adaptations, environmental factors, and evolution, that is, to be a physiological ecologist.

Some underlying assumptions

Before considering adaptations of organisms to extreme environments and comparisons between species, I shall identify a series of assumptions that are the foundation for my view of biology. My confidence in them is sufficient for me to treat them as axioms. I present them here because they underlie the rest of this essay.

1. Two of the salient characteristics of "good" science are originality of conception and generality of application. The example par excellence is Darwinian natural selection; it offers a mechanism so widely applicable that it is almost coexistent with reproduction and so innovative that it shook and continues to shake people's perception of causality and of themselves. The characteristics of innovation and generality have contrasting relations to different disciplines of biology. The simpler (i.e., more basic) the level of integration in the discipline, the more widely applicable (i.e., more general) its findings are apt to be. Consider molecular biology, cell biology, and genetics; the early findings in these fields can have broad relevance because

they apply to a wide variety of organisms. (However, as these fields mature, investigators have an increasing tendency to focus on minor variations and special cases, just as do some organismic biologists.) The more complex the level of integration in the discipline, the greater the diversity and the greater the opportunities for discovering new mechanisms and developing innovative ideas – consider behavior and comparative physiology – but the greater the difficulty in finding generalities that are not so obvious as to be trivial.

2. At present, only two known processes lead to the evolution of adaptive change. These are chance and natural selection, and natural selection is itself a stochastic process. Chance plays a substantial role both in generating the genetic variability on which natural selection operates and in shaping the ecological arena in which it acts.
3. Natural selection increases fitness but it produces systems that function no better than they must. It yields adequacy of adaptation rather than perfection. The adaptive changes that result from its actions are adjustments to local and immediate conditions. It is blind to the long-term consequences of the biological changes that it produces. The chances that short-term adaptations will meet long-term challenges are necessarily slight. Consequently, for a given population or a given species, the long-term probability of extinction is high.
4. Despite the high long-term probability of extinction, every organism alive today is a link in a chain of parent-offspring relationships that extends back unbroken to the beginning of life on earth. Every organism is a part of an enormously long success story – each of its direct ancestors has been sufficiently well adapted to its physical and biological environments to allow it to mature and reproduce successfully. Viewed thus, adaptation is not a trivial facet of natural history, but a biological attribute so central as to be inseparable from life itself.
5. All sexually reproducing animals exist as parts of breeding populations. From the point of view of the evolution of adaptations, the breeding population is the key unit of biological organization. All the taxonomic categories in the modern Linnean system can be interpreted as phylogenetic inferences. However, under some circumstances a subspecies, and more rarely a species, can exist as a breeding population. Because species can have functional coherence as breeding populations, in this essay I shall speak as if they ordinarily do.
6. Diversity of form and diversity of function are salient features of organismic biology. One of the most impressive contrasts in all of biology is the uniformity of the physiological machinery at the cellular level and the diversity of the adaptive patterns into which this machinery is assembled at the organismic level.
7. The responses of organisms ignore the categories of scientific specializa-

tion into which we biologists have divided ourselves. Organisms are functionally indivisible and do not fall neatly into the conventional compartments such as physiology, morphology, behavior, and genetics. It is the intact and functioning organism on which natural selection operates. Organisms are therefore the central element of concern to the biologist who aspires to a broad and integrated understanding of biology.

8. For a person interested in physiological ecology, it is operationally effective to think of an organism not just as an animal or plant housed in the laboratory or seen in the field, but as an interaction between a complex, self-sustaining physicochemical system and the substances and conditions which we usually think of as the environment (Bartholomew, 1964). As Claude Bernard pointed out a century ago, organism and environment form an inseparable pair. Each can be defined only in terms of the other.
9. A final axiom is biological only to the extent that it involves an attribute of the human nervous system. For a human being, unstructured detail is chaos. The complexity of the world in which we live, and also the complexity of the science that we use to interpret it, requires us to create structures of logical relationships on which to attach our observations and interpretations. This platitudinous statement is worth emphasizing in the present context, because students of physiological ecology risk being overwhelmed by the diversity of the systems they study. Fortunately, biological diversity has pattern. One of the more obvious elements of this pattern is supplied by species. Consequently, it is extremely difficult for a student of organismic biology to avoid comparing species, either explicitly or implicitly. Not to do so is to ignore one of the central features of the system the student is attempting to understand.

Some themes in physiological ecology

Composing a definition for a field in biology can be a sterilizing effort because definitions delimit, and biology is functionally indivisible. However, it should be possible to identify a discipline operationally without confining it. I shall present my view of physiological ecology by enumerating some of the topics subsumed by the phrase "ecologically relevant physiology." It is obvious that the integration that follows is a personal view that grows out of the circumstances of my scientific career. It is incomplete, but it is coherent – tug on one thread and the entire fabric vibrates.

Because organism and environment represent inseparable elements in a single system, all aspects of physiology (or for that matter all aspects of organismic biology) are in some cosmic sense "ecologically relevant." To reduce the situation to reasonable dimensions, we can propose that the physiological processes that are most apt to be ecologically relevant are not those that deal with internal integration, but rather those that deal with (1) exchanges of food, water, energy, and metabolites between organisms and their environments; (2) the exchanges of information (social signals, pheromones) and gametes between organisms; (3) the acquisition of information about the environment by organisms; and (4) the effects of the physical environment on physiological capacity and performance. Many of these processes are conveniently studied in terms of rates and limits (e.g., energy metabolism, physiological constraints). Others are effectively studied as rhythms or cycles (e.g., activity, migration). Still others are effectively examined in the context of mechanisms of communication and orientation (e.g., social signals, navigation). Chapter 1 enumerates some of the achievements of the field in these various contexts.

By assembling a nonexclusive list of topics that most students of physiological ecology will recognize as being appropriate to the field, we have identified an intellectual enterprise with indefinite boundaries that is given coherence by its concern with the functional dynamics of whole organisms existing in the natural environment, having long evolutionary histories, and sustained by complex interrelations in both time and space with the rest of the natural world.

Extreme environments and species comparisons

Two of the early, persistent, and interlocking themes in physiological ecology are the adaptation of organisms to extreme environments and comparisons between species. I shall discuss the former briefly and the latter more extensively. I shall develop these themes by describing several general research formats that rely on interspecific comparisons. It is noteworthy that each of them is derived from the same major paradigm that underlies the entire comparative approach to biology. That paradigm is neither complex nor obscure; it proposes that each species represents a coherent solution to the problem of living. Each species can therefore reveal to us by its adaptations something of biological interest that will increase our understanding of the biological world.

Adaptations to extreme environments: rationale and philosophy

The study of physiological adaptations to extreme environments – the polar regions, the tops of high mountains, low-latitude deserts, saline lakes, hot springs, the depths of the sea, bottom sediments – has the attraction of allowing an investigator to focus on those aspects of an organism's physiology that allow it to cope with overt, clearly definable challenges such as extremes of temperature, lack of free water, low partial pressures of oxygen, trace element deficiency, high pressures, intense radiation, high osmotic pressures, and extremes of pH. These quantifiable environmental challenges are potentially limiting factors and are so overriding that they can appear to overshadow other ecological considerations.

Another incentive to study adaptations to extreme environments is the widely held perception that animals occupying physically difficult environments offer especially attractive opportunities for the study of ecologically relevant aspects of physiology. An analysis of the mechanisms by which animals have successfully adjusted to the rigors of their habitats can provide the investigator with unusually revealing insights concerning adaptive modification of the fundamental patterns of physiology characteristic of the group as a whole. Furthermore, it can afford particularly favorable opportunities for the delineation of aspects of the fundamental patterns themselves that are often not readily observed or measured in animals occupying relatively undemanding situations.

Early in the history of physiological ecology, this point was made in a more restricted context by Cowles and Bogert (1944) in their "Preliminary study of the thermal requirements of desert reptiles," probably the most influential paper ever published on the physiological ecology of reptiles:

> The great advantage inherent in studies of desert reptiles lies in the extraordinarily high maximum temperatures, as well as the greatly exaggerated temperature changes, so characteristic of desert climates. Such environmental conditions may amplify subtle details of thermal relationships that would otherwise escape notice. The resulting accentuation of temperature responses throws into relief temperature relations which in the equable climate of the tropics might otherwise remain as imperceptible, or at least unperceived, nuances in thermal adaptations.

This general perception has been widespread among most students of animal physiology, not just among physiological ecologists, and has been an important factor in the study of environmentally relevant physiology. The pioneering and exciting studies of Knut and Bodil Schmidt-Nielsen on desert rodents and of Per Scholander and associates on Arctic vertebrates come immediately to mind.

Over the years it has become clear that adjustments to the physical environment are behavioral as well as physiological and are inextricably intertwined with ecology and evolution. Consequently, a student of the physiology of adaptation should not only be a technically competent physiologist, but also be familiar with the evolutionary and ecological setting of the phenomenon that he or she is studying. This in turn requires that the student be cognizant of aspects of organismic biology that lie outside the boundaries of traditional physiology.

Physiological adaptations are accessible to dissection into components and to analysis of mechanisms, but they are also attributes of organisms and products of evolution. To be effective they must be adequately supported by behavior and, of course, behavior is powerfully constrained by morphology.

Moreover, an organism must not only be adapted to meet major environmental challenges, it must also be capable of carrying out essential functions throughout its entire life cycle, including embryonic development, growth, and reproduction. Even those adaptations that meet the most extreme challenges of the physical environment are constrained by essential functions that may be quite unrelated to these challenges.

When one is analyzing physiological adaptations to an extreme environmental condition, it is essential to evaluate the comparative setting. This is necessary not only because different kinds of animals have different ways of meeting a given environmental challenge, but also because such a challenge can be adequately specified only in terms of the animal that deals with it. What can be a severe physiological challenge to one type of organism may have negligible impact on another. Sometimes this variation is due to physiological differences, but in terrestrial habitats it often occurs because behavioral avoidance, not physiological adaptations, is an organism's primary response to an environmental challenge (Chapter 1). This point is elementary, but it is by no means trivial.

The terrestrial environment is far from monolithic. In fact, in terrestrial habitats there is no such thing as "the environment." Instead, there are virtually as many environments as there are species of animals. A few large animals such as plains-dwelling ungulates, cowboys, and ostriches live, at least part of the time, in an environment that in terms of air temperature, humidity, radiative flux, and air movements approximates the environment described by meteorologists. However, such is not the case for most animals most of the time. Most terrestrial vertebrates weigh less than 100 g, and few arthropods weigh more than a gram. For these creatures the physical environment consists of cracks and crevices, sheltered nooks and tunnels. For them, distances are measured in meters, not kilometers, and the difference between sun and shade or surface and burrow can mean the difference between life and death. A terrestrial animal can pick and choose among the different environmental conditions that occur from place to place and time to time in its habitat. By its behavior it can assemble its own environment. This behaviorally generated physical microenvironment is the one with which the organism's physiological capacities can cope. It is this species-specific environment, not the one that physiologists would face were they in the same part of the world, to which the organism's physiology is adjusted.

Environments are classified as "extreme" by us physiologists, not by the animals that live in them. The number of factors potentially contributing to the presence or absence of species in an extreme environment is almost limitless, and many, probably most, of them are unrelated to those aspects of the local environmental situation that a physiologist may judge to be difficult. Consider the Adelie penguin, which is immersed for days at a time in ice water and breeds under conditions that are always near and frequently below freezing. Its insulation is so effective that while foraging in the water it must continuously maintain the thermal windows of its feet and underside of its wings in a state of vasodilation. Low temperatures present no physiological challenge; however, on clear days, even with air temperatures near freezing, these birds must maximize their respiratory evaporative water loss to prevent overheating even while remaining motionless.

The preceding considerations can lead us to a biologically realistic and operationally reasonable approach to the study of physiological adjustments to the physical environment. If we base our physiological measurements on the environmental factors with which the organisms actually interact rather than on the ones with which we interact (or can conveniently measure), we can minimize the chance that our analysis will be devoted to experimental artifact rather than to significant physiological variables. Making environmental measurements in biologically realistic situations can be difficult. It requires substantial knowledge of the natural history of the animal being studied, and it usually requires either the design and fabrication of special instruments or extensive alteration of existing devices.

Comparisons between species

I shall examine this topic by responding to a series of questions concerning the relevance of interspecific comparisons to physiological ecology. Does the species-comparison approach have a central theory? Is it just a collection of adaptive stories? Does it do more than demonstrate that animals live where they can? What are the analytical paradigms of the comparative approach? Does the comparative approach have a useful future? How can it profitably be modified? Does the species-comparison approach have a central theoretical structure? My answer to this last question is, "Of course." However, its theoretical structure has a number of components and cannot be summarized in a single sentence. It has operational constraints and cannot stand alone. I shall first deal with this central structure.

Most students of organismic diversity have been either systematists or students of evolution. They have been mainly concerned with speciation, the evolutionary derivation of one group of organisms from another, or the analysis of the patterns of interactions between species. Students of ecologically relevant physiology approach animal diversity from a different point of view, namely, by analyzing the physiological (and behavioral and ecological) adjustments that allow animals to accommodate themselves to their environments. Not only are these adjustments almost infinitely diverse, they presumably also have strong genetic and phylogenetic components and must, therefore, be dealt with in an evolutionary context. Animal diversity, expressed in terms of functional differences between species, offers an investigator an opportunity to examine quantitatively the physiological mechanisms underlying responses that can be of acute ecological and evolutionary relevance. However, it is easy to be overwhelmed by the diversity of physiological responses. As pointed out above, the human mind requires some logical structure within which to organize the diversity.

In retrospect, physiological ecology appears to have found a substantial component of its initial logical structure in the New Synthesis that dominated evolutionary and organismic biology during the 1940's and 1950's. From the perspective of the New Synthesis, physiological diversity could be viewed as the result of natural selection, itself a stochastic process, acting in diverse environments, and operating at different rates and starting from different temporal and taxonomic points. Each species could be treated as a coherent but temporary solution to the general problem of living, that is, maintaining organismic integrity by the regulated processing of energy and materials and at the same time remaining functional long enough to reproduce. By examining many species, one could hope to discern patterns of physiological adaptation that were correlated with measurable environmental variables. Deviations from patterns and free-form adaptive solutions could be readily accommodated because of the stochastic nature of natural selection. Many of the persistently followed lines of research in physiological ecology were started by attempts to use physiological characteristics to answer evolutionary questions that were important at the time of emergence of the New Synthesis (convergence, divergence, and parallelism, local optima in fitness, the factors determining clines, and the factors limiting geographic distribution).

Thus the central focus of the species-comparison approach is not (or should not be) to compile an encyclopedia of the physiological differences between species. Rather, it is a way of studying patterns of adaptation. It does not have a single paradigm. It can be employed in a variety of ways and can be used to address a variety of questions. Its versatility is due in part to the fact that species can be used to exemplify various levels in the taxonomic hierarchy. Every kind of animal that has been described is assigned to a species and this in turn assigns it to a genus, a family, an order, and so on. Consequently, when one compares physiological functions of animals belonging to different species, one can use those species as exemplars of categories ranging from breeding population to phylum. Obviously, when species are used as exemplars of higher categories, it is prudent to make measurements on as many carefully selected species in those categories as feasible.

Types of interspecific comparisons

Four examples of applications of species comparisons will be considered. Documentation will depend mostly on studies with which I have been either directly or indirectly associated, not only because of my familiarity with them, but because I know the questions to which they were addressed.

Convergence and divergence in physiological functions: flapping flight

Animals belonging to very distantly related taxa may evolve adaptations that allow them to have ecologically similar modes of life, even though the physiological machinery and morphological structures that support their performance are totally different. Flight powered by wing flapping occurs only in vertebrates and insects. Every student of elementary zoology learns that the wings of birds and insects have analogous locomotor functions, but are totally different in embryonic derivation and structural organization, and have totally different mechanisms for linking muscle contraction to wing movement.

That flapping flight evolved separately in birds and insects cannot be seriously challenged. However, it is of physiological, ecological, and evolutionary interest to compare the energy costs for flapping flight in birds and insects. Although the flapping flight of insects and vertebrates are separate evolutionary events, with insect flight much the more ancient, a substantial convergent evolution occurred once flying vertebrates appeared in the later Mesozoic. A particularly striking example is offered by hummingbirds and sphinx moths. Members of these two families (Trochilidae and Sphingidae) belong to different classes and different phyla. Sphinx moths include some of the largest insects. Hummingbirds are the smallest birds, and the two groups overlap in body mass (Bartholomew, 1981). They share the remarkable habit of feeding on flower nectar while hovering. Sphinx moths and hummingbirds are aerodynamically similar and resemble each other when in flight despite their totally different morphological organization.

Has natural selection resulted in patterns of flight with similar energy efficiency in hummingbirds and sphinx moths? This question is of both theoretical interest physiologically and direct relevance ecologically. Hovering flight is energetically the most expensive mode of locomotion, and the energy to support it has to be supplied by the flowers on which the animals feed. This question can be approached by scaling the energy cost of flight in birds and sphinx moths and by comparing the flight energetics of moths and birds of similar mass. The energy costs of hovering flight in these two structurally and physiologically different types of organisms are remarkably similar (Figure 2.1).

In this example, species are being used as exemplars of major taxonomic categories, and the research question being addressed pertains to physiological convergence in members of different phyla. Consequently, a comparison of a single species of hummingbird with a single species of moth, while amusing, would not be scientifically convincing. However, by directly comparing several species from each of the families and also by fitting these data

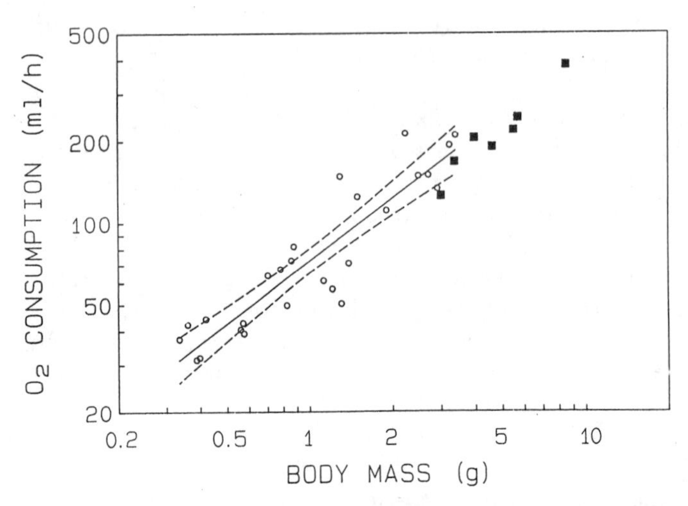

FIGURE 2.1 Oxygen consumption during hovering flight in relation to body mass in 26 species of sphinx moths and 6 species of hummingbirds. The regression line is fitted only to the data for the sphinx moths. The equation for the line is $Y = 72.3\ X^{0.77}$; $r^2 = .83$. The dashed lines enclose the 95% confidence interval for the line. Data for sphinx moths are from Bartholomew and Casey (1978). Data and sources for the hummingbirds are given in Bartholomew and Lighton (1986).

into an allometric analysis, one can arrive at a reasonably satisfactory answer. One can also translate this similarity into a general statement of evolutionary interest. The aerodynamic constraints imposed by the combination of forward flight, while searching for flowers, and sustained hovering, while feeding on nectar, are associated with the evolution of similar body size, body shape, and wing shape in sphingids and hummingbirds despite fundamentally different patterns of morphological organization. The energy expenditures during hovering flight in these convergent insects and birds are virtually identical and extremely high. The mass-specific oxygen consumption of hovering hummingbirds represents the maximum metabolic rate known among vertebrates. From this one can infer that sphingids and hummingbirds are approaching the biological limit of aerodynamic performance for animals of their size, and further that the arthropod pattern and the vertebrate pattern of structure and function support similar aerodynamic efficiencies in the size range from 1 to 10 g.

Adjustments that allow closely related species to live in dissimilar environments: equatorial, temperate, and Arctic pinnipeds

Most species of the order Pinnipedia (seals, sea lions, and walruses) occur in cool temperate and polar regions. Various combinations of large size, thick layers of subcutaneous fat, and dense pelage make them essentially indifferent to low environmental temperatures. Neither prolonged submersion in ice water nor sustained exposure to air temperatures far below freezing seriously challenges their thermoregulatory capacities. However, the effectiveness of their heat retention often makes it difficult for them to avoid overheating while they are on land, even under relatively cool conditions.

All members of the family Otariidae (sea lions and fur seals) breed on land, where the very effectiveness of their adaptations for heat retention while in the water can present them with acute physiological problems. This is shown with particular clarity by the northern fur seal *Callorhinus ursinus*. Young male *Callorhinus* frequently die of overheating when forced to walk only a few hundred yards even though the sky may be overcast and the air temperature no more than 10 °C. They are unable to lose the heat they generate by terrestrial locomotion. Their body temperatures rise to 43 °C or more, and they collapse of heat prostration (Bartholomew and Wilke, 1956).

The fur seals of the southern hemisphere belong to the genus *Arctocephalus*. They have a circumpolar distribution on the islands of the subantarctic, and substantial numbers occur on the Antarctic Peninsula. Like the northern fur seal, they are extremely well insulated by fur and blubber. Nevertheless, one species in the genus, *Arctocephalus galapagoensis*, breeds in the Galapagos Islands, which are located on the equator off the west coast of South America. Compared with most equatorial regions the water around the Galapagos is relatively cool. As long as the fur seals remain in the water, they can easily avoid overheating. The temperature of the sea is usually 12 to 18 °C below body temperature, and the seals can dump heat to the water through their naked and highly vascular flippers. However, the islands on which *A. galapagoensis* come ashore and breed are characterized by intense solar radiation and tropical air temperatures. How do they manage? They do so by behavioral rather than physiological thermoregulation (Bartholomew, 1966; Trillmich and Moren, 1981). They select areas where they can avoid direct solar radiation. During the warmer parts of day, they typically frequent wave-worn caves where they are completely sheltered from the heat of the sun, or they occupy talus slopes with massive boulders among which they can find cool and shaded retreats. During the tradewind season they haul out on low rocky ledges on the windward side of islands where they are continuously drenched with spume and spray from the heavy seas.

Otariids have evolved a highly polygynous breeding organization (Bartholomew, 1970). The principal populations of the California sea lion, *Zalophus californianus*, occur on the islands off the coast of California and northern Mexico. These islands are bathed by cool, upwelling water and are characterized by persistent summer overcast. *Zalophus californianus* breeds during the summer. On these islands the rookeries are divided into persistently maintained territories by the males, and virtually all the breeding behavior takes place on land (Peterson and Bartholomew, 1967).

A substantial population of *Z. californianus* also occurs in the Galapagos Islands where conditions are of course much warmer than in the coastal islands of California, too warm to support a strictly terrestrial breeding pattern. The Galapagos population of *Zalophus* solves this thermoregulatory problem by changing its breeding behavior from the terrestrial pattern characteristic of otariids. During the daylight hours, instead of holding territories exclusively on land, the breeding bulls maintain aquatic territories while remaining totally or partially submerged in tidal pools and channels. They restrict their terrestrial activity to the nighttime. Here again we see a thermoregulatory problem resolved by behavioral rather than by physiological adaptation.

The relevance of these two examples to the species-comparison approach is two fold. In the fur seals, behavioral thermoregulation involving escape from solar radiation allows a member of a subantarctic genus to extend its range to include an equatorial archipelago. In *Zalophus*, occupancy of the same equatorial archipelago by a population of a predominantly cool-temperate species also involves behavioral rather than physiological changes, but these changes require that the animals spend most of the daylight hours in the water and haul out on shore only at night. This behavioral adjustment, although superficially simple and direct, requires a drastic reorganization of a breeding structure that is exclusively terrestrial in other members of the family.

The response of the Galapagos population of *Zalophus californianus* is also of interest because these animals are an isolated tropical population of species which is otherwise Holarctic. Hence the comparison is *intra*specific, rather than *inter*specific.

Do related species living in the same difficult environment have similar responses to it? Reproductive timing of desert rodents

The demanding physical and biotic conditions associated with extreme aridity profoundly affect the reproductive performance of desert rodents. Erratic rainfalls followed by transient pulses of plant growth often result in irregular breeding schedules. A widespread pattern in rodents involves a seasonally flexible schedule characterized by opportunistic breeding triggered by the unpredictable availability of fresh green vegetation, which is important for reproduction (Beatley, 1969; Reichman and van de Graaff, 1975).

Animals, including desert animals, interact not with the entire environment available to them, but only with selected elements of it. Because of their mobility and capacity for selection and rejection, they can pick and choose among the different physical and biotic conditions that occur from time to time and place to place in their habitat. Thus, in a given habitat there is more than one solution for any given biological problem. Potentially the solutions are as numerous as the species in that habitat. In practice, of course, phylogenetic constraints are inescapable, and closely related species may be able to solve similar problems only in similar ways. Desert animals differ in size, patterns of locomotion, and food habits. These differences are accompanied by different environmental constraints and different opportunities. Moreover, even severe deserts are not environmentally monolithic. Consequently, it should be possible for closely related species to have different reproductive schedules in what is superficially the same simple desert environment. But do they?

G. J. Kenagy and I examined this possibility by undertaking a long-term laboratory and field study of the reproductive timing of five locally abundant, coexisting rodent species in a desert community in the Owens Valley, Inyo County, California (Kenagy and Bartholomew, 1985). The species studied included four nocturnal heteromyids (two kangaroo rats and two pocket mice) and one diurnal sciurid (a ground squirrel).

Dipodomys microps, the Great Basin kangaroo rat (mass 50 to 65 g) is unique among heteromyids in that it feeds mostly on the leaves of a perennial shrub, the salt bush *(Atriplex confertifolia). Dipodomys merriami*, Merriam's kangaroo rat (mass 35 to 40 g), is primarily granivorous, but for reproduction it needs some green vegetation in its diet. *Perognathus longimembris*, the little pocket mouse, a granivorous species with a mass of about 8 g, is among the smallest of rodents. *Perognathus formosus*, the long-tailed pocket mouse, weighs about 20 g and also is primarily granivorous. Both species of pocket mouse hibernate and remain dormant underground during fall and winter. *Ammospermophilus leucurus*, the antelope ground squirrel, whose mass is about 100 g, is the most widespread desert ground squirrel in North America. It is diurnal and omnivorous. Unlike most ground squirrels, it neither hibernates nor estivates.

In the Owens Valley, plant growth is highly seasonal and is largely dependent on irregular winter precipitation that produces pulses of growth of moisture-containing vegetation in the late winter and spring. The five sympatric rodent species accommodate their reproduction to this environmental schedule in a variety of ways. Their patterns of reproduction correlate with differences in body size, food habits, and duration of hibernation. In all five species, breeding begins in late winter and early spring, and reproductive readiness of the males precedes that of the females by six to eight weeks. Among the heteromyids, the onset of breeding is in the same sequence as body mass, with the largest species breeding first. *Dipodomys merriami* has only two young per litter, and the young are weaned in less than three weeks. In good years a female may produce several litters; males produce sperm all year long; and the juveniles often reach reproductive maturity before the breeding season ends. This high reproductive potential can result in a large population increase when rains are appropriately spaced. *Dipodomys microps* also has two quickly weaned young per litter. Its breeding season rarely lasts more than six weeks. It produces only one litter per year, and its young never mature sexually in the season of their birth. Because of its diet of salt bush leaves, it rarely fails to breed even in poor years. Saltbush is a perennial plant with a highly predictable spring growth of leaves, which supply the lactating females with a dependable source of green and succulent food that is largely unaffected by the vagaries of the local winter precipitation. As a result, despite its modest reproductive potential, the breeding success of *D. microps* exceeds that of *D. merriami.*

Prolonged hibernation of the pocket mice restricts their breeding season, as compared with the kangaroo rats. *Perognathus formosus* hibernates for 3.5 months, *P. longimembris* for 6.5 months; both species breed after hibernation ends. Their periods of dormancy overlap the period of winter rainfall and may extend beyond it. As a result they have a very restricted annual breeding season and are prone to complete reproductive failure. However, they produce about five young per litter, and, when the growth of annual vegetation continues into late spring, they can produce two litters in a single breeding season. Consequently, over the long haul their total annual reproductive potential approximates that of the two species of kangaroo rat.

The breeding schedule of the ground squirrel is quite different from that of the heteromyids. Its breeding season lasts only two weeks. It occurs always at the same time in the early spring and is unaffected by the erratic annual variations in rainfall. The ground squirrels produce large litters (mean 8, range 5 to 14), and the young grow slowly (8 weeks to weaning). They adjust the magnitude of their breeding effort after conception. If conditions are poor during pregnancy, some of the embryos are resorbed. If conditions are poor late in the breeding season, the weaker young die. When conditions are good throughout the season, all or most of the large litter are successfully weaned.

All five of the coexisting species are long-lived and maintain relatively stable populations. *Ammospermophilus leucurus* is a "pulse averager" with a slow, prolonged reproductive effort that matches the historical probability distribution of rainfall and plant production. *Dipodomys merriami* is a "pulse matcher," responding directly and rapidly to the occurrence of pulses of food production. *Dipodomys microps* has a dependable food supply and is a "pulse ignorer," indifferent to variations in rainfall and pulses of production of annual plants. The two species of pocket mice are "pulse gamblers," producing large litters in a narrow time window. Sometimes they succeed in bringing off a litter; sometimes they fail completely.

With regard to the reproductive timing of these coexisting species of desert rodents, the answer to the question, "Do related forms living in the same difficult environment have similar adaptations to it?" is negative. They face the same environmental challenge, but in their reproductive schedules they respond to it very differently.

How do distantly related forms adapt to the same or similar difficult physical environments? Flying insects that breed in the winter

For me the interest in this particular question is that it causes one to focus attention on the phylogenetic constraints to physiological adaptation. Every major adaptive change inevitably constrains the nature of the subsequent adaptations of all later members of the lineage in which it occurred. Every organism is thus in a kind of phylogenetic trap. Some patterns of adaptation are available to it; others are not. The frequency of parallel evolution of functions and structures documents this point. An organism can most easily change by modification of mechanisms that are already in place. What happens when species in different orders or families of the same class adapt to similar difficult environments? Clearly, there is no single answer, but physiological comparisons of species can be instructive.

At middle and high latitudes, most insects overwinter in a dormant state (usually as eggs or pupae, but sometimes as adults). However, a few species carry out their breeding flights during times of year when air temperatures are at or below 0 °C. Although the selective factors favoring this unusual timing have never been analyzed in detail, the most obvious inference is that reduced predation pressure for flying insects during fall and winter offers a substantial advantage. Although other factors are probably involved, the absence of migratory birds and bats, as well as the dormancy of predatory insects, reptiles, and nonmigratory bats, virtually eliminates predation as a source of adult mortality for flying insects. In any event, at least one scarab beetle and several geometrid moths engage in mating flights under conditions of cold that are totally immobilizing for most insects. This remarkable performance involves profoundly different physiological adjustments in the scarab and the geometrids.

Most insects are ectothermic. However, several groups, including some moths and some beetles, have evolved the capacity to sequester some of the heat produced by the contraction of flight muscles and to use it to regulate body temperature, particularly in the thorax and the head (see Heinrich, 1981, for a review). This endothermic capacity is well developed in large beetles of the family Scarabaeidae (Bartholomew and Casey, 1977; Bartholomew and Heinrich, 1978; Morgan and Bartholomew, 1982). Rain beetles (*Pleocoma*) are large, heavily wing-loaded scarabs (mass of males about 2 g) that are common in the mountains of southern California. The larvae remain underground for many years. Females are flightless and about twice as large as the males. As adults, males do not feed. The breeding flights, during which the males search for the females that stand in the entrances to their burrows, coincide with the occurrence of storm fronts. These fronts bring snow to the local mountains, and air temperatures are often substantially below freezing. Mating flights of *Pleocoma australis* occur at dusk and at dawn. The males cannot fly when their thoracic temperature is less than 35 °C. Even in freezing weather they can generate enough heat by quivering their wing muscles to raise the thorax to flight temperature (Morgan, 1987) and can successfully carry out their breeding flights in the dark during snowfalls or freezing rains.

Although moths belonging to a number of families are endothermic during flight, some large-winged, light-bodied moths such as members of the family Geometridae are ectothermic and fly with slowly beating wings and with thorax temperatures that are essentially the same as that of the air (Bartholomew and Heinrich, 1973).

In the northeastern United States, a few species of geometrids emerge in the fall. Like rain beetles, the females are flightless, and the adult males do not feed. Two of these geometrids (*Operophtera bruceata* and *Alsophila pometaria*) have been studied (Heinrich and Mommsen, 1985). Beginning in late November and continuing until the first heavy snows, these small (less than 10 mg) moths perform their breeding flights both during the day and at night. They neither bask nor shiver. Thoracic temperatures remain within 1 °C of air temperature. They can fly at air temperatures as low as −3 °C. Their body temperatures during flight (as low as −3 °C) are the lowest so far measured. The data on the biochemistry of insect thermoregulation are extremely scanty. However, it is noteworthy that these two winter-flying geometrid moths have extremely low wingbeat frequencies and the lightest wing-loading of any moths known.

In this example of interspecific comparisons, distantly related insects (a scarab beetle and geometrid moths) have evolved patterns of winter breeding flights, presumably in part as a method for reducing mortality from predation. Some aspects of their breeding performances are similar (flightless females, flying but nonfeeding males). However, the mechanisms that allow effective flight at low temperatures in the scarab and the geometrids differ completely. The beetles are heavily wing-loaded, fly with rapidly beating wings, and expend large amounts of energy to maintain thorax temperatures 35 °C or more above air temperature. The geometrid moths manage to fly with extremely low rates of energy expenditure, with thoracic temperatures at or below 0 °C by virtue of being extremely lightly wing-loaded and beating their wings very slowly.

Both the scarab beetle and geometrid moths are constrained by the morphological and functional organization of the families to which they belong. All scarabs are heavily wing-loaded, and all having a mass of more than 1 g are strongly endothermic. All geometrids have large wings, small bodies, and slow wing beats; none is endothermic. Despite these morphological and physiological differences, the beetles and the moths have evolved surprisingly similar patterns of winter reproductive behavior.

Specific questions, general answers

Most scientists, including me, hope to establish general relationships that will answer or at least help to answer many specific questions. As pointed out earlier, the process of establishing satisfactory generalities becomes more difficult as one approaches the ecological and evolutionary levels of biological organization. Of course, anyone can propose a general relationship, but it is impossible to know whether or not it is really general until many relevant cases have been evaluated. Such evaluations necessarily involve species comparisons, either explicitly or implicitly.

Let me cite an example from my own experience. A quarter of a century ago, Vance Tucker and I discovered that, under controlled conditions, the bearded dragon (*Amphibolurus barbatus*), a large Australian lizard of the family Agamidae, heated more rapidly than it cooled. We found that this physiological control of body temperature was primarily dependent not on metabolic heat production but rather on the modulation of rates of heat exchange with the environment by control of heart rate and probably by vasomotor control of the circulation of blood in the integument. We knew immediately (Bartholomew and Tucker, 1963) that we had demonstrated the first example of physiological control of body temperature in a reptile, but we did not know whether or not this unanticipated capacity was unique to *Amphibolurus barbatus* or was widespread among reptiles. We immediately looked for and found the same capacity in several monitor lizards and a skink. A year or two later, I found it in a member of the family Iguanidae. Clearly, physiological control of rates of change in body temperature was widespread among lizards. Subsequently, other investigators demonstrated its existence in snakes, turtles, crocodilians, and sphenodontids (see Bartholomew, 1982, for a review).

In this case, a series of comparative studies of different species by many investigators has gradually produced a general understanding of one aspect of body temperature control in reptiles. Generalizations about the ecological importance of this physiological capacity are still limited, and the specifics of its physiological mechanisms in different groups remain to be established. The point of this example is that species comparisons should be more than just encyclopedic compilations of the diversity of form and function among organisms; they should be an integral part of the process of determining the generality of adaptive patterns.

Do species comparisons produce merely a series of "adaptive stories?"

The species-comparison approach clearly does produce "adaptive stories," but these need not be trivial results. As previously emphasized, adaptations are a vital part of organismic biology. When the word "merely" is introduced into any scientific question, it casts a pejorative light on the related scientific enterprise. The compilation of an encyclopedia of randomly selected, unrelated "adaptive stories" is obviously an inappropriate goal. In contrast, the rigorous analysis of thoughtfully selected cases that can serve as examples that help to identify mechanisms and patterns of evolutionary adaptation is a commendable goal for any biologist.

Do species comparisons just demonstrate that animals can live where they do?

The comparative approach necessarily shows that animals can live where they do. This self-answering question becomes more attractive when it is rephrased as: "What physiological mechanisms allow animals to live where and how they do? In its original form, the question trivializes the study of adaptation, and for that matter the study of evolution.

Does the approach based on species comparisons have adequate analytical paradigms?

This is a challenging question. Faced with an overwhelmingly complex universe, all scientists need paradigms. As understanding increases, the paradigms change. After all, one of the central factors in the success of science is that it is self-correcting.

I find paradigms more useful for communicating with other investigators than for guiding my own research. In this essay I have indicated several formats within which species comparison is an effective analytical tool. In a sense each of these formats is a paradigm, or at least it allows a paradigm to be stated if one is so inclined. In the hands of a resourceful investigator, these formats can be used effectively. However, it should be possible to improve them or to create new ones.

General questions, specific answers

We come now to a matter of scientific style, or perhaps more accurately, a matter of personal scientific taste. For me, looking for generality by documenting the breadth of occurrence of some physiological response in many species is not the most rewarding or productive way to increase scientific understanding of the ways in which organisms function (see section above, Specific questions, general answers). I find it more interesting to attempt to answer general questions about basic biological functions than to propose general answers to questions of physiological mechanisms. Students of organisms are especially favored by the nature of the entities they study.

Every species has evolved a specific set of solutions to general problems that all organisms must address. By the very fact of its existence a species demonstrates that it is able to carry out effectively a series of general functions. Most of these functions are thoroughly identified, and offer a framework within which one can integrate one's view of biology and ensure that one does not become lost in a morass of unstructured detail, even though the ways in which different species perform these functions may differ widely.

A few obvious examples will suffice. All successful animals must remain functionally integrated. All must obtain materials from their environments, and process and release energy from these materials. All must remain adequately hydrated. All must reproduce. All must differentiate and grow. By focusing questions on these obligatory and universal capacities, one can ensure that one's research will not be trivial and will have some chance of achieving general significance. As a physiological ecologist, I have tried to ask questions that are not more than a step or two away from these universals of organismic biology. What are the effects on rates of energy metabolism of size, of taxon, of ontogenetic state, of environmental temperature, of body temperature, of food availability, of activity? What are the effects of environmental factors on reproductive schedules? What determines patterns of water flux? How is body temperature controlled?

These questions can be addressed cogently to virtually any species. If patterns of physiological function at the level of the whole organism are to be found, questions such as these should reveal them. In fact, having directed these very general questions to a variety of species, I find that I have received mostly specific answers. Every species does things slightly differently. However, these specific and different answers are ordered by the fact that all are answers to a few general questions. Frequently, one can organize the specific answers into variations on the central theme of the general question. It helps to begin one's inquiries with broad general questions and then, after reconnoitering the situation, to pose more limited and more pointed questions that address critical elements of the general question. It is almost inevitable in organismic biology that one will come up with highly specific solutions to general biological problems, but that does not mean that one need lose sight of the patterns of adaptive response. Indeed, the questions, if appropriately selected and phrased, can help to identify patterns.

What, if any, are the advantages of working at the species level rather than at higher or lower levels?

Unfortunately, but understandably in the context of physiological comparisons, the word "species" is often quite loosely and uncritically interpreted. Ecological physiologists usually make measurements on individual organisms and then, from similar measurements on a number of individuals, make some inference about the average characteristics of the species to which someone else has assigned these organisms – a somewhat uncomfortable process when one stops to consider it. Sometimes the individuals that are measured are from different populations. Sometimes they are measured at one season, sometimes at another. They are frequently measured at different times of day. Often they have different histories of captivity.

It is important to emphasize that assigning functional characteristics to an entire species is a typological inference, just as is assigning common functional characteristics to taxonomic categories higher than species. The more remote the category from the breeding population, the more tenuous is the inference. It is not possible to make physiological measurements on a taxonomic category, only on some of the individuals in it, and occasionally on groups of animals included within it. Aside from feasibility, which is often operationally paramount, natural history is the best guide to the particular biological entities most appropriate for measurement. Knowledge about the natural history of a species enables the proposition of meaningful questions and also suggests the component of a population on which to make measurements.

Individual organisms are not necessarily the units of choice. For breeding birds exposed to heat or cold, the critical unit may be the brood, plus parent, plus nest. For small mammals in the cold, it may be a huddled group of adults. Under some circumstances for ants, bees, and wasps, it may be an entire colony. From measurements on each of these groups, one can make inferences at the species level which are as valid physiologically and sometimes more valid ecologically than inferences based on measurements of individual organisms. Obviously for some situations the species is not an adequate functional frame of reference; consider a coral reef, for example. Sometimes when one makes a measurement on what is ostensibly a single individual, one is actually measuring the performance of a complex assembly; consider the rate of heat production of a ruminant or a rabbit. How much of the heat comes from the tissues of the mammal? How much, from the microorganisms in its gut?

When one finds that two or more species differ with regard to some physiological function, the measurements usually record differences that evolved at some past time and in response to past selective pressures. Although one can sometimes give one's comparative data a limited amount of contemporary relevance by attempting to identify selective factors that are presently contributing to the maintenance of differences between subspecies or breeding populations, by and large when one compares species one is making historical interpretations.

Differences between individuals are the raw materials for evolutionary change and for the evolution of adaptations, yet of course most physiologists treat these differences as noise that is to be filtered out. From the standpoint of physiological ecology, the traditional emphasis of physiologists on central tendencies rather than on variance has some unhappy consequences. Variation is not just noise; it is also the stuff of evolution and a central attribute of living systems (Chapter 7). The physiological differences between individuals in the same species or population, and also the patterns of variation in different groups, must not be ignored (Bucher and Bartholomew, 1984). We should not allow the traditions of physiological measurement to cause us to overlook the obvious fact (Mayr, 1982, p. 47) that differences between individuals can be real, and that the mean values for the same characteristics in several individuals are only inferences about a population.

Should the comparative approach be modified?

Of course the comparative approach should be modified, but it should not be modified by attempting to force it into a mold shaped by the traditions of other kinds of biological investigation. The traditional approaches of any discipline are, after all, anthropogenic. Few things are more sterilizing than allowing one's research approach to be determined by the traditions of the field rather than the nature of the system being studied. As knowledge increases, one's perception of the natural world changes, and traditional approaches lose their effectiveness. It is from this changing perception rather than from the traditions of the field that changes must come.

Species comparison has been one of the useful approaches available to the student of physiological ecology. It clearly has limitations, but it also has virtues. It does not focus on techniques, or on intellectual fads. It deals with whole organisms existing in breeding populations and recognizable as species, and these are fundamental components both of the biological world and of the subject matter of ecology. However, as pointed out in the previous section, species are not always the most appropriate unit for biological comparison.

As long as there are students of organisms, they will make comparisons, because comparing is an essential element in establishing relations between phenomena and thus in escaping the chaos of unstructured detail. At this stage in the development of physiological ecology, efforts should be made to deal critically with intraspecific comparisons. Rather than just comparing different species, one should adopt some of the formats developed for interspecific comparisons in an attempt to compare breeding populations within species (Chapter 5) and individuals within the same breeding population (Chapter 7). Such efforts should allow investigators to approach more closely the dynamics of evolutionary change, and, in appropriate situations, to integrate the findings of physiological ecology with those of population genetics.

References

Bartholomew, G. A. (1964) The roles of physiology and behaviour in the maintenance of homeostasis in the desert environment. In *Homeostasis and Feedback Mechanisms*, ed. G. M. Hughes, pp. 7–29. *Symp. Soc. Exp. Biol.,* vol. 18 Cambridge University Press.

Bartholomew, G. A. (1966) Interaction of physiology and behavior under natural conditions. In *The Galapagos*, ed. R. I. Bowman, pp. 39–45. Berkeley and Los Angeles: University of California Press.

Bartholomew, G. A. (1970) A model for the evolution of pinniped polygyny. *Evolution* 24:546–559.

Bartholomew, G. A. (1981) A matter of size: an examination of endothermy in insects and terrestrial vertebrates. In *Insect Thermoregulation*, ed. B. Heinrich, pp. 1–78. New York: Wiley.

Bartholomew, G. A. (1982) Physiological control of body temperature. In *Biology of the Reptilia*, ed. C. Gans and F. H. Pough, pp. 167–211. New York: Academic Press.

Bartholomew, G. A., and Casey, T. M. (1977) Endothermy during terrestrial activity in large beetles. *Science* 195:882–883.

Bartholomew, G. A., and Casey, T. M. (1978) Oxygen consumption of moths during rest, pre-flight warm-up, and flight in relation to body size and wing morphology. *J. Exp. Biol.* 6:11–25.

Bartholomew, G. A., and Heinrich, B. (1973) A field study of flight temperatures in moths in relation to body weight and wing loading. *J. Exp. Biol.* 58:123–135.

Bartholomew, G. A., and Heinrich, B. (1978) Endothermy in African dung beetles during flight, ball making, and ball rolling. *J. Exp. Biol.* 73:65–83.

Bartholomew, G. A., and Lighton, J. R. B. (1986) Oxygen consumption during hover-feeding in free-ranging Anna hummingbirds. *J. Exp. Biol.* 123:191–199.

Bartholomew, G. A., and Tucker, V. A. (1963) Control of changes of body temperature, metabolism, and circulation by the agamid lizard, *Amphibolurus barbatus. Physiol. Zool.* 36:199–218.

Bartholomew, G. A., and Wilke, F. (1956) Body temperature of the northern fur seal, *Callorhinus ursinus. J. Mammal.* 37:327–337.

Beatley, J. D. (1969) Dependence of desert rodents on winter annuals and precipitation. *Ecology* 50:721–724.

Bucher, T. L., and Bartholomew, G. A. (1984) Analysis of variation of gas exchange, growth patterns and energy utilization in a parrot and other avian embryos. In *Respiration and Metabolism in Embryonic Vertebrates*, ed. R. S. Seymour, pp. 359–372. The Hague: Dr W. Junk.

Cowles, R. B., and Bogert, C. M. (1944) A preliminary study of the thermal requirements of desert reptiles. *Bull. Am. Mus. Nat. Hist.* 83:265–296.

Dobzhansky, T. (1937) *Genetics and the Origin of Species.* New York: Columbia University Press.

Heinrich, B. (1981) *Insect Thermoregulation.* New York: Wiley.

Heinrich, B., and Mommsen, T. P. (1985) Flight of winter moths at 0 °C. *Science* 228:177–179.

Huxley, J. S., ed. (1940). *The New Systematics.* Oxford: Oxford University Press (Clarendon Press).
Huxley, J. S. (1942) *Evolution, the Modern Synthesis.* London: Allen and Unwin.
Kenagy, G. J., and Bartholomew, G. A. (1985) Seasonal reproductive patterns in five coexisting California desert rodent species. *Ecol. Monogr.* 55:371–397.
Mayr, E. (1942) *Systematics and the Origin of Species.* New York: Columbia University Press.
Mayr, E. (1982) *The Growth of Biological Thought.* Cambridge, Mass.: Harvard University Press.
Morgan, K. R. (1987) Temperature regulation, energy metabolism, and mate-searching in rain beetles (*Pleocoma* spp.), winter-active endothermic scarabs (Coleoptera). *J. Exp. Biol.* 128:107–122.
Morgan, K. R., and Bartholomew, G. A. (1982) Homeothermic response to reduced ambient temperature in a scarab beetle. *Science* 216:1409–1410.
Peterson, R. S., and Bartholomew, G. A. (1967) *The Natural History and Behavior of the California Sea Lion.* Special Publ. no. 1, American Society of Mammalogy.
Reichman, O. J., and van de Graaff, K. M. (1975) Association between ingestion of green vegetation and desert rodent reproduction. *J. Mammal.* 56:503–506.
Simpson, G. G. (1944) *Tempo and Mode in Evolution.* New York: Columbia University Press.
Trillmich, F., and Mohren, W. (1981) Effects of the lunar cycle on the Galapagos fur seal, *Arctocephalus galapagoensis. Oecologia* 48:85–92.

Discussion

FEDER: You mentioned something about "ground squirrelness" that lets these animals inhabit all sorts of environments, which is unusual among animals. This seems relevant to the concept of a "key adaptation," advanced by Karel Liem [Harvard University]. In certain groups, one finds special characters that provide entrée to a new adaptive zone, a new way of doing things that permits a group to establish itself in a wide range of environments. Liem's specific example was the evolution of pharyngeal jaws in cichlid fishes, which enabled them to speciate widely. George Lauder's criticism of key adaptations [cited in Chapter 3] is that such adaptations arise typically only once in a given line. There is no rigorous way to test whether the expansion of a group such as ground squirrels is due to the evolution of this presumed key adaptation, or if the key adaptation is a chance correlate. Can you see any way to distinguish between these alternatives rigorously?

BARTHOLOMEW: Frankly, you ask more of physiological data than I have been inclined to do during most of my life. I am willing to ask more, but I am a little more modest about the quality of the inference that I can draw. I was impressed by the fact that whether you see ground squirrels in the Kalahari, or in the Arctic tundra, or in the dry savannahs of southern Mexico, they really represent a coherent adaptive type. My approach to identifying the features of this adaptive type has not been particularly complicated. One thing that ground squirrels, with the exception of one subgenus, seem to have in common is a temporal scheduling of their energy expenditures, which permits them to adjust their annual cycle so that they are active only at the appropriate times of year. To do that requires ground squirrels to remain hypothermic for a long period. It seemed to me that this must mean a central adjustment.

FLORANT: Are behavioral changes one way to overcome physiological problems?

BARTHOLOMEW: They are probably a paramount way, certainly the first way. I can recount an anecdote that really impressed me. I went to central Africa specifically to study the renal function of flamingos, which filter unicellular algae out of highly saline lakes. It was soon clear that this was no problem. To be sure, the flamingos did filter unicellular algae out of hypersaline lakes, but every few hours or so they would fly to a fresh water spring some place and flush themselves out. Consequently, electrolyte burdens were not persistent. I never published these observations. In summary, I would say that, surely, the first place to look is always behavior; it is quick, flexible, and a long-term trend that has certainly existed in vertebrates and presumably in some other groups that have a rapidly responding nervous system in which learning or imprinting can occur.

FLORANT: This brings to mind your and Dawson's work on water birds.

BARTHOLOMEW: Yes, that work in the extremely hot Gulf of California is a useful example. There were pelicans, gulls, and blue herons, totally different animals, physiologically at least. Yet they were nesting within meters of each other in exactly the same environment, which was extremely demanding. They did it by differences in behavior, but more than their nesting behavior differed.

POWERS: Do you have any general feeling about what fraction of these behavioral responses you see in these different adaptations are a function of learning versus genetic background?

BARTHOLOMEW: I have pondered that endlessly, but I am unable to handle it. The physiologist, or the behaviorist for that matter, studies the phenotypic expression. Consequently we seldom can specify the genetic underpinning, or whether this underpinning differs for two similar phenotypes.

POWERS: In at least some of these cases, culturing the animals involved and determining whether key behavior patterns are inherited or learned might be useful. We have done a little of this, and we find that some things are inherited and some are learned.

BARTHOLOMEW: The numbers of things that are learned continuously surprise me. When I started out, I assumed that most of these things were hard-wired, in the sense of the circuitry. The more animals I studied the less impressed I was by hard wiring, and now I think perhaps that the burden on the investigator is the other way, for taking learning into account.

BENNETT: There is an old evolutionary saw that an animal will respond first with behavior, because it is more plastic than physiology, which comes next, then morphology. And, ignoring the difficulty in stipulating whether a certain activity is behavior, physiology, or morphology, do you think that kind of progression has validity?

BARTHOLOMEW: On the basis of my work with vertebrates, I assume that the animal's first level of response to any environmental medium is apt to be behavioral, and that this imparts a capacity allowing use of their morphology and physiology in a variety of ways in the short haul. Certainly behavioral plasticity is striking among vertebrates.

FUTUYMA: Perhaps a clarification is in order regarding Bennett's question. The evolutionary saw, as I understand it, suggests that, given an environmental change, an essentially immediate behavioral response of individual organisms may occur without any genetic change. This is soon followed by physiological acclimation. Subsequently, genetic changes affecting morphology, behavior, or whatever, occur. I was not sure if Bennett was asking whether the genetic changes in the behavior – that is, the evolution of the behavior – would precede evolution of physiological or morphological traits. Is that what you meant? I am not sure why behavior should evolve any faster than morphology.

BARTHOLOMEW: My interpretation was based on the sequence that you outlined.

BENNETT: I think that previous presentations have been unclear concerning whether this sequence is supposed to involve phenotypic or evolutionary responses. It is almost certainly true phenotypically, but it may also be that behavior evolves faster than physiology or morphology.

3 The analysis of physiological diversity: the prospects for pattern documentation and general questions in ecological physiology

MARTIN E. FEDER

Introduction

Ecological physiology historically has emphasized the demonstration of pattern rather than the testing of hypotheses. Initially these demonstrations focused on extreme environments (e.g., patterns of physiological adaptation in deserts, high elevations, and cold environments), major taxa (e.g., responses to cold in insects vs. fishes vs. mammals), and the limits of physiological performance. More recently, a focus on less extreme environments and on animals undergoing routine behaviors has burgeoned alongside the initial foci. As outlined in the preceding chapters, the collective elaboration of pattern in ecophysiological attributes has been both a productive and a scientifically successful enterprise in its own right. Moreover, it has established a firm foundation for the effective proposition and testing of general hypotheses.

The discussion of goals for ecological physiology accordingly has usually focused on whether patterns are sufficiently documented rather than what major questions should be answered (but see Prosser, 1975, 1986b). Clearly, the documentation of ecophysiological pattern is still incomplete. Many unexamined species and populations remain to be reconciled with already recognized patterns, and perhaps novel patterns remain to be recognized. The future demonstration of pattern in ecophysiology, however, is justifiable only if the scientific advances it promises are commensurate with the effort expended in the process. The purpose of the first part of this essay is to consider whether the documentation of pattern in ecological physiology has reached the point of diminishing returns, and whether further case studies of physiological adaptation to the environment will really more firmly establish the conclusions upon which these patterns bear. My analysis suggests that the further elaboration of pattern in ecophysiological attributes is not a sufficient agenda for the future, although it will continue as a natural consequence of any ecophysiological investigation. Therefore, the field should also emphasize the proposition and solution of general questions, some of which are outlined in the second part of this essay.

The status of pattern documentation

What questions have historically justified the demonstration of pattern in ecophysiological attributes? Will elaboration of these patterns; study of additional attributes, species, and environments; or the discovery of novel patterns yield more robust answers to these questions?

1. *Do physiological characteristics vary among organisms, or are physiological characteristics similar in different organisms?*
2. *Is the variation in physiological attributes random, or does it show pattern?*

A first important documentation is that variation in physiological attributes is near universal. Ecological physiologists have recognized that many (if not all) physiological functions (e.g., metabolic rate, enzyme-substrate affinity, lethal temperature) vary among populations of a species, species of a genus, genera of a family, and so on (e.g., Table 3.1) (Prosser, 1955, 1973, 1986a), although physiological variation within populations is less well characterized (Prosser, 1955; Chapter 7). A second important documentation is that variation in physiological attributes is not random, but evidences pattern, structure, or regularity (e.g., see Figure 3.1). We do not see all possible combinations of physiological characteristics or unbounded variation in physiological attributes.

These findings are now second nature to ecological physiologists. That these conclusions are taken for granted does not minimize their significance. Diversity in physiological attributes is not a necessary outcome of evolution. It could be, as is somewhat more the case at the cellular level, that all organisms are similar in physiological attributes, with interindividual, interpopulational, and interspecific differences manifested solely in terms of gross behavior, population structure, and so forth. It could be, just as all animals contain carbon and water, that the inner workings of all animals are the same as in the laboratory rat. The finding of physiological diversity tells us something nonobvious but fundamental about the nature of life.

Recognition of pattern in physiological diversity is significant for several reasons. First, it is a necessary condition for the emergence of a "nomothetic" ecological physiology, that is, a field concerned with the lawlike properties reflected in repeated events, as opposed to the idiographic description of unique unrepeated events (Gould, 1980c). Second, the occurrence of pattern makes the field tractable. One need not study every species or higher taxon individually to understand its ecophysiological attributes; a smaller number of carefully chosen exemplars will suffice. Third, the recognition of pattern enables distinction between those explanations with which pattern is consistent and those explanations of which pattern is exclusive. At least in

TABLE 3.1 An example of variation in a physiological characteristic, the hydrogen ion concentration in the blood

Blood $[H^+]$ (nmol/L)[a]	Species	Reference[b]
6.03	*Salmo* (rainbow trout)	75
6.76	*Salmo* (rainbow trout)	75
8.13	*Rana* (bullfrog)	46
8.13	*Cryptobranchus* (salamander)	[c]
8.32	*Rana* (bullfrog)	46
8.51	*Carcinus* (crab)	48
9.12	*Salmo* (rainbow trout)	75
9.33	*Callinectes* (crab)	48
9.55	*Chelydra* (snapping turtle)	46
10.23	*Bufo* (toad)	46
10.96	*Rana* (bullfrog)	46
11.22	*Salmo* (rainbow trout)	75
12.02	*Pseudemys* (red-eared turtle)	60
12.30	*Scyliorhinus* (dogfish shark)	41
12.59	*Cryptobranchus* (salamander)	[c]
12.88	*Uca* (fiddler crab)	48
13.49	*Rana* (bullfrog)	46
13.80	*Carcinus* (crab)	48
13.80	*Scyliorhinus* (dogfish shark)	41
14.45	*Bufo* (toad)	46
15.14	*Pseudemys* (red-eared turtle)	60
15.14	*Salmo* (rainbow trout)	75
15.85	*Callinectes* (crab)	48
16.22	*Carcinus* (crab)	48
16.22	*Scyliorhinus* (dogfish shark)	41
16.98	*Carcinus* (crab)	48
16.98	*Callinectes* (crab)	48
17.38	*Chelydra* (snapping turtle)	46
17.38	*Uca* (fiddler crab)	48
17.78	*Rana* (bullfrog)	46
17.78	*Cryptobranchus* (salamander)	[c]
20.42	*Bufo* (toad)	46
23.99	*Pseudemys* (red-eared turtle)	60
25.70	*Pseudemys* (red-eared turtle)	60
26.92	*Bufo* (toad)	46
28.84	*Chelydra* (snapping turtle)	46
31.62	*Pseudemys* (red-eared turtle)	60
31.62	*Gallus* (chicken)	46
32.36	*Uca* (fiddler crab)	48
35.48	*Pseudemys* (red-eared turtle)	60
39.81	*Homo* (humans)	47

[a]As do most physiological characteristics, the hydrogen ion concentration in the blood varies considerably. The data are arranged from the lowest to the highest concentrations of hydrogen ions, without respect to phylogenetic affinity. Variation is evident among individuals of the same species, among species, and among higher taxa. Overall, variation is approximately sixfold.

[b]References correspond to reference numbers in Reeves (1977).

[c]Moalli, Meyers, Ultsch, and Jackson (1981).

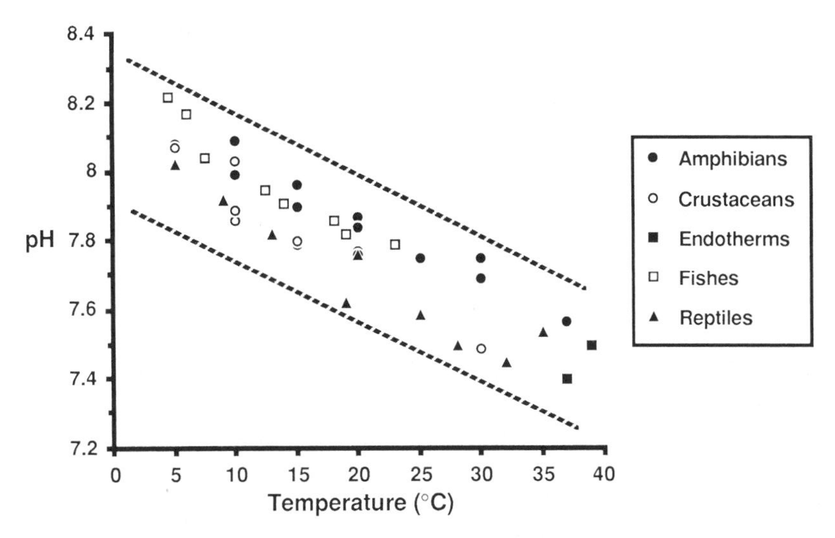

FIGURE 3.1 An example of pattern in physiological diversity. When the data of Table 3.1 are expressed as pH and then plotted as a function of the body temperature at which the measurements were made, a regular pattern emerges: pH falls as body temperature increases. This pattern is evident in numerous phylogenetic groups. The similarity between the slope of this relationship and the slope of the relationship between the pN of water (i.e., the pH at which pH = pOH) and temperature suggests that animals regulate their blood pH at a constant "relative alkalinity" (Rahn, 1967), perhaps to maintain a constant protein net charge state (Reeves, 1977). One insight to emerge from this analysis is that ectotherms do not lack the ability to regulate blood pH (as was previously thought), but are able to regulate blood pH at whatever level their current body temperature dictates.

theory, we have a means by which to determine the "general gas laws" by which species are governed (Schopf, 1979). Finally, the patterns that have been recognized point out several important general questions, which are discussed in the latter portion of this chapter.

Having granted that physiological diversity and pattern in physiological diversity are significant findings that have justified ecophysiological analysis in the past, is either sufficiently inconclusive to merit more substantiation? With the exception of variation within populations (Chapter 7), the answer is obviously no.

3. *Is the pattern of physiological diversity consistent with the effects of natural selection?*

A third (if not the central) contribution of ecological physiology is its relating variation in physiological characters within populations, species, or higher taxonomic levels, to the environments in which organisms live. In a synthetic sense, what is being demonstrated is that the physiology of organisms is in "equilibrium" (Lewontin, 1969; Eldredge and Gould, 1972) with their environment.[1] The physiological characteristics of individuals in a population are adequate to support maintenance, growth, and reproduction in the particular environment in which the population occurs. Thus, as environments vary from place to place, the physiological characteristics of their inhabitants vary correspondingly (e.g., Figures 3.2 and 3.3). As Dobzhansky (1951) put it: "The enormous diversity of organisms may be envisaged as correlated with the immense variety of environments and of ecological niches which exist on earth."

The demonstration of equilibrium between physiological and environmental variables has long held the highest of priorities for ecological physiologists; indeed, the field has invested much of its energies into assembling the best possible case for equilibrium. The resultant advances have been conceptual, technical, and empirical. We have, for example, realized that physiology-environment correlations should be sought at the molecular and cellular levels as well as at higher levels (Prosser, 1986a; Chapter 5), that "behavior" and "morphology" should be considered coequal with "physiology" in our analyses (Bartholomew, 1958; Gans, 1986; Chapters 1 and 2), that microclimate may be more meaningful than gross climate in characterizing physiology-environment correlations (Bartholomew, 1958), that organisms from extreme environments may exhibit very obvious physiology-environment correlations (Chapter 2), that function of a part in the context of a whole organism may yield different insights than function of a part in isolation in an experimental preparation (Chapter 1 in Gans, 1974; Huey and Stevenson, 1979), and so on. The issue at hand, however, is whether additional documentation of "equilibrium" will advance the field conceptually. First, let us consider the conceptual advances that documentation of equilibrium has afforded.

Gould (1980a) has characterized the seminal contribution of George Gaylord Simpson to the biological sciences as the demonstration that the fossil record is consistent with the major features of evolutionary theory. Lest this seem trivial, imagine if Simpson had found that the fossil record were inconsistent with evolutionary theory! In much the same sense, the documentation of equilibrium between physiological attributes and environmental variables is of fundamental importance because it is consistent with how Darwinian natural selection ought to work: natural selection, acting in diverse local environments, is expected to result in changes in physiological variables that enhance the Darwinian fitness of each physiological variant in the particular environment in which it finds itself. In evolutionary time, the cumulative

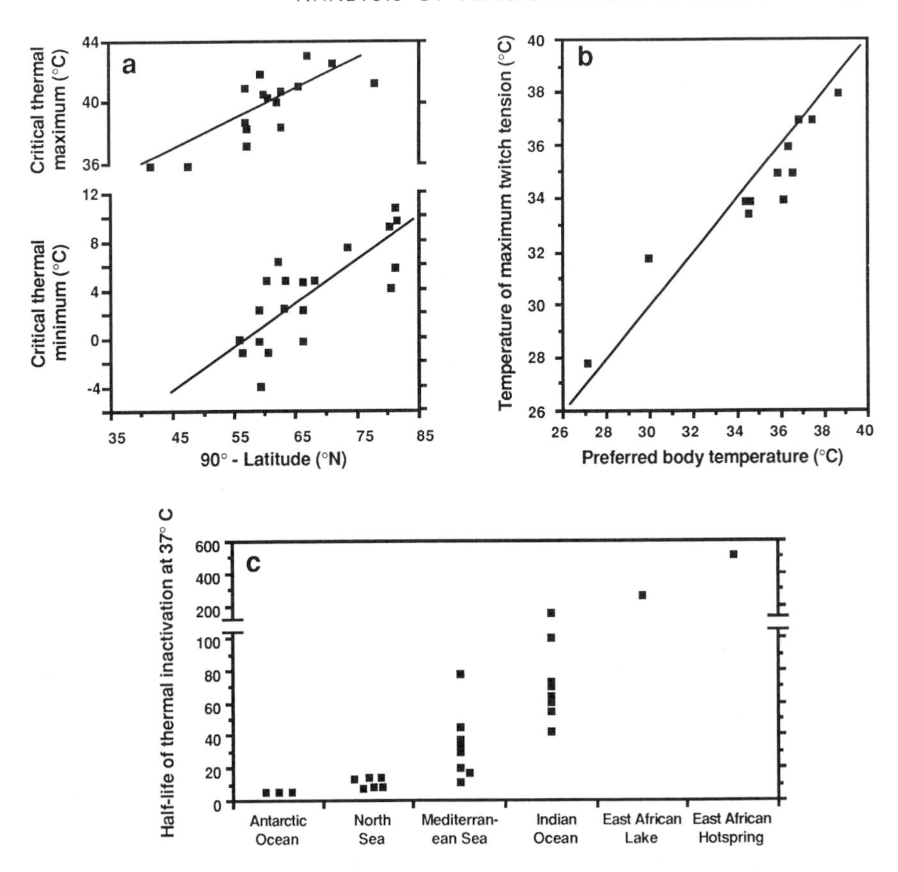

FIGURE 3.2 Three examples of "equilibrium" between different physiological characteristics of species and an environmental variable, temperature. (a) Indices of thermal tolerance, the critical thermal maximum and the critical thermal minimum, are correlated with latitude in amphibian species. Tropical species are able to tolerate warmer temperatures than are temperate species, whereas temperate species are able to tolerate cooler temperatures than are tropical species. (From Snyder and Weathers, 1975.) (b) The temperature at which skeletal muscle develops maximum isometric twitch tension is positively correlated with the preferred body temperature in lizard species. (From Licht, Dawson, and Shoemaker, 1969.) (c) The time required for exposure to a warm temperature (37 °C) to inactivate myofibrillar ATPase is positively correlated with the thermal environments of fish species. (From Johnston and Walesby, 1977.)

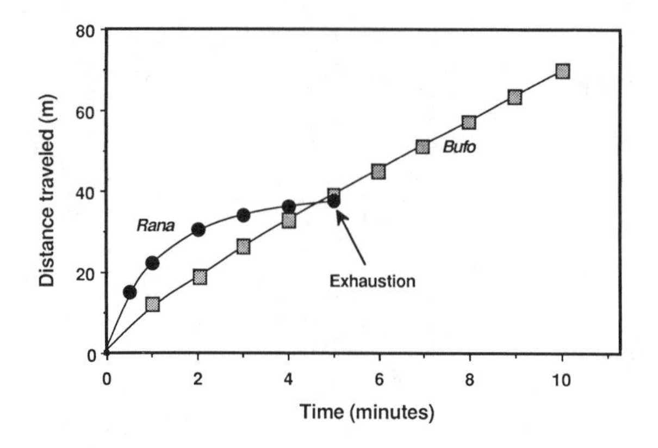

	Bufo [Toad]	*Rana* [Frog]	Reference
ATP generated during intense activity [μmoles ATP/(g mass • min)]			
Aerobic	3.7	1.8	b
Anaerobic	4.0	13.7	b
Aerobic [% of total]	48%	12%	
Blood oxygen capactiy (Vol %)	15.0	10.5	c
Blood volume (% of body mass)	15.7	7.0	c
Ventricle mass (% of body mass)	0.311	0.118	c
Enzyme activity [μmoles product/(g muscle•min)]			
Citrate synthase	20	7	d
Lactate dehydrogenase	75	108	d
Phosphofructokinase	0.69	1.21	a
Contractile properties of skeletal muscle			
Contraction time (msec)	71	54	d
Maximum rate of tension rise [kN/(m^2•msec)]	3.13	4.25	d

FIGURE 3.3 An example of "equilibrium" between the physiological characteristics of species and the ways in which they exploit their environment. Toads (*Bufo*) respond to threat or stimulation with relatively slow and steady movement or static defense; frogs (*Rana*) respond with intense activity, including powerful leaps (Bennett, 1974). Toads may forage widely for mates or prey, traveling up to 40 m in an hour (Wells and Taigen, 1984); frogs are more sedentary, often sitting and waiting for food to come to them. This difference is reflected in laboratory measurements of locomotor performance (shown in graph from Putnam and Bennett, 1981): *Rana* exhaust in five minutes whereas *Bufo* do not fatigue. Physiological characteristics that favor sustained activity and aerobic metabolism have greater values in *Bufo* than in *Rana*. Physiological characteristics that favor rapid or intense activity have greater values in *Rana* than in *Bufo*. [Sources: a, Bennett (1974); b, Bennett (1980); c, Hillman (1976); d, Putnam and Bennett (1983).]

result of this natural selection should be equilibrium between physiological and environmental variables. Insofar as the major empirical finding of ecological physiology is not inconsistent with this prediction, ecological physiology "supports" natural selection. Indeed, the phrases "not inconsistent" and "supports" are too weak, given the variety and intricacy of cases in which "physiology" and "environment" are correlated.

Consistency of ecophysiological patterns with Darwinian theory was not always as obvious as it seems today. The New Synthesis, which explicitly set forth the ways in which the biology of populations could be used to examine the predictions of Darwinian theory, emerged in the 1930's and 1940's. Most of the now-senior practitioners of ecological physiology were at relatively early stages of their professional careers at this time (Chapter 2). For example, George Bartholomew (Chapter 2) recollects discussion of the then-new New Synthesis while an undergraduate, a time at which the concept of a cline was novel and geographic variation in physiological characteristics relatively unexplored. Importantly, the extent to which analyses of physiological diversity could shed light on evolutionary mechanisms, and vice versa, was unknown. Thus, at the time a general exploration of ecophysiological variation represented a significant and fundamental expansion of knowledge with important implications for both physiology and evolutionary biology. Moreover, a physiological approach to adaptation appeared to offer special promise in that the functional consequences associated with variation in organismal characteristics could be observed directly. The question of consistency with natural selection was thus a more than adequate justification for past documentations of "equilibrium." Is it an adequate justification for the future?

Ecological physiologists have assembled an enormous number of case studies that establish "equilibrium" beyond any reasonable doubt. Bartholomew (Chapter 2) has outlined some of the variants of this pattern: related species that achieve different physiological solutions in dissimilar environments, unrelated species that achieve common physiological solutions to similar problems or in similar environments, and so on. This is a significant achievement, but also a significant problem: the field is a victim of its own success. It can continue to document physiology-environment correlations with as yet unexamined variables or in as yet unexamined species, but further studies will not appreciably augment the overwhelming mass of evidence already assembled to demonstrate "equilibrium." We know in advance what the outcome of additional studies will be: some feature of organismal function will be correlated with some environmental variable. If the documentation of equilibrium is a foregone conclusion, it is time to emphasize other issues.

4. *Does variation in physiological characteristics provide useful insights into the phylogenetic relationships of animals?*

Ecological physiology, it is sometimes claimed, may offer us key insights into the phylogenetic relationships of organisms by furnishing useful systematic characters (Ross, 1981). Ross (1981) questions this claim because of the high frequency of convergence evident in ecophysiological attributes. Similar variation in physiological characters in distinct taxa (i.e., convergence) will yield the appearance of phylogenetic affinity where none exists. Clearly, all physiological characters do not show convergence, and therefore could be of use in constructing phylogenies. Difficulties of convergence aside, the literature of systematics suggests that physiological characters have seldom, if ever, been of real value in resolving phylogenies; indeed, most phylogenies are now so well established that even contradictory physiological characters are unlikely to prompt their revision (Ross, 1981). Thus, the likely provision of systematic insight is an insufficient justification for the continued documentation of ecophysiological patterns in the future.

5. Does the analysis of physiological diversity disclose general principles of adaptation?

Natural selection, acting upon unrelated organisms facing similar environmental challenges, sometimes has resulted in similar modifications of physiological characteristics. This modification is not random or unstructured, but itself follows certain rules or patterns that may be defined as "general principles" of ecological physiology. For example, one nonuniversal but recurrent feature of organisms is countercurrent exchange; another is the pH-temperature relationship depicted in Figure 3.1; a third is the energetic consequences of hovering flight (Chapter 2). A final justification for the continued documentation of ecophysiological patterns is that it will disclose such "general principles" of physiological adaptation, which are held to have inherent value as unique and important contributions to the body of scientific knowledge.

A major difficulty of general principles of ecophysiology is that their definition is so inclusive. The minimum conditions that must be satisfied to establish a general principle are two fold: the physiological attribute it concerns (1) must have been derived independently in at least two lineages; and (2) must be recognizable as shared by these lineages. Thus, it is difficult to name a physiological attribute that does not reflect or define a general principle. Unlike mechanics, in which a small number of laws suffice to describe diverse phenomena, the general principles of ecological physiology are as numerous as instances of convergence in ecophysiological traits. Although describing all "general principles of ecological physiology" may provide considerable grist for the mills of more mechanistic physiologists, it is a never-ending process with a probable low yield of conceptual advances for our understanding of the process of physiological adaptation and its consequences.

Conclusion

Darwin (quoted in Medawar, 1969) belittled unbounded inductivism by likening it to going into a gravel pit, counting the pebbles, and describing their colors. Although Darwin's point is well taken, it is also true that wholesale inspection of pebbles will sometimes reveal diamonds or gold nuggets. In the final analysis, the effort expended and the likelihood of reward will jointly determine whether the inspection of pebbles is a worthwhile activity; consulting a geologist beforehand may increase the likelihood of reward and is therefore justifiable. Ecological physiologists need to consult one another as to whether the likelihood of future scientific reward justifies the continued description of adaptive patterns, which strategies bring large rewards most rapidly, and how limited resources should be apportioned among alternative investigative strategies.

Opinions are mixed as to whether the mine of ecophysiological patterns is played out and whether mining is a rewarding industry. At one pole, Carl Gans (1978) has argued that "all animals are interesting," that in the hands of a competent scientist even the most mundane of species can reveal important insights, and that therefore any unexamined species is an appropriate subject for investigation. At the other, Robert Platt (1964) has held that a field advances most quickly by the structured proposition and solution of general questions that transcend particular cases, and that studies not addressing such questions directly are wasted effort. While he agrees that the field should focus on general questions, George Bartholomew (1982, 1986, personal communication) feels that general answers are seldom forthcoming and that the best that can be expected is "a variety of different and highly specific answers to any given general biological question" (1986, p. 328). Each of these viewpoints is meritorious. Nonetheless, the research traditions of ecological physiology are such that Platt's approach has been relatively unexploited, which is why I wish to advocate it here.

Traditional pattern documentation, which has made numerous valuable contributions to our understanding of life, may be approaching the point of diminishing returns because the objectives that justify its continuance either have been met or cannot be met. By contrast, unsolved general questions abound concerning the nature of adaptation. A focus on these general questions may do much to invigorate the field. To this end, I have attempted to identify some questions from evolutionary organismal biology that may have particular relevance for ecological physiology and may serve as a complement to the elaboration of ecophysiological pattern. These questions uniformly predate modern ecological physiology in general and this essay in particular; Simpson (1953), for example, outlined many of the same questions, as did Darwin before him. Ecological physiology as a field has seldom addressed these questions directly, however. Because ecological physiologists already have immense knowledge of ecophysiological diversity and are able to quantify organismal function, they may be in a unique position to make substantive contributions to the solution of these general questions.

General issues for the future

The physiological systems that ecological physiologists study have numerous components and intricate regulatory mechanisms. All components and controls must work together adequately for organisms to grow and reproduce. Complexity is the first of three interrelated issues embodied in the questions that follow. If general conclusions in ecological physiology are to be realistic, they must account for complexity and its evolution. The challenge, for example, is not only to explain how a particular enzyme facilitates speed in a particular species, but to explain how natural processes have yielded a coadapted complex of traits involving multiple systems, which as a group facilitate locomotion. Such processes obviously involve the transformation of numerous interacting physiological components from a primitive to a derived state. Can we realistically endorse classical microevolutionary processes as a sufficient explanation of physiology-environment equilibrium (Eldredge, 1985)?[2] Can a reductionist approach suffice to explain the evolution of complexity?

Insofar as "physiological adaptations" are complex and involve interacting systems, a second general issue is the constraint of each interacting component on the form and function of all others. Constraints can take several forms: acute conflicts between interacting components of physiological systems (Chapter 15); mechanical or structural constraints (Gould and Lewontin, 1979); linkage of physiological characteristics to ongoing processes of growth, development, and reproduction (Chapters 10 and 16); and genetic linkage that may prevent the independent evolution of physiological traits (Chapter 9). As a result of such constraints, the physiological characteristics of animals may be more appropriate for ancestral environments than extant ones, or may be prevented from approaching "optimal" function (Chapter 4). A general question concerns the extent to which constraint limits equilibrium between physiological variables and environmental ones, or retards the rate of equilibration between "physiology" and "environment."

Ecological physiologists usually attribute physiological diversity to the equilibration of organism and environment. A major implication of constraint, of course, is that not all variation in physiological characteristics is due to adaptation to the environment; some is surely due to other factors, which in turn may have either strong or negligible links to the environment. This implication does not obviate ecological physiology, but necessitates the elevation of phylogeny, history, ontogeny, and size to full equality with environment as potential explanations for physiological diversity. We need to know how important each of these additional variables has been in the determination of physiological diversity before we can accurately attribute physiological variation to adaptation. What follows are several specific questions or issues that touch upon these general issues of complexity, constraint, and attribution.

Equilibrium: testing the key assumptions

The consistency of equilibrium with its usual explanation, adaptation to the environment via natural selection, implies only that the data do not preclude the explanation; that is, the explanation is a plausible one (Gould, 1980c). Consistency does not imply that the data require such explanation or that alternative explanations can be excluded, nor does it demonstrate that physiological diversity has arisen through natural selection or even that the necessary conditions have been met for the origin of physiological diversity through natural selection. Ecological physiologists know far more about the potential results of natural selection than about how (and if) it actually modifies physiological features. A reasonable starting point for addressing these concerns would be to examine the key assumptions upon which rest the implied evolution of physiological characteristics. The assumptions are as follows:

For evolution of physiological traits to occur by natural selection, physiological traits must vary among individuals within populations and this interindividual variation in physiology must be heritable. Are these conditions met?

Although we know much about physiological variation among species, higher taxa, and (to a lesser extent) populations within species (but see Chapter 5), we know comparatively little about physiological variation among individuals within populations (Chapter 7). A key assumption underlying the implied evolutionary origin of physiological diversity is unsubstantiated.

Through electrophoresis and other techniques, population geneticists have revealed an unexpected degree of diversity in protein and gene structure in natural populations. Why not simply assume that physiological traits are similarly diverse? Most of the traits discovered through modern biochemical techniques have not yet been assigned any particular function. Indeed, a whole school of evolutionary theory has grown up about the assumption that most enzyme polymorphisms are selectively neutral (Chapters 5 and 8). If many or most traits are selectively neutral, then the amount of variability in these traits will be related to stochastic factors (e.g., the mutation rate) and have little bearing on "equilibrium." We know that this is not always the case. For example, we know that alternative forms of hemoglobin profoundly affect organismal function (Chapter 5), and hence hemoglobin phenotypes are not as variable as they might otherwise be. We need to characterize physiological variation in enough populations to know whether in general it is large and random (which might indicate the absence of stabilizing selection), small (which might indicate stabilizing selection, limited genetic variability for physiological traits, "constraint," or some other explanation), or nonexistent (which would have profound implications for the origin of physiological diversity). Chapter 7 considers progress in this area.

Few would be surprised by findings of physiological diversity within populations, and the general description of this diversity may be just a matter of time. By contrast, the genetic basis (or, more specifically, the heritability) of the traits that ecological physiologists typically think about may prove more elusive. For obvious reasons, we know most about relatively simple traits encoded for by a small number of genes with a limited number of alleles; the further we depart from this condition, the less we understand. The traits ecological physiologists think about are, unfortunately, often complex. The expression of complex traits may represent the control of numerous genes (Chapters 5 and 8) or relatively few (e.g., Alberch, 1980,[2] and Chapter 9). The interaction of environmental factors with gene expression, especially during development, may play a major role in expression of complex traits (Alberch, 1980). The sorting out of these possibilities, and an understanding of the relative importance of each alternative means of expression of complex traits remain tasks for the future. To accomplish these tasks, we must know the mechanisms by which simple changes in DNA base-pair sequences are translated into complex organismal traits. Such knowledge exceeds the grasp of modern molecular biologists (at least for the moment), and yet we are faced with extrapolating this knowledge to the characters of interest to us as they appear in natural populations. Characterization of heritability, fortunately, requires "only" the breeding of the animals of interest or knowledge of their pedigrees. Some (but not enough) heritabilities have been determined for complex traits of interest to ecophysiologists in both wild and domestic species (Arnold, 1986). However, heritability is but a partial answer to a larger question.

Is variation in physiological characters related to variation in fitness?

Physiological variation must affect fitness for evolution to occur by natural selection. Given their focus on adaptation, ecological physiologists have assembled relatively few data bearing on whether variation in physiological traits within natural populations actually affects fitness (Arnold, 1983; see also Arnold's concluding remarks in Chapter 17). We typically do not know, for example, whether an animal with a greater than average metabolic rate or a lower than average cost of transport contributes more offspring to its population than an average conspecific.

The available data are mixed in their support of a close linkage between physiological variation and Darwinian fitness. Several exemplary efforts have demonstrated the fitness consequences of allelic variation at loci coding for metabolic enzymes (Koehn, 1984; Watt, Carter, and Donohue, 1986; see also Chapters 5 and 8). These studies are of special interest in light of arguments that much isozyme variation is selectively neutral (Chapter 5). However, the traits these studies examine are not the complex, multi-component "adaptations" or performance traits (e.g., salt glands, the fish gill, sprint speed, maximum oxygen consumption) of interest to many ecophysiologists. More studies have failed than have succeeded in demonstrating a relationship between variation in such complex physiological characters and fitness. For example, relationships are not evident between sprint speed and survivorship in juvenile lizards (Raymound Huey, personal communication); between metabolism, calling, and courtship success in toads (Wells and Taigen, 1984); or between either metabolism or locomotor performance and foraging success within a toad population (B. Michael Walton, personal communication).

Do no such relationships exist, or are our analytical procedures inadequate to resolve them? In his recent review of selection in natural populations, Endler (1986) cited direct demonstrations of selection for 314 different traits in 141 species. Thus, demonstration of relationships between variation in complex physiological traits of animals and their Darwinian fitness (if any relationships exist) is possible. It may, however, be very difficult. Endler (1986) enumerates ten classes of reasons for failing to detect selection when it is present, and seven classes of reasons for mistaken detection of selection when none is present in natural populations. This enumeration specifies two general problems for the demonstration of physiology-fitness relationships.

First, programmatic demonstrations may well require larger sample sizes, more extensive fieldwork, more regular sampling, longer-term studies, and greater background knowledge of natural history than most physiologists are willing to provide or granting agencies are willing to support. As Endler (1986, p. 98) has stated: "There are no shortcuts in demonstrating natural selection." Indeed, many of the investigations listed above may have failed to detect physiology-fitness relationships because of inadequate sampling. The stringent requirements for characterizing such relationships raises an important point: if the failure to demonstrate a relationship between variation in complex physiological traits and fitness can always be attributed to inadequate methodology as opposed to obvious alternative explanations (e.g., that of no relationship), then the search for such relationships becomes a self-fulfilling prophecy rather than a scientific enterprise.

Second, variation in the complex physiological trait of interest must be large in relation to the breadth of the physiology-fitness function (Figure 3.4). Many ecophysiologists, I think, infer from demonstrations of physiology-environment equilibrium that the physiological traits they study are under strong stabilizing selection such that any variation in these traits is tanta-

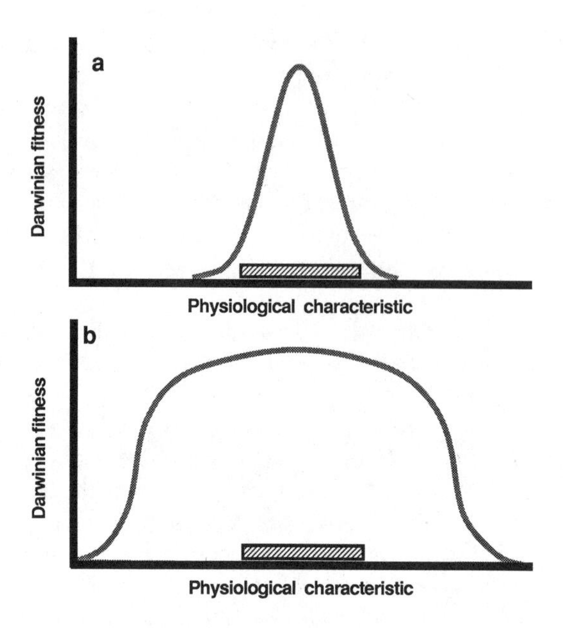

FIGURE 3.4 Two possible extreme forms of the relationship between physiological variation and fitness. For a given amount of variation in a physiological attribute (indicated by the horizontal bar), stabilizing selection would be strong in (a) ("peak" model) and weak in (b) ("plateau" model). An open question for ecological physiologists is how the actual relationships between physiological variation and fitness are distributed between the two extremes shown.

mount to a decrease in fitness (Figure 3.4a). Too high a hemoglobin oxygen affinity for a given elevation or too low a gill surface area for a given body size, for example, would compromise fitness according to this logic. If, however, interindividual variation is either minor or negligible with respect to the breadth of the physiology-fitness function (Figure 3.4b), detection of a relationship between physiological variation and fitness will be difficult or impossible. Because we have so little knowledge of the actual shape of the physiology-fitness relationship for complex traits, we cannot predict how serious this problem may be.

Given these problems, it may make sense for ecological physiologists to avail themselves of natural populations whose demography and natural history are already understood in detail (e.g., Clutton-Brock, Guinness, and Albon, 1982), and only then to perform physiological studies on characteristics of likely importance during actual differential mortality or reproduction in the field. In addition, ecological physiologists should be alert for unusual situations in which differential mortality or differential reproduction is clearly in evidence (e.g., Lande and Arnold, 1983). Such instances may not be representative ones, but they may be productive. Ecological physiology badly needs more studies of whether variations in physiology among individuals are correlated with fitness.

Do the physiological characteristics that promote Darwinian fitness of individuals within a population also promote the persistence of species through evolutionary time?

Ecophysiologists often ascribe the persistence of species and higher taxa in evolutionary time to the various physiological (and other) adaptations in the individual organisms that comprise the species. However, the nexus of recent interests in mass extinctions, macroevolution, punctuated equilibria, and clade selection suggests that features other than adaptations in the traditional ecophysiological sense may be important in determining the evolutionary persistence of species and higher taxa, as follows (Stanley, 1979; Gould, 1982; Vrba and Eldredge, 1984; Jablonski, 1986): the persistence of a taxon will be related to the difference between the speciation rate and the extinction rate. Characteristics that promote speciation include physiological (and other) adaptations in the traditional sense; but also include a population structure of small isolated populations, a low capacity for dispersal, extinction of major competitors, and heterogeneity of a taxon's habitat through evolutionary time. The latter two "characteristics" are functions of biotic, abiotic, geological, and stochastic processes unrelated to the taxon's adaptations. In addition to traditional adaptation, geographic range may affect extinction rate. All else being equal, widespread taxa are less likely to become extinct, apparently because the chance of some populations not being affected by a given extinction event is maximized (Figure 3.5). From study of Cretaceous marine bivalves and gastropods, Jablonski (1986) has suggested that during catastrophic mass extinctions, these rules change (p. 132):

> Many traits of individuals and species that had enhanced the survival and proliferation of species and clades during background times become ineffective during mass extinctions, and other traits that were not closely correlated with survivorship differences become influential.

In other words, the present array of physiological types may have little to do with the long-term adaptation of their ancestors to ancestral environments, and may reflect mainly that their ancestors by chance were at the right place at the right time during mass extinctions.

These suppositions have not been examined exhaustively and are strenuously disputed by some evolutionary biologists, in part because relevant data are difficult to assemble from the fossil record. Because of their familiarity with extreme environments (Chapter 2), ecological physiologists may be in a unique position to assess the relative importance of adaptedness in the tra-

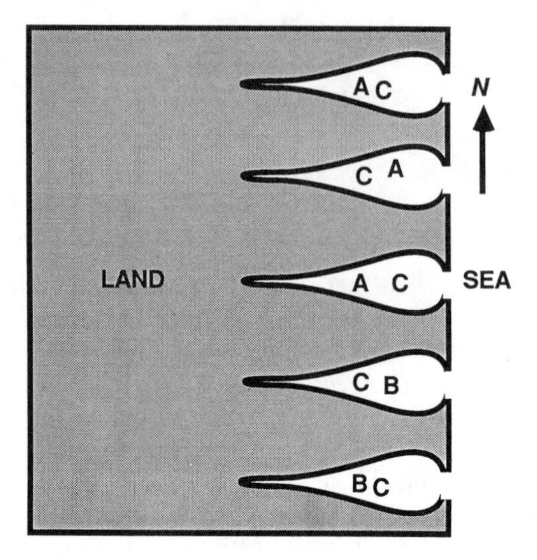

FIGURE 3.5 A hypothetical example of "species selection" by differential extinction rates. Suppose three species of flying insects whose aquatic larvae are intolerant of seawater are distributed in estuaries along a north-south transect as shown. Suppose occasional storm fronts moving from east to west inundate the estuaries with seawater. Storms affecting the southernmost two estuaries are likely to exterminate all populations of species B but only some populations of species C. Storms affecting the northernmost three estuaries are likely to exterminate all populations of species A but only some populations of species C. All else equal, species C is likely to persist longer than the other species through the years. The physiological attributes that engender the persistence of C are those promoting dispersal of adults, not those promoting salinity tolerance of larvae.

ditional sense versus the other characteristics enumerated above in determining species' persistence in the face of local extinctions. Are species with otherwise equivalent physiologies but different geographic ranges and population sizes differentially prone to extinction? To what extent are physiological features that promote dispersal or hinder gene flow key to a taxon's survivorship? Can one quantify extinction rates in extant populations by comparing the performance of species in replicates of transient environments (e.g., drying vernal ponds) or habitat islands (e.g., mountain tops or rare host plants) (e.g., see Smith, 1983)? The implications of these suppositions, if proven, for ecological physiology are enormous: physiological adaptation in the traditional sense may be of secondary importance in promoting the persistence of species and higher taxa. The implication of physiological findings for the hypotheses of "species selection" may be equally large.

Peaks or plateaus: How is variation in physiological attributes related to variation in fitness?

Assuming that physiological variation and fitness are related, many ecological physiologists discuss physiological adaptations as if attributes of an organism typically represent the result of natural selection for the "correct" solution to pressing problems posed by the environment, such that deviation from the "correct physiology" in any particular environment will have dire consequences for fitness. This is typified in Figure 3.4a: either too high or too low a level of a physiological attribute will reduce fitness in a given environment or situation. A sharp peak is certainly implicit in textbook discussions that stress the importance of spectacular adaptations (e.g., salt glands, countercurrent exchangers, unusual metabolic pathways), which presume that animals lacking such adaptations would soon succumb to their environments. A sharp peak is also implicit in the numerous ecophysiological studies demonstrating a close correlation between the physiological attributes of animals and the environments from which they come (e.g., hemoglobin P_{50} vs. altitude, water vapor conductance vs. numerous biotic and abiotic variables in bird eggs, preferred body temperature vs. thermal performance optimum in lizards); animals often appear not to express too high or too low a P_{50}, water vapor conductance, thermal tolerance range, thermal optimum, etc. for the specific circumstances in which they find themselves.

An alternative view (Figure 3.4b) is that, within limits, variation in any given physiological attribute is relatively unimportant in its consequences for fitness; only very large deviations from the "correct" physiology (or gross environmental change) should reduce fitness. As long as gross limits are not exceeded, fitness should be unaffected if an Arctic fox's coat were 10% less insulative than normal, if a lizard were to regulate its body temperature with 5 °C less precision, and so on. Thus, according to this view, the shape of the fitness function is a plateau and not a peak. Why might this be so? In real environments, stochastic variation in abiotic factors, in motivation, and in the distribution and abundance of predators, prey, or potential mates may be so large as to confound the importance of any particular "physiological adaptation" in a given instance. Real animals have multiple redundant safety systems; other components (e.g., behavior; see Chapter 2) can often compensate for deleterious variation in any particular physiological component. Physiological traits may be maintained when not essential (or maintained in excess) either because selection for their removal is slight or because their expression is linked to or constrained by the need to express some essential characteristic (Chapter 9).

A question for ecological physiologists is how actual physiology-fitness relationships are distributed between the extremes represented by the peak model and the plateau model. If most relationships more closely resemble the plateau model than the peak model, then natural selection should maintain not a close match, but only a loose correlation between physiological characteristics and environmental ones. As the environments of populations change, selection might permit physiological evolution to lag well behind environmental change. Dissimilarity in physiological attributes could not be equated with dissimilarity in fitness.

How is this question to be answered? Although direct characterization of physiology-fitness relationships in the field would be preferable, the difficulties with this procedure are numerous, as outlined above and by Endler (1986). The few direct characterizations that have been attempted for complex physiological or performance characters thus far support the plateau model better than its alternative. The fineness or the coarseness of the equilibria between physiological and environmental variables constitutes a less direct examination of these alternatives. The data, however, need to be examined carefully for bias; demonstrations of poor relationships or no relationships may seldom be published (but see McNab, 1971). A largely unexploited tactic in animal ecological physiology for distinguishing between these alternatives is the field experiment; application of this tactic to plants has already yielded significant insights (e.g., Clausen, Keck, and Hiesey, 1940). One variant is to alter a physiological characteristic experimentally, release the experimental subjects, and observe if and by how much fitness is affected. Another is to alter *environmental* variables in the field (e.g., by moving animals to novel environments), follow free-ranging experimental subjects, and observe by how much fitness changes and how rapidly a new equilibrium is achieved between physiological and environmental characteristics (e.g., Berven, 1982). According to the "peak model," even small changes in physiological characteristics or environmental variables should cause large changes in fitness; according to the "plateau model," only large changes in physiological or environmental variables should cause changes in fitness. For example, Silberglied, Aiello, and Windsor (1980) removed presumptive protective coloration from the wings of butterflies, released and recaptured the insects, and surprisingly found that the absence of protective coloration did not affect the vulnerability of insects to natural predators. Closer approximations of fitness are essential. We may already know, for example, that an inability to regulate body temperature precisely causes a 10% decrement in locomotor speed in a foraging lizard; we need to know whether the 10% decrement in speed translates into a negligible change, a 10% change, or a 100% change in fitness (Huey, 1982).

The historical trajectory of physiological complexity

The evolution of any complex physiological adaptation involves the historical transformation of numerous interacting components from a primitive to

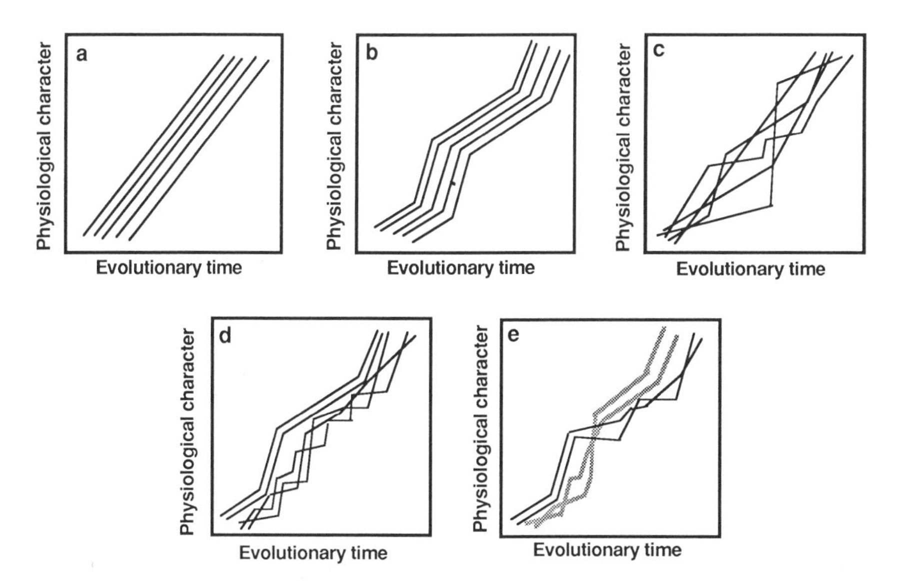

FIGURE 3.6 Possible patterns of change in the components of a complex physiological trait through evolutionary time. All component traits might change in parallel, at a constant rate (a) or at a variable rate (b). Each component trait might change at its own rate, with extreme variations prevented by selection (c). Some component traits may be rate-limiting, with others free to vary as long as they do not depart drastically from the rate-limiting components (d). If so, some components may be rate-limiting early in an evolutionary transformation [solid lines in (e)], whereas others may be rate-limiting later in a transformation [shaded lines in (e)].

a derived state. The patterns of such transformations are of interest because they may inform us of both the mechanisms of the transformation and the constraints upon them (Simpson, 1953; Lauder, 1981, 1982).

Consider the tranformation from a salivary gland of an ancestral reptile to a salt gland, from a reptilian kidney to a mammalian one, from the unshelled egg of an amphibian or fish to the shelled egg of a reptile or bird, or from the respiratory system of an inactive mammal species to that of an active mammal species. Transformation could occur in at least five ways (Figure 3.6): (a) parallel and gradual evolution of component traits; (b) parallel but punctuated evolution of component traits; (c) independent evolution of component traits; (d) rate of evolution limited by a single component or set of components; (e) rate of evolution limited first by some components and then by others.

Parallelism among the evolutionary trajectories of the components of a complex trait should bear some relationship to the control of expression of these components. If, for example, expression of each of a series of components shares a common genetic control, a common biochemical precursor, or a common developmental pathway (Alberch, 1980; Lauder, 1981), the components might be expected to change at similar rates through evolutionary time. Lauder (1981) has proposed that major coordinated changes in the phenotype may occur by the dissolution of functional links between components or the duplication of components (Lauder, 1981). Such changes would be consistent with rapid alterations in the phenotypic trajectory interspersed with periods of relative stasis (Figure 3.6b).

By contrast, "independent" (i.e., nonparallel) evolutionary trajectories of components of a complex trait (Figure 3.6c) might reflect (1) the absence of common control of expression of the components and (2) selection. Complex traits presumably play some functional role in their primitive state and while evolving as well as in their derived form. Variation in the components of a complex trait thus should be constrained during evolution by the need to maintain the coordinated function of the complex trait. If common control of expression of the components is absent, then each component should exhibit some variation independent of the other components' variation; however, selection should operate against any animal in which an important component varies so much that function of the complex trait is compromised. Additionally, if components differ in their importance to the function of the complex traits, less important components may be less constrained and more important components may be more constrained in their evolutionary trajectory (Figure 3.6d and e).

Distinction among these models is possible by tracing the evolutionary trajectories of components of complex traits in independently derived lineages of organisms (Lauder, 1982). In many ways, transformational analysis may be a simpler task for paleontologists than for ecological physiologists. Paleontologists can measure directly the rate of change of components as expressed in the fossil record, whereas ecological physiologists are limited to inferring trajectories from their end points. This is, however, not an impossible task for ecological physiologists, particularly if the analysis is conducted in a phylogenetic context [see Lauder (1981, 1982) and Chapter 4]. Although few ecological physiologists have attempted to distinguish among these models of evolutionary trajectory directly, many have an excellent intuitive "feel" for patterns and mechanisms of variation within individuals. For example, the ever-expanding literature on acclimation clearly documents the large extent to which components of an organism can vary independently of one another, and the control of acclimation (i.e., central vs. local) is an emerging theme in this area. The controversy regarding "symmorphosis" (Chapter 14) essentially concerns the extent to which selection maintains parity among components of a functional complex during evolutionary time. Distinction among these and other models of evolutionary trajectories is thus a significant problem that is already within the purview of ecological physiology, but has seldom been recognized as such.

The relative importance of adaptive and nonadaptive mechanisms in generating or constraining physiological diversity

Ecological physiologists often invoke only the cumulative effects of natural selection to explain the physiological diversity of organisms (Gould and Lewontin, 1979; Gould, 1980c). However, as discussed above, mechanisms other than "typical" natural selection may account for some fraction of physiological diversity. These mechanisms are nonadaptive in that selection for increased fitness is not involved in their action. Thus, an important question is, What is the relative importance of natural selection and various nonadaptive mechanisms in producing physiological evolution?

A second question concerns the limits of physiological diversity. We see repeated instances of well-defined physiological types, generally recognized along taxonomic lines: a mammalian physiology, an insect physiology, an elasmobranch physiology, and so on. Although we can recognize obvious physiological variants within each major group, we do not see all possible combinations of physiological traits: no moist-skinned terrestrial animals (e.g., slugs or amphibians) that routinely are endothermic, for example, and no healthy mammals with sharklike concentrations of urea in their blood. To paraphrase Gould (1980b), vast regions of potential physiology are unoccupied, whereas others are swarming with minor variations on common physiological themes. Why is this so (Simpson, 1953; Gould, 1980b; Wake, 1982), and what is the relative importance of natural selection and nonadaptive mechanisms in determining the commonness or rarity of physiological types?

Natural selection, of course, can almost always be invoked to explain both the amount and the limits of physiological diversity (Gould and Lewontin, 1979): physiological characteristics are so diverse because organisms evolved in so many diverse environments. The stochastic nature of natural selection is likely to promote multiple solutions to any given environmental challenge (Bartholomew, 1986; Chapter 2), thereby amplifying physiological diversity. The unexploited regions of physiological space are vacant because they represent unworkable solutions to environmental challenges (e.g., a terrestrial endotherm with the skin of an amphibian would either incur an intolerable level of water loss or expend energy uneconomically in behavioral hydroregulation). However, nonadaptive mechanisms may also account for physiological diversity and may be as consistent with patterns of physiological diversity as are adaptive explanations:

1. Phylogenetic inertia (Wilson, 1975; Ridley, 1983). Whether due to limited genetic variability, difficulty in simultaneously modifying the various components of a complex trait, a "plateau-shaped" fitness function, or simply the slow pace of natural selection, physiological characteristics may change very slowly in the face of environmental change or not change at all. If so, then some physiological space may be empty because selection and other evolutionary processes have not yet hit upon the improbable combinations of physiological characteristics necessary to occupy such empty regions. In theory, a marine mammal that osmoregulates in the manner of a shark is not impossible, but its evolution would entail numerous physiological adjustments whose joint probability is vanishingly small. Also, as a population begins to exploit a new environment, it will doubtless retain many characteristics it had in its previous environment. Although retained features may improve fitness in the new environment, they would not have originated in response to selection by the new environment (Gould and Vrba, 1982).
2. Genetic correlation. As detailed in Chapter 9, natural selection on one trait may lead to correlated changes in all other traits to which it is genetically linked. This phenomenon may cause nonadaptive (or even maladaptive) changes in physiological traits, or may promote changes that fortuitously enable an animal to occupy a new environment or exploit its current environment better. If this process is important and widespread, then physiological space may be filled as it is largely due to physiological traits that have been "hitchhiking" on other (potentially nonphysiological) traits, which were the actual object of selection. For example, Kingsolver and Koehl (1985) have proposed that the evolution of increased body size (and a correlated response in the size of wings that served as thermoregulatory structures) led to wings large enough to support flight in insects.
3. Developmental constraints. Canalization during development may not allow enough unbounded phenotypic variation for evolution to occur in any direction of change; instead, evolution may likely proceed only along a relatively small number of paths specified by "developmental constraints" (Alberch, 1980). Although this viewpoint remains a controversial one, it suggests that some vacant regions of physiological space may be unoccupied because developmental constraints preclude evolution in those directions and that some well-occupied regions of physiological space might represent evolution that has been canalized by developmental constraints.
4. Constraints and chance. At any point in evolution, natural selection may act along several paths dictated by the variation in an evolving population (Wright, 1932; Bock, 1959). In some cases, only one "choice" is feasible (e.g., only increased thermal tolerance, and not decreased thermal tolerance, will improve the survivorship of a sessile organism faced with stead-

ily increasing environmental temperatures), but in many cases several options are equally viable in terms of their immediate utility. The choice between equally viable options is largely a matter of chance. For example, in suspension-feeding brachiopods, in which the lophophore is the food entrapment organ, the evolution of increased lophophore size occurred along two coequal pathways (LaBarbera, 1986). In "plectolophes," lophophore size increased by expansion of its median portion; in "spirolophes," the lophophore formed a spiral and its size increased by expansion of its margins. As far as can be determined, both alternatives were equally advantageous at their origin.

Because selection is a response to the immediate utility of a trait rather than a trait's long-term consequences, chance "choice" of a pathway early in evolution may pose unforeseen opportunities or constraints later in evolution. For example, in salamanders the loss of lungs may have allowed the loss of tongue elements used in breathing, which in turn may have allowed the development of a projectile tongue and hence novel modes of foraging (Roth and Wake, 1985). As brachiopods evolved larger body sizes, the allometry of "plectolophe" lophophores proved inadequate to support function; lineages of plectolophe brachiopods were constrained to small size (LaBarbera, 1986). Comparable phenomena may result from extinction, which, like selection, is in part a stochastic process (Jablonski, 1986; also see above). We do not know to what extent diversity (or lack thereof) may stem from a species being poorly or well adapted, or may simply represent the luck of the draw.

How are we to assess the relative importance of natural selection and these alternative mechanisms in determining physiological diversity? First, we must eschew an exclusive focus on natural selection as the mechanism generating physiological diversity, and view selection as one of several potential mechanisms that may have been at work in any given case. Second, we must incorporate adaptation, phylogenetic inertia, correlated responses to selection, developmental constraints, chance, and other potential mechanisms as formal and coequal variables in our analysis. This will require different kinds of data sets and different sorts of analyses. In essence, what is needed is a way of examining the effects of one potential mechanism of physiological diversity while the effects of all others are either held constant or removed from the analysis. Both statistical techniques (e.g., partial correlation analysis, path analysis, multivariate analysis of variance, canonical correlation analysis) and "comparative techniques" (Ridley, 1983; Clutton-Brock and Harvey, 1984; Huey and Bennett, 1986; see also Chapter 4) are available to achieve this end. To examine the importance of constraint (developmental, mechanical, on one component by another, etc.), we need analyses of contingency. In how many instances have physiological characteristics evolved independently of one another? How frequently have two or more traits evolved only in the presence of one another? To what extent is the expression of one physiological trait necessary, sufficient, both necessary and sufficient, or neither necessary nor sufficient for the expression of a second trait?

What are the general consequences of physiological diversity for the ecological and evolutionary properties of animals?

Ecological physiologists may never be able to predict which physiological specializations will evolve in response to a given environmental challenge (Bartholomew, 1986). Once a specialization is in place, however, can we determine what its subsequent consequences are likely to be for a species's ability to exploit various environments and to persist in evolutionary time (Pough, 1978)?

A common theme in ecological physiology concerns how the evolution of physiological innovations enables species or higher taxa to invade previously unoccupied "adaptive zones" (Simpson, 1953; see also Mayr, 1960; Liem, 1973; and Lauder, 1981): new environments, vacant niches, or new ways of exploiting already occupied environments. For example, some ecological physiologists view the evolution of aerial respiration as entrée to terrestriality, the evolution of the cleidoic egg as a key to the amniotes' invasion of xeric environments, and the evolution of salt glands as allowing those birds and reptiles that possess such glands to exploit saline environments. In abstract terms, ecophysiologists' recognition of key innovations in physiological evolution implies that *general* ecological and environmental consequences attend physiological characteristics. Much as a single electron in the outermost shell of an atom will confer certain chemical properties regardless of the identity of the element in question, these *general* consequences of physiological characteristics are, by implication, more or less independent of the particular species in which the physiological characteristics occur. The admission of "general ecological-evolutionary consequences of physiological traits" has several important implications. First, it suggests that independent evolutions of a physiological characteristic can be viewed as independent trials of a natural experiment (Ridley, 1983); that the presence of a given physiological characteristic ought to have regular and repeatable consequences; and that the historical and phylogenetic idiosyncrasies of individual species cannot wholly confound recognition of general consequences. Much as ensembles of gas molecules behave in accordance with the general gas laws, ensembles of species ought to behave in ecological and evolutionary time according to their physiological characteristics, although individual molecules or species can and do behave eccentrically (Schopf, 1979; Gould, 1980b, pp. 112–116; Ridley, 1983, pp. 40–43). Second, if species can be expected to behave in lawlike fashion according to their physiological characteristics, ecological physiologists can pose hypotheses concerning general ecological-evolutionary consequences of physiological characteristics and can falsify these hypotheses with established methodologies and statistics (Lauder, 1981; Ridley, 1983; Clutton-Brock and Harvey, 1984). Given this perspective, the emergent questions are: Can we recognize general behavioral, ecological, and evolutionary consequences of major physiological characteristics; that is, consequences that are independent of the species in which the physiological trait is expressed? If so, what are these general consequences of major physiological characteristics?

There is a tension between this perspective and another mode of ecophysiological study, which considers each species an irreducible adaptive case unto itself that can only be compared to (but never equated with) other unique species. This tension is in part a matter of intellectual preference (inductivism vs. other modes of investigation, the idiographic vs. the nomothetic approach; see Gould, 1980c), and in part a legitimate disagreement about the prospects for recognizing general consequences in light of several biological issues (Bartholomew, 1986; Chapter 2):

1. Each species has its own unique evolutionary history with its own set of selection pressures and its own stochastic responses in evolutionary time. Accordingly, detection of general consequences, necessarily by examination of multiple species, may be impossible or unlikely.
2. Natural selection can yield multiple solutions to any given environmental challenge. Not only one physiological innovation, but scores or hundreds of possible innovations may give entrée to a new adaptive zone. The importance of any one physiological characteristic as an evolutionary breakthrough diminishes accordingly, and the prospects for recognizing the general consequences of any particular feature against the background of alternative solutions may be dim.
3. Key physiological innovations may come to bear on fitness only during instances of intense selection (e.g., broad-scale extinctions). Between incidents of intense selection, more or less any suite of physiological characters might suffice. If intense selection is rare, the identification of key innovations and their consequences may be impossible in practice.

These challenges may ultimately defeat most attempts at ecophysiological generalization. Yet, even if general ecological-evolutionary consequences of physiological attributes cannot be recognized, this failure would be of profound interest.

Recent awareness of general consequences and their implications has highlighted several tactics for their detection:

1. Establish the independence of repeated trials in natural experiments (Lauder, 1981; Ridley, 1983; Chapter 4). One way to avoid problems of nonindependence is to include phylogeny as an explicit factor in analyses of

physiological diversity; Chapter 4 discusses how this can be done. Another is to examine taxa that are as distantly related as possible. For example, in plants, insects, fishes, birds, mammals, and possibly dinosaurs, endothermy is associated with high rates of growth and sustained activity; only in birds and mammals is endothermy usually associated with routine thermoregulation. This observation suggests that a general consequence of endothermy is the capability for sustained high rates of growth or activity, a conclusion that would be confounded had the analysis been limited to birds and mammals (Bennett and Ruben, 1979).

2. Search for and focus upon instances in which putative "general consequences" are not supported by the data. An excellent example is Kramer's (e.g., 1983) work on the general consequences of aerial gas exchange for aquatic vertebrates. This work began on a field expedition to document adaptations for aerial respiration in fishes, when Kramer noticed that the majority of fish species in hypoxic Amazonian swamps did not breathe air despite a potentially large selection pressure to do so. He subsequently began descriptive and experimental work to test explicitly the presumed general benefits of aerial respiration. Similarly, my own work on cutaneous gas exchange began when I learned that tropical plethodontid salamanders both attain large body sizes and experience high body temperatures, conditions from which they were thought to be excluded by the inadequacies of cutaneous gas exchange. Subsequent work has discounted many a priori expectations regarding cutaneous gas exchange (Feder and Burggren, 1985).
3. Distinguish among multiple competing explanations of general consequence, one of which is that no general consequences exist, rather than defend or refute a single hypothesis (Platt, 1964; for physiological examples, see Kramer, 1983, and Feder, 1984). Defense of a single hypothesis is too often self-fulfilling, and the textbooks of ecological physiology are replete with "general consequences" that have been subjected to repeated substantiation but never falsification. Perhaps it is time to declare open season on such "general consequences."

Toward a nomothetic ecological physiology

An increased emphasis on the solution of general questions in ecological physiology has the potential to invigorate the field in many ways. Numerous ecological physiologists perceive a lack of direction in the field, or perceive their work as adding minor entries to a vast encyclopedia of adaptations whose broad outlines are already well established. For these people, general questions can serve as a novel and exciting focus. Numerous workers outside of the field perceive ecological physiology as a curiosity or an anachronism. To paraphrase a comment directed at a sister field (Gould, 1980c, p. 101):

"The flowering of ecological physiology has yielded a panoply of elegant individual examples and few principles beyond the unenlightening conclusion that animals work well." Answering the questions posed herein clearly has the potential to alter such perceptions, and, as detailed throughout this volume, has done so already. We have direct evidence that such invigoration is possible from the field of paleobiology. Only seventeen years ago, paleontology was characterized as a moribund field whose major objective was the encyclopedic accumulation of fossils and the description of their relationships to one another (Gould, 1980c). Today, paleobiology is a thriving field whose general questions (e.g., punctuated equilibria vs. gradualism; periodicity and importance of extinction; macroevolution vs. microevolution) are regularly debated in the journals *Science* and *Nature*. What achieved this transformation was an emphasis on general questions (Gould, 1980c), brought about in no small part by a volume on which this one was patterned (Schopf, 1972) and the founding of the journal *Paleobiology* (Gould, 1980c). This is not to say that the fostering of a nomothetic ecological physiology is a trivial matter. But how might it be achieved?

To establish a focus on general questions alongside the traditional focus on ecophysiological patterns, we need an increased awareness of general questions, and of their significance, and debate on both the general questions themselves and the most expeditious means of answering them. This chapter was written in hopes of stimulating such debate and awareness. The general questions it includes are not the only ones or even novel ones, but they can serve as points of departure for discussion. Another means of increasing awareness of general questions is to implement Platt's (1964, p. 352) procedure on a daily basis:

> It consists of asking in your own mind, on hearing any scientific explanation or theory put forward, "But, sir, what experiment could *dis*prove your hypothesis?"; or, on hearing a scientific experiment described, "But, sir, what hypothesis does your experiment *dis*prove?"

A focus on general questions will also require officers of scientific societies, journal editors, and symposium conveners to encourage the nontraditional, often theoretical discourse that general questions will entail. A focus on general questions cannot succeed, however, by excluding the traditional foci on ecophysiological patterns, natural history, and case studies of unusual species or populations. These traditional endeavors both complement and potentiate the solution of general questions, for general questions can be neither framed nor solved rigorously without accurate information on the natural history and systematics of their subjects (Gans, 1978; Bartholomew, 1986; Feder and Lauder, 1986; Greene, 1986; Chapter 4).

Shortly before I began this essay, I received a set of the journal *Physiological Zoology* dating back to 1956. As I scanned the titles of papers in chronological order, it became obvious that, technical advances aside, many of the 1956 papers would not have been out of place in the 1986 volume, and vice versa. This is an interesting statement about the rate of conceptual progress in ecological physiology. However, I think the prospects of a similar occurrence thirty years hence are minimal. The field must change or it will disappear. On the negative side, ecological physiology is under increasing financial pressure and its academic positions are at risk due to both general budgetary constraints and competition with disciplines with seemingly greater biomedical relevance, particularly molecular biology. On the positive side, this is a time of considerable conceptual ferment in organismal biology. Many of the developments considered in this volume are quite recent and are yet to be assimilated into ecological physiology. Bridges are being built to conceptual advances in both molecular biology and evolutionary biology. That this volume appeared speaks to a growing and active concern about the degree of conceptual progress in ecological physiology. Thus, prospects for a nomothetic ecological physiology are clearly visible. However, these prospects require active support if they are to flourish, and it is up to us to develop them.

Acknowledgments

Several papers were especially insightful to me in assessing the future rewards of pattern documentation and especially influential in their advocacy of general questions: Platt (1964), Schopf (1972, pp. 3–6), Gould and Lewontin (1979), Gould (1980c), Ross (1981), and Wake (1982). I am grateful to the following colleagues, who were willing to be distracted from more data-intensive enterprises for seemingly interminable conversations in which the ideas presented herein were (hopefully) refined: G. Bartholomew, W. Burggren, J. Edwards, G. Florant, R. Full, R. Huey, D. Jablonski, G. Lauder, P. Licht, J. Markin, K. Miller, A. Pinder, H. Pough, G. Roth, P. Ulinski, and D. Wake. In addition, C. Gans, R. Full, T. Garland, Jr., L. Houck, R. Huey, M. LaBarbera, K. Miller, A. Pinder, H. Pough, L. Prosser, and D. Wake made numerous helpful and instructive comments on the manuscript. Writing was supported by NSF Grant DCB 84–16121.

Notes

1. Throughout this chapter, the word "environment" is used with very broad meaning (cf. Ridley, 1983). It may refer to the ensemble of abiotic and biotic factors that an animal encounters where it lives, a species's niche, or a species's way of life (e.g., herbivory vs. insectivory, wide foraging vs. sit-and-wait predation, semelparity vs. iteroparity).
2. Throughout this section, reviews and examples have been cited to lead the interested reader into the primary literature, not to credit the priority of ideas.

References

Alberch, P. (1980) Ontogenesis and morphological diversification. *Am. Zool.* 20:653–667.

Arnold, S. J. (1983) Morphology, performance and fitness. *Am. Zool.* 23:347–361.

Arnold, S. J. (1986) Laboratory and field approaches to the study of adaptation. In *Predator-Prey Relationships: Perspectives and Approaches from the Study of Lower Vertebrates*, ed. M. E. Feder and G. V. Lauder, pp. 157–179. Chicago: University of Chicago Press.

Bartholomew, G. A. (1958) The role of physiology in the distribution of vertebrates. In *Zoogeography*, ed. C. L. Hubbs, pp. 81–95. Washington, D.C.: American Association for the Advancement of Science.

Bartholomew, G. A. (1982) Scientific innovation and creativity: a zoologist's point of view. *Am. Zool.* 22:227–235.

Bartholomew, G. A. (1986) The role of natural history in contemporary biology. *Bioscience* 36:324–329.

Bennett, A. F. (1974) Enzymatic correlates of activity metabolism in anuran amphibians. *Am. J. Physiol.* 226:1149–1151.

Bennett, A. F. (1980) The metabolic foundations of vertebrate behavior. *Bioscience* 30:452–456.

Bennett, A. F., and Ruben, J. A. (1979) Endothermy and activity in vertebrates. *Science* 206:649–654.

Berven, K. A. (1982) The genetic basis of altitudinal variation in the wood frog, *Rana sylvatica*. II. An experimental analysis of larval development. *Oecologia* 52:360–369.

Bock, W. J. (1959) Preadaptation and multiple evolutionary pathways. *Evolution* 13:194–211.

Clausen, J., Keck, D. D., and Hiesey, W. W. (1940) Experimental studies on the nature of species. *Carnegie Inst. Washington Publ.* 520:1–152.

Clutton-Brock, T. H., and Harvey, P. H. (1984) Comparative approaches to investigating adaptation. In *Behavioural Ecology: An Evolutionary Approach*, 2nd ed., ed. J. R. Krebs and N. B. Davies, pp. 7–29. Sunderland, Mass.: Sinauer.

Clutton-Brock, T. H., Guinness, F. E., and Albon, S. D. (1982) *Red Deer: The Behavior and Ecology of Two Sexes.* Chicago: University of Chicago Press.

Dobzhansky, T. (1951) *Genetics and the Origin of Species*, 3rd ed. New York: Columbia University Press.

Eldredge, N. (1985) *Unfinished Synthesis: Biological Hierarchies and Modern Evolutionary Thought.* New York: Oxford University Press.

Eldredge, N., and Gould, S. J. (1972) Punctuated equilibria: an alternative to phyletic gradualism. In *Models in Paleobiology*, ed. T. J. M. Schopf, pp. 82– 115. San Francisco: Freeman, Cooper.

Endler, J. A. (1986) *Natural Selection in the Wild.* Princeton, N.J.: Princeton University Press.

Feder, M. E. (1984) Consequences of aerial respiration for amphibian larvae. In *Respiration and Metabolism of Embryonic Vertebrates*, ed. R. S. Seymour, pp. 71–86. Dordrecht: Dr W. Junk.

Feder, M. E., and Burggren, W. W. (1985) Cutaneous gas exchange in vertebrates: design, patterns, control, and implications. *Biol. Rev.* 60:1–45.

Feder, M. E., and Lauder, G. V. (1986) Commentary and conclusion. In *Predator-Prey Relationships: Perspectives and Approaches from the Study of Lower Vertebrates*, ed. M. E. Feder and G. V. Lauder, pp. 180–189. Chicago: University of Chicago Press.

Gans, C. (1974) *Biomechanics: An Approach to Vertebrate Biology.* Philadelphia: J. B. Lippincott.

Gans, C. (1978) All animals are interesting! *Am. Zool.* 18:3–9.

Gans, C. (1986) Functional morphology of predator-prey relationships. In *Predator-Prey Relationships: Perspectives and Approaches from the Study of Lower Vertebrates*, ed. M. E. Feder and G. V. Lauder, pp. 6–23. Chicago: University of Chicago Press.

Gould, S. J. (1980a) G. G. Simpson, paleontology, and the Modern Synthesis. In *The Evolutionary Synthesis: Perspectives on the Unification of Biology*, ed. E. Mayr and W. B. Provine, pp. 153–172. Cambridge, Mass.: Harvard University Press.

Gould, S. J. (1980b) The evolutionary biology of constraint. *Daedalus* 109:39–52.

Gould, S. J. (1980c) The promise of paleobiology as a nomothetic, evolutionary discipline. *Paleobiology* 6:96–118.

Gould, S. J. (1982) Darwinism and the expansion of evolutionary theory. *Science* 216:380–387.

Gould, S. J., and Lewontin, R. C. (1979) The spandrels of San Marco and the Panglossian paradigm: a critique of the adaptationist programme. *Proc. R. Soc. Lond.* [*B*] 205:581–598.

Gould, S. J., and Vrba, E. S. (1982) Exaptation – a missing term in the science of form. *Paleobiology* 8:4–15.

Greene, H. W. (1986) Natural history and evolutionary biology. In *Predator-Prey Relationships: Perspectives and Approaches from the Study of Lower Vertebrates*, ed. M. E. Feder and G. V. Lauder, pp. 99–108. Chicago: University of Chicago Press.

Hillman, S. S. (1976) Cardiovascular correlates of maximal oxygen consumption rates in anuran amphibians. *J. Comp. Physiol.* 109:199–207.

Huey, R. B. (1982) Temperature, physiology, and the ecology of reptiles. In *Biology of the Reptilia*, vol. 12, ed. C. Gans and F. H. Pough, pp. 25–91. New York: Academic Press.

Huey, R. B., and Bennett, A. F. (1986) A comparative approach to field and laboratory studies in evolutionary biology. In *Predator-Prey Relationships: Perspectives and Approaches from the Study of Lower Vertebrates*, ed. M. E. Feder and G. V. Lauder, pp. 82–98. Chicago: University of Chicago Press.

Huey, R. B., and Stevenson, R. D. (1979) Integrating thermal physiology and ecology of ectotherms: a discussion of approaches. *Am. Zool.* 19:357–366.

Jablonski, D. (1986) Background and mass extinctions: the alternation of evolutionary regimes. *Science* 231:129–133.

Johnston, I. A., and Walesby, N. J. (1977) Molecular mechanisms of temperature adaptation in fish myofibrillar adenosine triphosphatases. *J. Comp. Physiol.* 119:195–206.

Kingsolver, J. G., and Koehl, M. A. R. (1985) Aerodynamics, thermoregulation, and the evolution of insect wings: differential scaling and evolutionary change. *Evolution* 39:488–504.

Koehn, R. K. (1984) *The Application of Genetics to Problems in the Marine Environment: Future Areas of Research.* Swindon, U.K.: Natural Environment Research Council.

Kramer, D. L. (1983) The evolutionary ecology of respiratory mode in fishes: an analysis based on the cost of breathing. *Environ. Biol. Fish.* 9:145–158.

LaBarbera, M. C. (1986) Brachiopod lophophores: functional diversity and scaling. In *Les Brachiopodes fossiles et actuels,* ed. P. R. Racheboeuf and C. C. Emig, pp. 313–322. *Biostratigraphie du Paléozoique,* vol. 4. Brest: Université de Bretagne Occidentale.

Lande, R., and Arnold, S. J. (1983) The measurement of selection on correlated characters. *Evolution* 37:1210–1226.

Lauder, G. V. (1981) Form and function: structural analysis in evolutionary morphology. *Paleobiology* 7:430–442.

Lauder, G. V. (1982) Historical biology and the problem of design. *J. Theor. Biol.* 97:57–67.

Lewontin, R. C. (1969) The bases of conflict in biological explanation. *J. Hist. Biol.* 2:35–53.

Licht, P., Dawson, W. R., and Shoemaker, V. H. (1969) Thermal adjustments in cardiac and skeletal muscles of lizards. *Z. Vergl. Physiol.* 65:1–14.

Liem, K. F. (1973) Evolutionary strategies and morphological innovations: cichlid pharyngeal jaws. *Syst. Zool.* 22:425–441.

Mayr, E. (1960) The emergence of evolutionary novelties. In *Evolution after Darwin,* vol. 2, ed. S. Tax, pp. 349–380. Chicago: University of Chicago Press.

McNab, B. K. (1971) On the ecological significance of Bergmann's rule. *Ecology* 52:845–854.

Medawar, P. B. (1969) *Induction and Intuition in Scientific Thought.* Philadelphia: American Philosophical Society.

Moalli, R., Meyers, R. S., Ultsch, G. R., and Jackson, D. C. (1981) Acid-base balance and temperature in a predominantly skin-breathing salamander, *Cryptobranchus alleganiensis. Respir. Physiol.* 43:1–11.

Platt, J.R. (1964) Strong inference. *Science* 146:347–353.

Pough, F. H. (1978) Review of *Biology of the Reptila,* vol. 5, *Physiology A. Copeia* 1978:370–372.

Prosser, C. L. (1955) Physiological variation in animals. *Biol. Rev.* 30:229–262.

Prosser, C. L., ed. (1973) *Comparative Animal Physiology,* 3rd ed. Philadelphia: W. B. Saunders.

Prosser, C. L. (1975) Prospects for comparative physiology and biochemistry. *J. Exp. Zool.* 194:345–348.

Prosser, C. L. (1986a) *Adaptational Biology: Molecules to Organisms.* New York: Wiley.

Prosser, C. L. (1986b) The challenge of adaptational biology. *Physiologist* 29:2–4.

Putnam, R. W., and Bennett, A. F. (1981) Thermal dependence of behavioural performance of anuran amphibians. *Anim. Behav.* 29:502–509.

Putnam, R. W., and Bennett, A. F. (1983) Histochemical, enzymatic, and contractile properties of skeletal muscles of three anuran amphibians. *Am. J. Physiol.* 244:R558–R567.

Rahn, H. (1967) Gas transport from the external environment to the cell. In *Development of the Lung, Ciba Foundation Symposium,* ed. A. V. S. de Reuck and R. Porter, pp. 3–23. London: J. and A. Churchill.

Reeves, R. B. (1977) The interaction of body temperature and acid-base balance in ectothermic vertebrates. *Annu. Rev. Physiol.* 39:559–586.

Ridley, M. (1983) *The Explanation of Organic Diversity: The Comparative Method and Adaptations for Mating.* Oxford: Oxford University Press (Clarendon Press).

Ross, D. M. (1981) Illusion and reality in comparative physiology. *Can. J. Zool.* 59:2151–2158.

Roth, G., and Wake, D. B. (1985) Trends in the functional morphology and sensorimotor control of feeding behavior in salamanders: an example of the role of internal dynamics in evolution. *Acta Biotheor.* 34:175–192.

Schopf, T. J. M., ed. (1972). *Models in Paleobiology.* San Francisco: Freeman, Cooper.

Schopf, T. J. M. (1979) Evolving paleontological views on deterministic and stochastic approaches. *Paleobiology* 5:337–352.

Silberglied, R. E., Aiello, A., and Windsor, D. M. (1980) Disruptive coloration in butterflies: lack of support in *Anartia fatima. Science* 209:617–619.

Simpson, G. G. (1953) *The Major Features of Evolution.* New York: Columbia University Press.

Smith, D. C. (1983) Factors controlling tadpole populations of the chorus frog (*Pseudacris triseriata*) on Isle Royale, Michigan. *Ecology* 64:501–510.

Snyder, G. K., and Weathers, W. W. (1975) Temperature adaptations in amphibians. *Am. Natur.* 109:93–101.

Stanley, S. M. (1979) *Macroevolution: Pattern and Process.* San Francisco: W. H. Freeman.

Vrba, E. S., and Eldredge, N. (1984) Individuals, hierarchies and processes: towards a more complete evolutionary theory. *Paleobiology* 10:146–171.

Wake, D. B. (1982) Functional and evolutionary morphology. *Perspect. Biol. Med.* 25:603–620.

Watt, W. B., Carter, P. A., and Donohue, K. (1986) Females' choice of "good genotypes" as mates is promoted by an insect mating system. *Science* 233:1187–1190.

Wells, K. D., and Taigen, T. L. (1984) Reproductive behavior and aerobic capacities of male American toads (*Bufo americanus*): is behavior constrained by physiology? *Herpetologica* 40:292–298.

Wilson, E. O. (1975) *Sociobiology: The New Synthesis.* Cambridge, Mass.: Belknap Press.

Wright, S. (1932) The roles of mutation, inbreeding, crossbreeding, and selection in evolution. *Proceedings of the Sixth International Congress of Genetics* 1:356–366.

Discussion

BENNETT: With respect to the final points of Feder's presentation, I found myself being philosophically in agreement, but wondering just how to investigate the questions discussed. For instance, one general question raised about "trajectory" was, "Do all traits comprising a complex adaptation evolve in parallel, or do some traits evolve more rapidly than others?" How would you resolve this?

FEDER: The most accessible answer to this question may lie in paleontology rather than in physiological ecology. Several instances exist of fairly good documentation of complex structures throughout the fossil record. According to James Hopson (University of Chicago), for example, the postcranial skeleton of mammals evolved in essentially parallel ways within several different lineages. It is as if there were only one way for the skeleton to evolve. One can, of course, observe the individual components of this complex structure, the postcranial skeleton, and see to what extent individual characters either maintain stasis or diverge with respect to one another. The consistency of trends in the evolution of the postcranial skeleton of mammals seems to me to imply three possible explanations: first, that natural selection is maintaining symmorphosis through time; second, that these characters are constrained during ontogeny, and it is essentially impossible for them to diverge from one another throughout the process of development; and third, that genetic control of expression of these characters is more or less fixed, thereby rendering it difficult to disentangle one particular regulatory gene, or complex of regulatory genes. The fossil record may provide insights into the relative likelihood of these alternatives.

BENNETT: But you can't dissect those. You have given three possible explanations and observed patterns, but no effective way to distinguish them.

FEDER: There is. A natural selection argument would imply a certain degree of variation or of discordance among the characters that might be inconsistent with either a developmental or a genetic explanation. A second way to approach the general problem involves looking at symmorphosis, but in a phylogenetic context as discussed by Huey. From a well-established cladogram, one might analyze the relationship of the component characters of a complex structure to one another along the array of different clades. That, I believe, will tell you something about whether individual characters can evolve independently or only linked with one another. George Lauder (University of California, Irvine) has addressed this issue in terms of constraint of one character on another during evolution. He calls this a "translational analysis," and gives a fairly explicit procedure for determining the relationship of one character to another, given a reasonable cladogram and the states of the components in extant organisms (Lauder, 1981, 1982).

ARNOLD: I agree that it would be good to focus on variation in physiological traits, but I want to comment on your expectation that much variation in physiological traits might signify weak stabilizing selection, whereas traits that are much less variable might reflect intense selection. Selection, particularly stabilizing selection, does tend to reduce variation in traits, and so you might expect this inverse relationship between phenotypic variation and intensity of selection. However other factors may be causing an increase in variation, including genetic recombination, mutation in each generation, and migration. Environmental effects – in other words, nongenetic differences among individuals – comprise another source of variation. The problem is that these other factors might confound your conclusion about the intensity of selection. There have been periodic attempts to relate variation and the intensity of selection in the field of morphology, and it has been a fairly disappointing exercise. Even if we had information on differences among traits in genetic variation, the mapping between a variation and selection would remain treacherous. There are other direct ways of studying selection. You could, for example, look for a correspondence between traits and fitness.

FEDER: That's a good comment. The major point that I wanted to make is that we currently have a rudimentary knowledge of variation in physiology. We need to know more about it, because it is a necessary condition for explaining equilibrium. Certainly, we must embrace the points that you have made. One point that you did not comment on is the inclusion of the notion of physiological constraint. Having one physiological character in place can limit the degrees of freedom of the others. Consequently, limited variability may reflect a physiological constraint rather than involve the other explanations we have considered.

DAWSON: I am pleased with this consideration of the need for interpreting the background of patterns of variation and the possible evolutionary meaning of these patterns. Variation often makes physiological ecologists uneasy in that they can take it to reflect unfavorably upon their experimental design and measurement techniques. While great care must be exercised in the application of these techniques, it is important to accept variation as a biological reality and to investigate with special care distinctive patterns.

BURGGREN: With respect to Feder's comments about the link between physiology and fitness, three studies, using two different traits, indicated that there did not seem to be any link. The apparent dismissal of any linkage between them on the basis of only those studies worries me. One advantage that encyclopedic knowledge of many aspects of ecological physiology gives us is a feel for the diversity of animals under different situations. For example, if comparison of the respiratory properties of the blood is confined to samples from llamas and humans, different perceptions of adaptations to high altitude are likely to result than if we use a broader assortment of montane species. It concerns me that, on the basis of a very small sample, the link between physiology and fitness is being questioned.

FEDER: You raise two important points. The first is also pertinent to Huey's presentation. One must gather data in a phylogenetic context. Man and llama are two very different clades, for which it is unclear whether the respiratory properties of the blood reflect differences in adaptation to the environment or phylogeny. As outlined in Huey's presentation, one might like to have a high-elevation species, a related low-elevation species, and an unrelated "outgroup" before reaching such a judgment. Your concern about inadequate sampling bothers me, too. The more species that one looks at, or the more populations that one looks at, the greater the likelihood of finding a positive correlation.

HUEY: All of the studies reported thus far are very short term. I think that this may tell us something about the nature of selection. In many instances the relationship between performance and fitness may in fact be variable over time. Certainly this is the lesson from the Galapagos finches. Moreover, our study on lizards was conducted over six weeks. That is hardly a definitive measure of the relationship between locomotor performance and fitness. Additionally, when we do interpopulation studies, as opposed to studies within populations, in most cases we do see correlates between, say, locomotor performance and apparent rates of predation. Perhaps by working from both the individual level up and the population level down, we can better assess the nature of selection on performance traits.

FEDER: We don't have a null model. Some biologists will automatically attribute the absence of a correlation between a physiological character and fitness to inadequacies in the search for correlation. However, perhaps there simply is no correlation. My preference would be to take the latter alternative as the typical null model. The burden of proof should lie on the investigator to demonstrate the presence of the correlation between performance and fitness.

KOEHN: I wanted to address the same connections from gene to physiology to fitness. You said that we have few data on that. Actually, there are quite a few results. There are many studies that have arisen primarily out of genetics, because of a great need to understand the possible adaptational significance of protein polymorphisms.

FUTUYMA: It remains interesting to assess the impact of variation on fitness. You will probably find some cases in which such an effect exists. However, it is interesting to ascertain the minimal difference necessary for a physiological function to exert an effect on fitness under various environmental conditions. This very much addresses the question of the slight, gradational steps by which Darwin's theory operates.

A general comment: I agree with much of what you have to say, but I am concerned at the apparent criticism of our predecessors. I wrote down some questions this morning, which I thought would be interesting contributions for me, an evolutionary biologist, to make to this session – interesting questions that physiologists ought to be able to address. I was greatly impressed this morning, to learn that Bartholomew was asking those questions ten or fifteen years ago, questions concerning such things as alternative adaptive solutions to the same problem. Maybe our predecessors were doing exactly what we should have been doing all along.

HUEY: Someone should give a brief defense of encyclopedic approaches. Feder's presentation reminded me of two things: one, a paper by George Bartholomew in *American Zoologist* a few years ago in which he said that we are our own worst judge of what is important; the other, a book by Louis Agassiz. In 1860, Agassiz was on a trip to Brazil, and he gave lectures on shipboard. He stated that basically we know everything there is to know, the great worlds have been discovered, and our task as scientists now is merely to fill in the gaps. At many points in history we may have felt as Agassiz did. We may, in fact, have sufficient knowledge about the adaptation of animals in diverse environments. However, sometimes we are wrong. A less complacent attitude can sometimes pay off.

BARTHOLOMEW: The encyclopedic detail that has accumulated over the years on aspects of physiology in relation to ecology is perhaps a consequence rather than a goal. One cannot always predict what is going to happen, but if one is a careful worker, one's data survive and may be of some use. In fact, I, somewhat reluctantly, have concluded recently in looking over my own publications, that ideas that guided me in gathering the data are probably much less permanent than the data themselves. The guiding force is the idea, but the monument is the volume of data.

FEDER: I will ask you to perform the following exercise, which in some ways led to the genesis of this conference. Go to a random sample of fellow ecological physiologists at a scientific meeting and ask them why they are working on their particular project. Usually, the answer is: "We know very little about this particular group. We need to know more; no one has yet studied the effect of environmental variable *X* on physiological function *Y* in species *Z*." There is indeed, among a significant fraction of the field, a matrix-filling, encyclopedia-writing mentality.

ARNOLD: But the way that I read your criticism of the encyclopedia is that it is insufficient justification for research to say that one is writing yet another paragraph in the encyclopedia of physiological ecology. Indeed this is a common problem in the rationales of grant proposals, and this is a common theme of criticisms that are leveled at them. Bartholomew is saying that the encyclopedia gets written not as a deliberate act but as a consequence of other activities. It is not a malevolent entity.

DAWSON: Let us accept the encyclopedia as a useful by-product accumulated in the course of testing hypotheses and interpreting relationships in physiological ecology.

POWERS: Just to get back to your point, I partially agree with both Feder and Bartholomew. Data are, and should be, sacrosanct. They should be collected very carefully so as to remain useful long after ideas have been disproven by us or others. The fact is that in physiological ecology, or any field of biology or of science as far as that goes, only a few people make the major contributions. Others fill in the gaps and do what I call "me-too biology," "me-too physics," or "me-too chemistry." Certain journals in this field embody this sort of approach. If "me-too" people are highly visible in physiological ecology, it can give the field a bad name.

FEDER: Bartholomew's statement and mine are complementary. The goal of this particular enterprise is to enhance the quality of research in physiological ecology.

4 Phylogeny, history, and the comparative method

RAYMOND B. HUEY

Comparative biology of contemporary organisms has three general goals: (1) to document the extent and patterns of existing organismal diversity, (2) to elucidate the mechanistic bases of how animals work, and (3) to deduce how and why those patterns evolved. The first two goals present few methodological challenges external to those of physiological expertise and adequate sampling, and comparative physiologists have been relatively successful at meeting these goals (e.g., Chapters 1 to 3). The third goal might present few external challenges as well, were the history of physiological change adequately preserved in the fossil record. However, because such direct historical markers are rare, the third goal in fact presents challenges with a vengeance, for the dynamics of historical pathways and causality must be deduced from indirect analyses of contemporary patterns. Such analyses are not easy, nor are they necessarily reliable.

I will argue here that recent developments of comparative methodology represent powerful – but underutilized – tools for reconstructing the dynamics of physiological, behavioral, and morphological evolution (Atz, Epple, and Pang, 1980; Ridley, 1983; Dobson, 1985; Liem and Wake, 1985) as well as for understanding the adaptive significance – or lack thereof – of physiological traits (Dumont and Robertson, 1986; Chapter 3). These developments are based on a phylogenetic perspective that explicitly considers the genealogical history and affinities of species under comparison. My chapter reviews some of these new phylogenetic approaches and illustrates ways in which these approaches might be applied in studies of ecological physiology. I hope these new developments challenge the assertion (Waterman, 1975, p. 313) that "phylogenetic explanations in comparative physiology must be rather nonrigorous and speculative."

The basic difficulty of deducing historical pathways from contemporary observations is readily demonstrated with a nonbiological example suggested by S. C. Stearns (personal communication). Consider how a person with no knowledge of history would classify the following countries as winners or losers of World War II: Germany, Japan, Great Britain, and France. Indices of current economic strength and international influence (our contemporary observations) would lead to the obvious but incorrect conclusion: Germany and Japan, winners; Great Britain and France, losers. When contemporary patterns depend on history, only a direct attempt to incorporate historical information is likely to yield trustworthy conclusions.

A phylogenetic perspective follows directly from the unarguable premise that species are not independent biological units that are devoid of history and of genealogical affinities. Quite the contrary – physiological traits of species reflect to varying degrees both their historical environments and their phylogenetic affinities. Inevitably, some characteristics of organisms are anachronistic (e.g., Darwin, 1859; Greene and Burghardt, 1978; Straney and Patton, 1980; Lauder, 1981, 1982; Janzen and Martin, 1982; Ridley, 1983; Stearns, 1983, 1984; Wanntorp, 1983; Dobson, 1985; Dumont and Robertson, 1986). These historical and phylogenetic considerations both complicate and make possible reconstructions of physiological evolution.

I will address five issues that relate phylogeny, history, and the comparative method. Specifically, I will show why phylogenetic information should be used to guide the choice of taxa for comparison, outline how phylogenies help circumvent some inherent statistical problems that arise from the nonindependence of species as sampling units, evaluate how phylogenetic data can promote the analysis of historical patterns, show how molecular-distance data serve as a basis for estimating rates of physiological evolution, and finally mention some limits and problems of phylogenetic approaches. (Some related issues are described in Chapters 3 and 5.) Overall, my intent is to show why a phylogenetic approach can be necessary and useful in many comparative physiological studies.

Many of the methods described here require a phylogenetic tree for the taxa being compared. For present purposes, we need not be concerned with the intemperate debate concerning how phylogenetic trees should be generated; rather we take a phylogeny as given, assume that it is valid (but see below), and use that phylogeny as a base for our analyses of physiological evolution. Suffice it to say that a variety of approaches are available (Mayr, 1981; Felsenstein, 1982; Sokal, 1985), of which "cladism" is probably dominant (Hennig, 1966; Eldredge and Cracraft, 1980; Wiley, 1981). Cladists attempt to deduce true phylogenetic relationships by clustering species according to patterns of traits shared in common, specifically shared derived traits (sometimes called "synapomorphies"). (Note: Systematists often distinguish between the true phylogeny and the deduced "phylogenetic hypothesis." In the interest of brevity, I will use "phylogeny" as a shorthand for phylogenetic hypothesis or estimated phylogeny.)

Phylogenies and the choice of species for comparison

Insights generated by comparative physiological studies often depend critically on the particular species selected for investigation. For many physio-

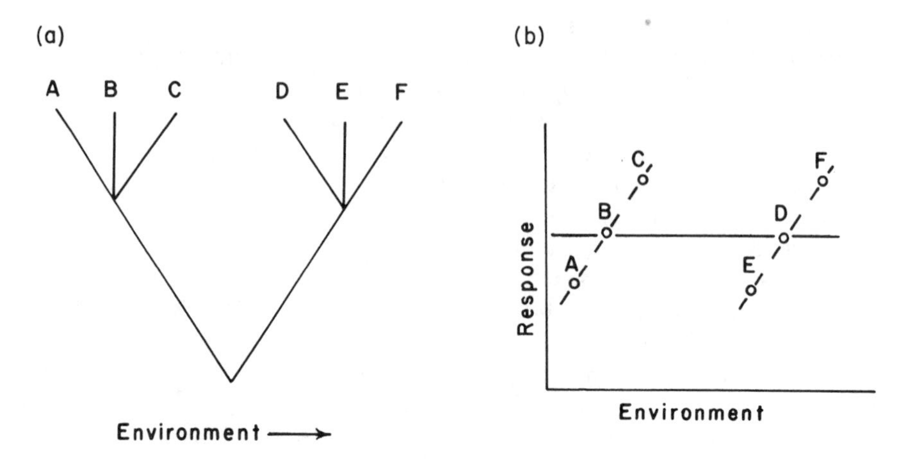

FIGURE 4.1 (a) A hypothetical phylogeny for two genera, each with three species, distributed along an environmental gradient. (b) Hypothetical physiological responses versus environmental position for the six species in (a). The solid horizontal line represents the regression of response on the environment if phylogeny is ignored, whereas the two dashed lines represent regressions for the genera analyzed separately. Failure to consider phylogeny would lead to the incorrect conclusion that the physiological response is unrelated to the environmental feature.

logical studies, especially those concerned primarily with mechanistic issues, the selection of species should be guided primarily by specific features (e.g, physiological or environmental) of the organism, by its tractability for physiological studies (the "August Krogh Principle"; Krebs, 1975), or by its accessibility. But for comparative studies that explicitly address adaptational or evolutionary issues, the selection of species should also be guided by phylogenetic considerations (Huey and Bennett, 1986). This assertion is justified below.

Phylogenetic considerations help answer the traditional question of whether comparative studies should select distant relatives or close relatives. For example, consider a set of species living in different parts of an environmental gradient (Figure 4.1a). In attempting to study physiological adaptation to that environmental feature, should one compare species a and b or species a and f? (Note: For reasons given later in this chapter, comparisons should involve at least three – not just two – species. Even then the statistical power is limited.)

The solution depends on the issue at hand. If one is interested primarily in discerning broad-scale patterns, then working with distant relatives is clearly appropriate (Chapters 2 and 3). For example, to contrast endothermy and ectothermy in vertebrates, one necessarily compares distant relatives (e.g., mammals and birds vs. reptiles, amphibians, and most fishes). Phylogenetic considerations are still useful here: they ensure that sampling *within* groups is representative. Similarly, in analyzing "key innovations" that may have promoted the radiation of certain taxa (for example, "relational analyses" of Lauder, 1981, 1982; also Chapter 3), wide taxonomic sampling is also necessary. On the other hand, if one is interested primarily in specific patterns (e.g., adaptation to extreme environments, Chapter 2), then comparing close relatives can be productive. Why?

A comparative study is, in effect, an experiment over historical time (Huey and Bennett, 1986). As in any experimental study in physiology, all extraneous variables should be rigorously controlled. However, if distantly related species are being compared, they may well differ in many features that reflect their independent evolutionary histories as well as the environmental correlate under investigation. For example, failure to consider phylogeny in Figure 4.1b would lead to the conclusion that hypothetical physiological response was unrelated to the environmental feature, when in fact that relationship is strong within the two genera. Consequently, any observed pattern – or the absence thereof – could be an artifact of phylogenetic heritage. Comparisons of distant relatives are therefore generally ambiguous (this is the familiar problem of comparing apples and oranges). In contrast, comparisons of close relatives are less likely to be confounded by phylogenetic artifacts, and observed differences are more likely to reflect adaptations to the environmental correlate.

In at least one case, however, comparisons of distant relatives enhance the analysis (see also Chapter 2). Imagine a comparative study of the scaling of some character (e.g., metabolic rate) on body mass. Measuring metabolic rate for a large size range of animals, which can be done only by gathering data from distant relatives, increases statistical power. This is useful as long as evolutionary changes in the dependent variable (metabolic rate) are closely coupled with those in the independent variable (body mass). When this is not true (e.g., Figure 4.1b), analyses of distant relatives may be very misleading.

This general problem is not merely academic. A failure to consider phylogenetic relationships can easily obscure a real pattern, or it can even create a false one. An extreme example comes from a review of the thermal dependence of sprint speed in twenty genera of lizards (Huey, van Berkum, Bennett, Hertz, unpublished data). We tested the hypothesis (Huey and Slatkin, 1976) that genera that are active in nature throughout broad ranges of body temperatures should be thermal generalists and thus should sprint well throughout a broad range of temperatures. This hypothesis predicts a positive correlation between the standard deviation of field body temperature and the "thermal performance breadth" (the range of body temperatures throughout which lizards' sprint speed exceeds some arbitrary criterion; see Huey and Stevenson, 1979).

If phylogeny is ignored and all genera are entered into the correlation, the traits are *negatively* correlated! In other words, genera that are active throughout the broadest ranges of body temperature have the narrowest performance breadths. If, however, closely related taxa are compared separately, the predicted positive pattern holds. For genera of the family Iguanidae, and also within the genera *Sceloporus* and *Anolis* (van Berkum, 1985, 1986), variability in body temperature is positively correlated with thermal performance breadth. This analysis suggests that evolution *within* families is very different from that *among* families, but why this occurs is obscure.

This example is extreme, but it underscores my point. By reducing phylogenetic artifacts, studies of closely related species reduce the risk of obscuring a real pattern or of even forcing a spurious one. As argued later in this chapter, comparisons should involve nested sets of relatives. This will permit physiologists to establish whether the observed patterns are general as well as to deduce evolutionary trajectories (Ridley, 1983; Felsenstein, 1985; Endler, 1986). For obvious practical reasons, such comparisons are unfortunately rare.

Application of the principle of studying close relatives sometimes runs into practical problems. Some important species (e.g., the tuatara, *Sphenodon*) have no living close relatives, and many taxa have limited physiological diversity. Little comfort can be provided to those who study these animals (but see Gans, 1983).

An additional problem arises when one attempts to determine the limits of "close relationships." Congeneric (or confamilial) comparisons are often used; but supraspecific taxonomic categories are arbitrary, such that taxonomy may be an unreliable guide for biological similarity (Simpson, 1944; Van Valen, 1973; King and Wilson, 1975). Moreover, some genera are young, whereas others are very old (e.g., Throckmorton, 1975; Bush, Case, Wilson, and Patton, 1977; and Stanley, 1979); and today's genus is often revised into tomorrow's genera. Some partial solutions are outlined below, but a working rule of thumb would be to compare the set of closest relatives that show adequate diversity for a given problem.

Phylogenies and statistical problems

Once species have been selected for comparison and studied, we must then proceed to an analysis of data. Here again phylogenetic considerations are important; unfortunately, they have usually been ignored. The problem arises from the fact that species are not statistically independent sampling units that lack history and genealogical affinities. In other words, ". . . species are part of a hierarchically structured phylogeny, and thus cannot be regarded for statistical purposes as if drawn independently from the same distribution" (Felsenstein, 1985, p. 1).

Consider the hypothesis that resistance to dehydration is greater for species from drier habitats (species A to C, Figure 4.1a) than for species from wetter habitats (D to F). Once data on dehydration resistance have been gathered, this hypothesis is easily tested by regression. But what are our taxonomic units? Typically, comparative biologists use *species*; but this choice assumes that species within each genus are evolutionarily and hence statistically independent, which frequently is not the case. This lack of independence artificially inflates the degrees of freedom in our comparisons, leading to an increased probability of a type I statistical error.

This is potentially a serious problem. Darwin (1859, p. 185) was aware of its importance to comparative studies, and it has been discussed extensively, especially in the behavioral literature (reviews in Ridley, 1983; Clutton-Brock and Harvey, 1984; Felsenstein, 1985). Currently, however, these discussions have rarely been incorporated into the literature of comparative physiology. Are there any solutions?

The first method, proposed by Crook (1965), is to make comparisons based on average values for an arbitrary taxonomic level (e.g., genus or family) above the species. This reduces, but does not remedy, the problem (Ridley, 1983): higher taxonomic categories are also part of the Linnean hierarchy and thus are not necessarily independent. For example, even though humans and chimpanzees are often placed in separate families, they are genetically very similar (King and Wilson, 1975; Sibley and Ahlquist, 1984), so that they are likely to be *physiologically* similar as well.

An extension of this method, developed by Clutton-Brock and Harvey (1984; see also Mace, Harvey, and Clutton-Brock, 1981; Stearns, 1983, 1984; Dunham and Miles, 1985), uses a nested analysis of variance to pinpoint the lowest taxonomic level at which the variance is large, and then makes comparisons based on trait values averaged at that taxonomic level. For example, in an analysis of thermal preferences and of optimal temperatures for sprinting in Australian skinks, Huey and Bennett (1987) found that 91% of the variance in thermal preference was among genera, suggesting that species averages should *not* be used as the basis for comparison.

All of these methods may reduce the problem. Nevertheless, all skirt the fundamental issue of phylogenetic independence as well as ignore the fact that the degree of independence can vary markedly among taxa even at the same taxonomic level. For instance, some genera are conservative for certain physiological traits, whereas others are not (van Berkum, 1986).

Ridley (1983) proposed a phylogenetically based solution for analyses of qualitative traits. Rather than trying to decide whether to compare the number of families or genera with certain traits, Ridley argued that it was more appropriate to focus on the number of independent evolutionary transitions between traits. Thus to test the hypothesis that trait X evolves to Y in habitat 1, one *counts the number of evolutionary transitions* that are associated with habitat 1. Ridley's method requires a phylogeny, a knowledge of the traits in the included taxa, and thus the use of phylogenetic techniques to identify evolutionary transitions (see below). Sadly, the resulting number of evolutionary events is almost certainly many fewer than the number of species studied.

Issues are somewhat more complicated for quantitative traits, which are the type of most interest to physiologists. Felsenstein (1985) proposed a method of studying correlations or regressions between traits (e.g., documenting allometric relationships) that corrects for lack of independence among taxa. His method requires knowing not only the topology of the tree, but also its branch lengths (i.e., times of separation); and it assumes that character change can be modeled as Brownian motion (i.e., directions of successive changes are independent; thus there is no evolutionary momentum). Starting with data on n species, Felsenstein's method generates $(n - 1)$ independent contrasts among species, which serve as the bases for correlations or regressions.

Felsenstein (1985, p. 13) also discussed a simpler method, which does not require detailed phylogenetic information. Independent contrasts are generated by comparing signs of correlations for pairs of closely related species: "... in a study of mammals we could use pairs consisting of two seals, two whales, two bats, two deer, etc. These contrasts would be independent ..." Correlations between traits can then be tested with nonparametric tests. Unfortunately, simplicity does not come cheaply: n species must be studied to generate only $n/2$ contrasts!

Finally, Cheverud, Dow, and Leutenegger (1985) developed a formal method of estimating the importance of phylogenetic effects in comparative analyses. Their method attempts to partition the total variance in traits into a component reflecting phylogeny (V_p, variance inherited from an ancestral species) and that reflecting specific adaptation (V_s, variance resulting from independent evolution of species). This method is conceptually similar to the familiar variance partitioning in quantitative genetics or in analysis of variance. When applied to an analysis of sexual dimorphism in body mass among primates, this method suggests that 50% of the variance among primates is associated with phylogeny; 36% reflects size (or scaling); and only 2% is due to habitat, mating system, and diet. Thus, to the question, Which primates should be highly dimorphic, one would respond: primates will generally be dimorphic, but large species should be especially dimorphic. Clearly phylogenetic conservatism is a dominant force in the evolution of sexual dimorphism in primates. [Using methodologies very different from those of Cheverud et al. (1985), Straney and Patton (1980), Stearns (1983, 1984), and Dunham and Miles (1985) also concluded that phylogeny is an important determinant of variation in morphology and life history traits.]

Reconstructing physiological evolution

We are now able to address some techniques that reconstruct the patterns of evolution of physiological or other traits. One starts with observed physiological diversity in some group, and then attempts to answer several questions: What were the directions, causes, and timing of evolutionary change within this group? Did the direction of change influence evolutionary outcomes? What were the sequences of change in complex physiological systems? What were the rates of change of physiological traits? These and related questions are central to historical analyses in comparative biology.

Several examples (see also Gans, 1970) have recently been described that demonstrate how an evolutionary perspective can clarify analyses of presumed physiological adaptations, specifically those of neuronal circuits (Dumont and Robertson, 1986). For instance, flightless grasshoppers possess flight motor neurons. Dumont and Robertson (1986) argue that these neurons presently have no functional significance but instead merely reflect evolutionary conservation of a trait that was functional in the flying ancestors of these grasshoppers (see also Chapter 3 for a general discussion of nonadaptive physiological traits). Here, knowing that flightlessness evolved from flying is the key to understanding an otherwise puzzling physiological trait.

Equilibrium analyses examine correlations between traits (Figure 4.2a) or between traits and environmental gradients (Figure 4.2c) of contemporary animals. ("Equilibrium" refers to the implicit assumption that traits are in evolutionary equilibrium with the environmental feature.) A positive correlation is interpreted as evidence that the traits are coadapted (Figure 4.2a), or that the trait is evolutionarily related to the environment (Figure 4.2c), in such a way that a change in the environment will lead to an evolutionary change in the trait. This is the traditional mode of analysis in comparative biology, and it does not utilize phylogenetic data.

Transformational analyses use phylogenetic data in an attempt to reconstruct the historical pathways of change that led to contemporary values of traits. Such analyses focus on dynamic patterns, not on static correlations. If transformational analyses are conducted for two (or more) traits (Figure 4.2b) or for traits and environments (Figure 4.2d), one can investigate whether changes in different traits are evolutionarily correlated (Figure 4.2b) or whether changes in the environment were accompanied by evolutionary changes in the trait (Figure 4.2d), respectively.

Lauder (1981) showed that equilibrium analyses sometimes provide misleading insights about the actual patterns of historical change. Moreover, equilibrium analyses yield less information than do transformational ones.

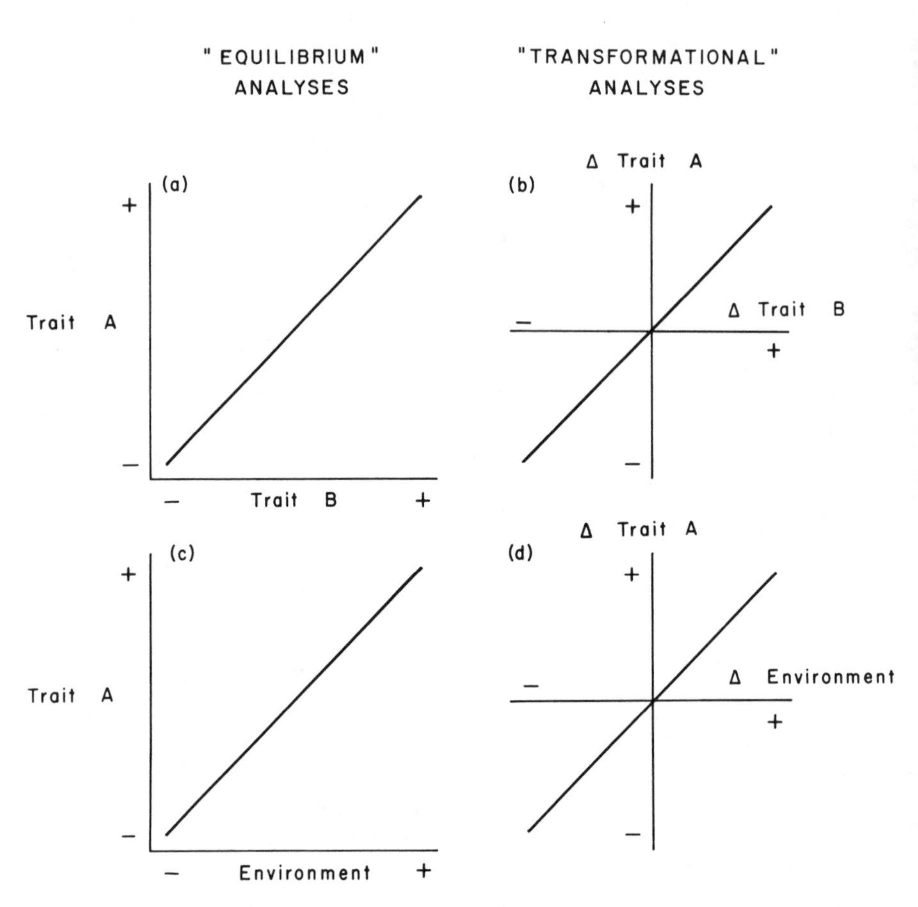

FIGURE 4.2 Equilibrium vs. transformational analyses (Lauder, 1981). Equilibrium analyses correlate two quantitative physiological traits in several species (a) or a quantitative trait and some environmental gradient (c). An extension of transformational analysis is shown in which the evolutionary *changes* in traits (b) and in environmental associations (d) are analyzed. Transformational analyses estimate these changes by deducing traits or environments of hypothetical ancestors (nodes in a phylogenetic tree) and then measuring the change between values for those ancestors and for contemporary species.

To demonstrate the relative power of both approaches, I will use data from a study of coadaptation between thermal preferences (T_p) and optimal temperatures (T_o) for sprinting in Australian skinks (Huey and Bennett, 1987). These skinks show exceptional differentiation in critical thermal maxima (Greer, 1980) and in thermal preferences (Bennett and John-Alder, 1986), and we wished to investigate whether the thermal dependence of sprinting shows equivalent and parallel differentiation. This might be expected; otherwise, a decoupling of thermal preferences and optimal temperatures would mean that lizards are routinely selecting temperatures at which they do not sprint well, which could decrease feeding success or increase risk of predation (Christian and Tracy, 1981; Webb, 1986). Sprint speed is a convenient organismal trait (Huey and Stevenson, 1979; Arnold, 1983), and its thermal dependence approximates those of several other performance traits in diverse reptiles (e.g., stamina, digestion, hearing efficiency; see Huey, 1982; Stevenson, Peterson, and Tsuji, 1985).

We first used an equilibrium analysis to investigate the degree of coadaptation between thermal preferences and optimal temperatures. Perfect coadaptation would be indicated if these traits shared a 1:1 relationship (dashed line in Figure 4.3). In fact, coadaptation is only partial: a 4 °C change in T_p is accompanied not by a 4 °C change in T_o, but rather by only a 1 °C change in T_o.

This equilibrium analysis seems to establish partial coadaptation, but fails to answer other questions. For instance, it does not tell us whether low or high thermal preferences are ancestral. This is a key issue if one wishes to understand the adaptive significance of traits (Dumont and Robertson, 1986; Huey and Bennett, 1986). If low thermal preferences were ancestral, then the evolution of genera with high thermal preferences has been accompanied by a "perfection" of coadaptation (Figure 4.3; i.e., by a "convergence" of T_p and T_o). But if high thermal preferences were ancestral, evolution has "decoupled" coadaptation (i.e., to a "divergence" of T_p and T_o). Such a decoupling would be an interesting evolutionary paradox, for it would imply that thermal preferences could evolve even at the cost of reduced locomotor performance.

Obviously, equilibrium analyses – even when correct (but see Lauder, 1981) – have limited power from the perspective of evolutionary reconstruction. To explore these issues more fully, we need transformational techniques. These usually require establishing ancestral versus derived states for traits (sometimes called "character polarity"), and they employ phylogenetic techniques.

Directions of evolutionary change

Deducing the direction of evolutionary change is a basic aspect of physiological reconstruction. Moreover, knowledge of directionality is sometimes useful in analyzing the adaptive significance of physiological traits (see the above discussion on flight motorneurons in flightless crickets) as well as in testing hypotheses concerning the ecological circumstances that might have selected for derived traits (Huey and Bennett, 1986).

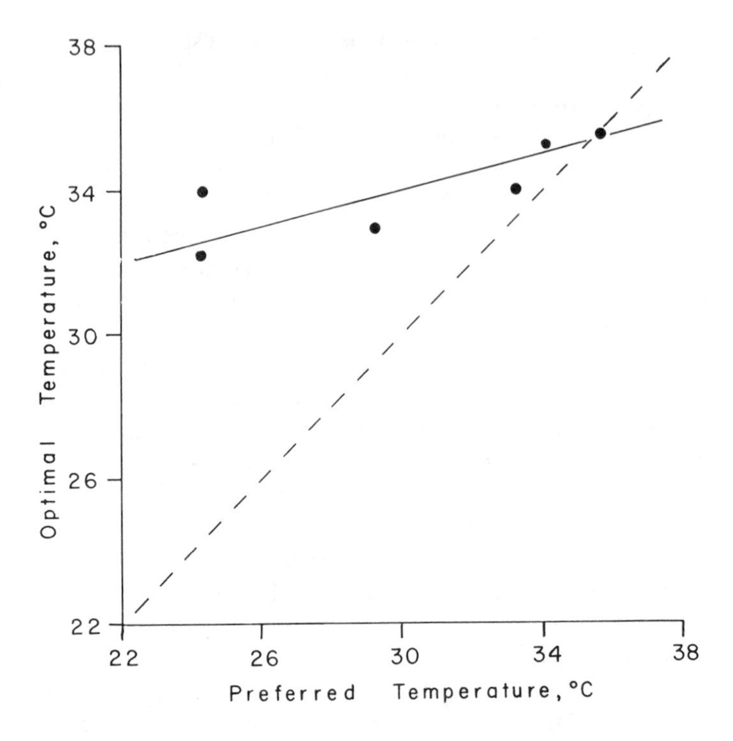

FIGURE 4.3 An equilibrium analysis of optimal temperatures for sprinting versus preferred body temperatures for genera of Australian skinks. The dashed line represents perfect coadaptation, whereas the solid regression line is the actual fitted regression. Clearly, coadaptation is only partial: optimal and preferred temperatures are closely coadapted for genera with high thermal preferences, but not for genera with low thermal preferences. This equilibrium analysis cannot determine whether low or high thermal preferences are derived. (From Huey and Bennett, 1987.)

The first method for deducing the direction of evolutionary change is simply to assume that the most common condition in a group represents the ancestral state. Thus, if most species in a given group have a low optimal temperature for some process, low optimal temperatures are assumed to be ancestral. Unfortunately this method ("ingroup" comparison) is unreliable: the current frequency of a trait might reflect its adaptive advantage or simply high speciation rates rather than ancestry (Crisci and Stuessy, 1980; see figure 4.1.5 in Ridley, 1983).

A more appropriate technique for deducing ancestral vs. derived states of discrete characters is called the "outgroup" method. This method predates (see Maslin, 1952) cladistic protocols (Hennig, 1966) but has been formalized

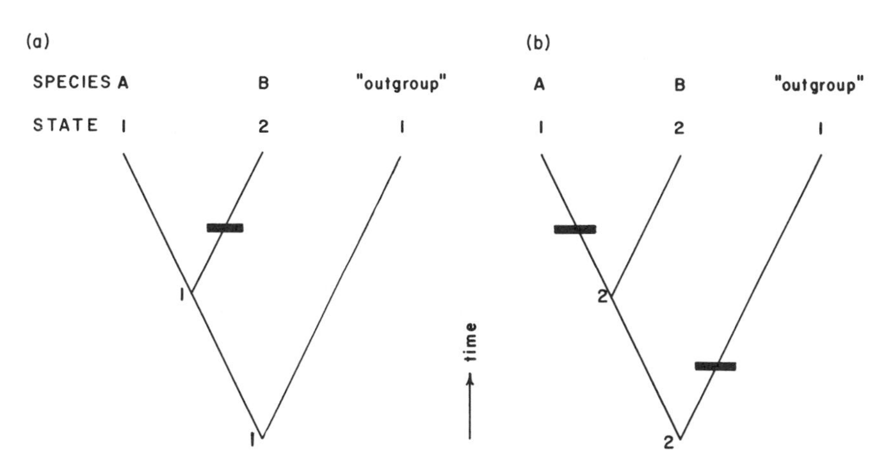

FIGURE 4.4 Outgroup method of deducing the directions of evolutionary change in character states (1 or 2) among a set of three species. We initially postulate that state 1 was ancestral (a) and then observe that only a single transition (horizontal bar) is required to account for the distribution of character states in the three species. We next postulate that state 2 was ancestral (b), but observe that at least two transitions are now required. Accordingly, state 1 is selected as ancestral, assuming parsimony.

and extended in the cladistic literature (e.g., Crisci and Stuessy, 1980; Watrous and Wheeler, 1981; Wiley, 1981). It assigns the ancestral state to be the character state in the hypothetical ancestor that minimizes the number of evolutionary transitions necessary to evolve the observed contemporary distribution of characteristics. Obviously, this technique assumes that major evolutionary change is rare – the assumption of parsimony. Accordingly, the method may run into difficulties where parallelism or convergence is rife (e.g., Felsenstein, 1983, 1985; Sokal, 1985).

Consider the hypothetical phylogenetic tree in Figure 4.4a, showing three species with associated character states (1 or 2). The most closely related species (A and B) are sometimes called "sister species," whereas their closest relative is called the "outgroup." If character states are known only for species A and B, the ancestral state cannot be deduced: ancestral states 1 and 2 are equally likely as either requires only (at a minimum) one transition to account for the distribution of states in species A and B. But if the character state is also known for a third species (the outgroup), the ancestral state can often be deduced. In the specific example in Figure 4.4a, the outgroup's state is assumed by appeal to parsimony to be ancestral, for this minimizes the number of evolutionary transitions (one, indicated by the horizontal bar in Figure 4.4a) necessary to account for the distribution of states in the three species. Were state 2 actually ancestral, this would require at least two transitions (e.g., Figure 4.4b). Note that the units of comparison need not be species – outgroup methods also deduce ancestral states among families, for example. Moreover, outgroup methods can be applied even when several character states occur in the outgroup (Maddison, Donoghue, and Maddison, 1984). Finally, in minimizing the number of evolutionary transitions, outgroup methods do not necessarily assign the outgroup's state to that of the ancestor (Figure 4.4a is thus a special case).

Most physiological studies deal with continuous rather than discrete characters, but two related methods are applicable here (see also Barrowclough, 1983). The median-rule algorithm (Kluge and Farris, 1969) proceeds by selecting the nodes of the tree ("hypothetical ancestors") from the median value among its three nearest neighbors (see Larson, 1984, for a nonphysiological example). This algorithm, which is the rule for assigning internal states in "Wagner networks" (Kluge and Farris, 1969), minimizes the absolute value of the total change in a given trait across the tree. The averaging-rule algorithm (Huey and Bennett, 1987; suggested by J. Felsenstein and based on Edwards and Cavalli-Sforza, 1964) "seeds" the nodes with arbitrary values and then substitutes the iterated averages (rather than the medians) of the three nearest neighbors. Several iterations are required to achieve convergence. This method minimizes the squared change in each branch of the tree, summed for all branches. (See Huey and Bennett, 1987, for a discussion of important statistical limitations of this algorithm.)

Huey and Bennett (1987), assuming a phylogeny based on Greer (1979), used the averaging algorithm to determine ancestral thermal preferences of Australian skinks. This analysis (Figure 4.5), with an accompanying sensitivity analysis, suggested that the ancestral thermal preference was probably 30 to 34 °C, such that the low thermal preferences (24 to 25 °C) of some genera are almost certainly derived. Evolution in these skinks appears to have led to a decoupling of coadaptation between thermal preferences and optimal temperatures (see also van Berkum, 1986). Such a decoupling, which was not evident from the above equilibrium analysis, is unexpected and calls for explanation (see below).

Change in multiple traits

A transformational approach for a single character can be extended for multiple characters, for environmental variables, or for geographic associations. If one wishes to determine whether the evolution of a character state was associated with a particular or extreme environment, one again uses the given phylogeny as a base and does multiple reconstructions – one for each character or environmental feature. Figure 4.6, for example, shows that the evolution of state 2 appears associated with a change in habitat from desert to forest. Demonstration of parallel trends in related lineages would support the hypothesis of a close evolutionary relationship between state 2 and forest

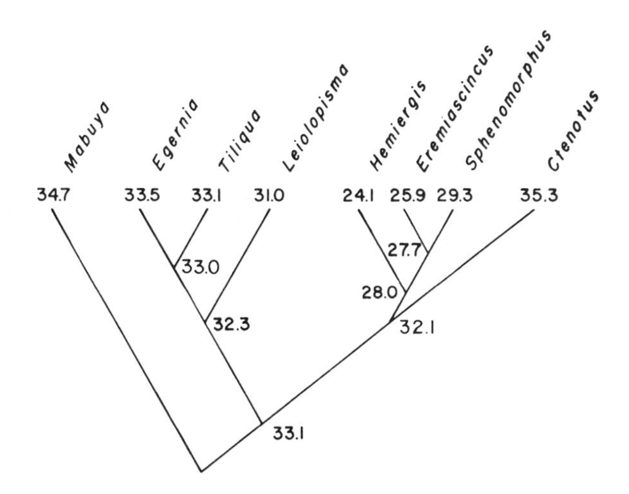

FIGURE 4.5 A phylogeny of the Australian skinks (based on Greer, 1979), with generic averages for thermal preferences and reconstructed ancestral thermal preferences using a minimum-evolution algorithm. A thermal preference of approximately 33°C appears to be ancestral.

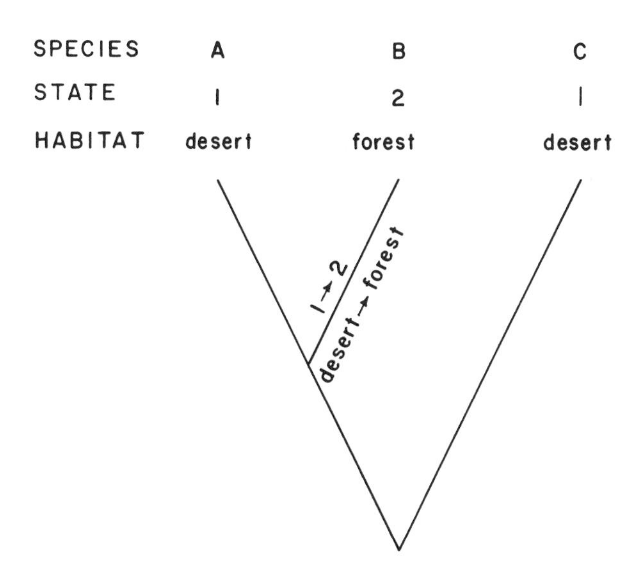

FIGURE 4.6 A phylogeny in which both character states and habitat associations are indicated. Given species C as the outgroup, the evolution of state 2 appears associated with the invasion of forest habitats by a desert-dwelling ancestor.

environments (Ridley, 1983; Greene, 1986). The mechanistic basis for that relationship could then be explored with traditional physiological studies.

If one is interested in coadaptation or in sequences of changes for complex physiological traits, similar analyses are run for each trait (Lauder, 1981). Such analyses may reveal whether changes in complex traits occur simultaneously ("saltatory" evolution) or sequentially ("mosaic" evolution; Simpson, 1944). Moreover, if transformational series are available in two directions (i.e., if association with desert habitats is ancestral to forest habitats in one group, but derived in a second), one can examine, for instance, whether the particular character states are dependent on the initial conditions (Dumont and Robertson, 1986). Thus, for example, does evolution follow a "last hired, first fired" sequence?

Data on Australian skinks (Huey and Bennett, 1987) exemplify the utility of a transformational approach to coadaptation. We used the averaging-rule algorithm (above) to estimate ancestral values of optimal temperatures, critical thermal maxima (CTMax), and thermal preferences (Figure 4.5). Then, for each genus, we estimated the changes in its optimal temperature, in its CTMax, and in its thermal preference from values for the nearest ancestral node (i.e., for the hypothetical ancestor). This enabled us to examine the change in optimal temperature or in CTMax (Figure 4.7) as a function of the change in thermal preference (ΔT_p). This analysis generally supports our equilibrium approach in showing that coadaptation is only partial: optimal temperature and critical thermal maxima (Figure 4.7) change slowly relative to thermal preferences. But the transformational analysis suggests one genus (*Eremiascincus*; circle in Figure 4.7) in which thermal preference has shifted to a low temperature, but the CTMax and perhaps the optimal temperature appear to have done just the opposite. In other words, this genus has evolved a thermal performance curve and preference that result in markedly reduced sprint speed at their thermal preferences. This result is unexpected and calls for explanation (see Huey and Bennett, 1987, for a discussion).

This example illustrates that equilibrium approaches can sometimes provide misleading (Lauder, 1981) or incomplete portraits of evolutionary change. Only transformational techniques identify the directions and sequences of change. In so doing, they should provide comparative physiologists with a powerful new tool for exploring the dynamics of evolutionary change.

Timing and rates of evolutionary change in physiology

A final and largely unexplored issue in comparative physiology concerns the timing and rates of evolutionary change. Physiology clearly evolves, but the temporal pattern of physiological evolution is rarely investigated. Useful questions include: (1) How do rates of physiological change vary among taxa or among traits? What accounts for relatively rapid physiological change in

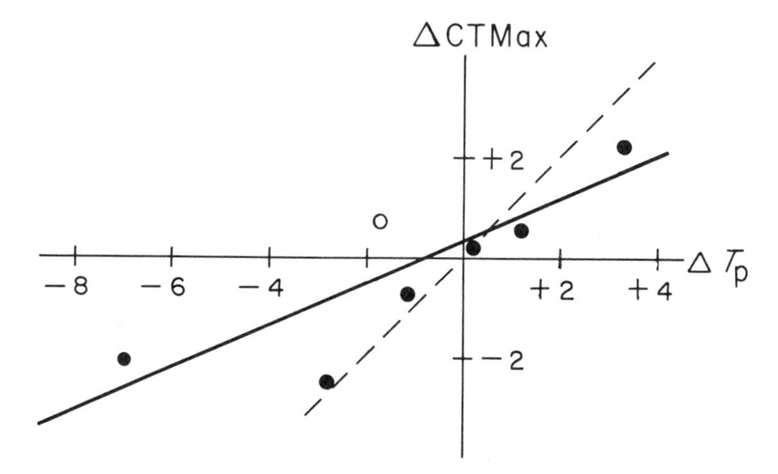

FIGURE 4.7 A transformational analysis (following Figure 4.2b) showing change in critical thermal maximum versus change in thermal preference for genera of Australian skinks. The data are the same as in the equilibrium analysis of Figure 4.3. The dashed line represents perfect coadaptation, whereas the solid line represents the fitted regression. Partial coadaptation is supported, as in Figure 4.3. The circle in the upper left quadrant represents *Eremiascincus,* which seems to have evolved a lower thermal preference but a slightly higher critical thermal maximum. See text. (From Huey and Bennett, 1987.)

certain taxa? (2) Do complex physiological pathways evolve more slowly than do simple pathways? Thus, is the resistance to evolutionary change proportional to the number of "links" in the physiological system or pathway? (3) Does the timing of physiological evolution correlate with major geological or biotic events (for a nonphysiological example, see Cadle, 1985)?

A recent example demonstrates how knowledge of the timing of physiological evolution can clarify analyses of the adaptive significance of physiological traits, specifically the significance of deafness in coleoid cephalopods (squids, cuttlefishes, and octopuses). Moynihan (1985) suggested that deafness could be an adaptation of these cephalopods to resist attacks from "booming" whales that use intensive bursts of sound to stun their prey (Norris and Møhl, 1983). A critical historical test refutes this hypothesis (Reid, Eckert, and Muma, 1986; Taylor, 1986): coleoid cephalopods apparently differentiated some 160 million years before "booming" whales, and they have neither vestigial ears nor acoustic centers in the brain.

Although information on the timing of physiological evolution is useful in theory, the rarity of physiological indicators in the fossil record (Thomas and Olson, 1980) has undoubtedly discouraged many attempts to gather such information. Moreover, alternative methods of estimating evolutionary rates are very indirect. For example, Bogert (1949) suggested that the rates of evolution of thermal preferences in lizards must be slow because activity temperatures of lizards from diverse environments showed limited intrageneric divergence.

Bogert's criterion for the rate of evolution – the magnitude of physiological divergence within genera – is unfortunately unreliable. Genera are not natural taxonomic units (Simpson, 1944); and, in any case, genera differ widely in age (references above).

An improved approach involves the integration of physiological and phylogenetic data. Specifically, biochemical data (e.g., electrophoresis, immunology, mitochondrial DNA), which are often used to generate a phylogenetic tree, can also be used to estimate divergence times for each branch in the tree, assuming the molecular-clock hypothesis (Wilson, Carlson, and White, 1977; Maxson and Maxson, 1979). By noting where evolutionary transitions occur in the tree, one can estimate the earliest time that the transition might have occurred. Moreover, rates of evolution of a trait can then be calculated as the magnitude of change in the trait divided by the time since divergence (Larson, Wake, Maxson, and Highton, 1981; Larson, 1984; see also Haldane, 1949). This general method specifies average rates of evolution but necessarily underestimates maximal (instantaneous) rates of evolution when evolutionary change is punctuational rather than gradual or when the transition occurred after, rather than simultaneously with, a branching event.

When applied to data on the thermal physiology of lizards of the *cristatellus* group of *Anolis* in Puerto Rico, Bogert's criterion and the molecular-clock approach lead to radically different conclusions. The lizards differ markedly in various physiological traits (critical thermal limits, thermal preferences, activity temperatures, desiccation rates). Adopting Bogert's criterion, Huey and Webster (1976) concluded that thermal physiology probably evolved rapidly in this species group. However, subsequent molecular data suggest that the *cristatellus* radiation is in fact old, perhaps dating to the Miocene (Gorman, Buth, Soulé, and Yang, 1980; Gorman, Buth, and Wyles, 1980), leading to the conclusion that average rates of physiological evolution might in fact have been slow (Huey, 1982).

The molecular-clock approach is probably the more objective of the two approaches, and only it (in the absence of fossil evidence) provides quantitative estimates of the timing and rates of evolution. Nevertheless, the approach is not without its problems (Gillespie, 1986; Vawter and Brown, 1986), and it makes several key assumptions. First, most biochemically detectable mutations are assumed to be selectively neutral. Second, rates of molecular evolution are assumed to be relatively constant among taxa and among molecules. Third, physiological evolution – the object of study here – is assumed to be independent of the molecular data that were used to generate the phylogenetic patterns (but see Chapter 5). Despite these assumptions, the phylogenetically based, molecular-clock approach seems to provide an opportunity to address unexplored issues in comparative physiology (see also Barrowclough, 1983).

Limits of a phylogenetic approach

Although I have argued that phylogenetic and historical information should be a fundamental part of comparative analyses that address evolutionary issues, I fully recognize that a phylogenetic approach is a double-edged sword. The benefits are clear, but the costs are often substantial. Moreover, some assumptions must be recognized.

First, the absence of a phylogeny for most taxa is the primary limitation on attempts to explore physiological evolution. Physiologists can circumvent this problem in two ways: (1) by studying only taxa for which phylogenies are already available, or (2) by developing phylogenies (probably in collaboration with a systematist) as an integral part of comparative physiological studies. The former is an unacceptably restrictive solution. The latter will often be the appropriate choice, but it requires an expansion of the scope of traditional studies in comparative biology. In either case, physiologists may find it convenient to develop strong links with systematists (Atz et al., 1980).

Second, even where estimated phylogenies are available, it is critical to appreciate that phylogenies are *hypotheses* of probable relationships. Phylogenies are often debated even for the best studied taxa (e.g., *Drosophila*, *Anolis*). Thus, phylogenies are not petrified trees of geological longevity. Instead, they are more like tinker toys: they can be assembled in multiple ways, and the constructed structure is often weak, full of holes, and tentative.

The consequences for comparative biologists are obvious – our reconstructions of physiological evolution are only as good as the phylogenies upon which they are based. We should consider the robustness of our conclusions under competing phylogenetic hypotheses (if available), explicitly mention the phylogeny selected for presentation, and be prepared to revise our conclusions if a better phylogenetic hypothesis is advanced subsequently.

Third, attempts to reconstruct evolutionary transformations (e.g., Figure 4.5) generally rely on the assumption that evolution is parsimonious – in other words, that evolutionary change (specifically convergence) is rare. This assumption need not always be true (Felsenstein, 1983), especially for the quantitative traits of interest to physiologists.

Fourth, because the reconstruction of physiological evolution is possible only by deduction, not by experimental manipulation, definitive answers to evolutionary questions will necessarily be evasive (Waterman, 1975; Huey and Bennett, 1986). (This problem is not unique to comparative biology – witness the problems faced by astrophysics.) Nevertheless, our "... potential to learn about historical phenomena seems inevitably limited. Given that evolutionary questions often provide the ultimate impetus for ecological and physiological research, this is frustrating" (Huey and Bennett, 1986).

Finally, because a phylogenetic approach is only beginning to penetrate the physiological literature, physiologists will need to make an effort to learn and to follow developments in this field. This will not be an easy task – the literature is scattered, and it is often contentious and filled with jargon.

Conclusions

I have argued that a phylogenetic perspective provides new directions and new opportunities for rigorous studies in evolutionary ecological physiology. I admit, however, that this perspective may create problems, not just solve them. As Felsenstein (1985, p. 14) wrote: "Comparative biologists may understandably feel frustrated upon being told that they need to know the phylogenies of their groups in great detail, when this is not something they had much interest in knowing." Nevertheless, two key benefits will result from efforts to use phylogenetic approaches. Not only will such a perspective help us select the most appropriate species, but it will also make possible analyses of historical patterns of physiological evolution. In so doing, the incorporation of a phylogenetic perspective may foster the exploration of certain fundamental questions: what are the patterns, directions, sequences, timing, rates, and adaptive bases of physiological change?

Acknowledgments

My interest in comparative approaches was stimulated by discussions initially with C. Gans, S. C. Stearns, D. B. Wake, and J. W. Wright and subsequently with A. F. Bennett, M. E. Feder, P. Harvey, C. Luke, H. W. Greene, T. Garland, Jr., M. Slatkin, and especially J. Felsenstein. I thank M. E. Feder, J. Felsenstein, T. Garland, Jr., and G. V. Lauder for very insightful comments on the manuscript. The writing of this paper was supported by NSF BSR 84-15855.

References

Arnold, S. J. (1983) Morphology, performance and fitness. *Am. Zool.* 23:347–361.

Atz, J. W., Epple, A., and Pang, P. K. T. (1980) Comparative physiology, systematics, and the history of life. In *Evolution of Vertebrate Endocrine Systems*, ed. P. K. T. Pang and A. Epple, pp. 7–15. Lubbock, Tex.: Texas Tech Press.

Barrowclough, G. F. (1983) Biochemical studies of microevolutionary processes. In *Perspectives in Ornithology*, ed. A. N. Bush and G. A. Clark, Jr., pp. 223–261. Cambridge University Press.

Bennett, A. F., and John-Alder, H. (1986) Thermal relations of some Australian skinks (Sauria: Scincidae). *Copeia* 1986:57–64.

Bogert, C. M. (1949) Thermoregulation in reptiles, a factor in evolution. *Evolution* 3:195–211.

Bush, G. L., Case, S. M., Wilson, A. C., and Patton, J. L. (1977). Rapid speciation and chromosomal evolution in mammals. *Proc. Natl. Acad. Sci. U.S.A.* 74:3942–3946.

Cadle, J. E. (1985) The Neotropical colubrid snake fauna (Serpentes: Colubridae): lineage components and biogeography. *Syst. Zool.* 34:1–20.

Cheverud, J. M., Dow, M. M., and Leutenegger, W. (1985) The quantitative assessment of phylogenetic constraints in comparative analyses: sexual dimorphism in body weight in primates. *Evolution* 39:1335–1351.

Christian, K. A., and Tracy, C. R. (1981) The effect of the thermal environment on the ability of hatchling Galapagos land iguanas to avoid predation during dispersal. *Oecologia* 49:218–223.

Clutton-Brock, T. H., and Harvey, P. H. (1984) Comparative approaches to investigating adaptation. In *Behavioural Ecology: An Evolutionary Approach*, ed. J. R. Krebs and N. B. Davies, pp. 7–29. Sunderland, Mass.: Sinauer.

Crisci, J. V., and Stuessy, T. F. (1980) Determining primitive character states for phylogenetic reconstruction. *Syst. Bot.* 5:112–135.

Crook, J. H. (1965) The adaptive significance of avian social organization. *Symp. Zool. Soc. Lond.* 14:181–218.

Darwin, C. D. (1859) *The Origin of Species.* New York: Penguin Books.

Dobson, F. S. (1985) The use of phylogeny in behavior and ecology. *Evolution* 39:1384–1388.

Dumont, J. P. C., and Robertson, R. M. (1986) Neuronal circuits: an evolutionary perspective. *Science* 233:849–853.

Dunham, A. E., and Miles, D. B. (1985) Patterns of covariation in life-history traits of squamate reptiles: the effects of size. *Am. Natur.* 126:56–72.

Edwards, A. W. F., and Cavalli-Sforza, L. L. (1964) Reconstruction of evolutionary trees. In *Phenetic and Phylogenetic Classification*, ed. V. H. Heywood and J. McNeill, pp. 67–76. Syst. Assoc. Publ. No. 6.

Eldredge, N., and Cracraft, C. (1980) *Phylogenetic Patterns and the Evolutionary Process.* New York: Columbia University Press.

Endler, J. A. (1986) *Natural Selection in the Wild.* Princeton, N.J.: Princeton University Press.

Felsenstein, J. (1982) Numerical methods for inferring evolutionary trees. *Q. Rev. Biol.* 57:379–404.

Felsenstein, J. (1983) Parsimony in systematics: biological and statistical issues. *Annu. Rev. Ecol. Syst.* 14:313–333.

Felsenstein, J. (1985) Phylogenies and the comparative method. *Am. Natur.* 125:1–15.

Gans, C. (1970) Respiration in early tetrapods – the frog is a red herring. *Evolution* 24:723–734.

Gans, C. (1983) Is *Sphenodon punctatus* a maladapted relict? In *Advances in Herpetology and Evolutionary Biology: Essays in Honor of Ernest E. Williams*, ed. G. J. Rhodin and K. Miyata, pp. 613–620. Cambridge, Mass.: Museum of Comparative Zoology.

Gillespie, J. H. (1986) Rates of molecular evolution. *Annu. Rev. Ecol. Syst.* 17:637–665.

Gorman, G. C., Buth, D. G., Soulé, M., and Yang, S. Y. (1980) The relationships of the *Anolis cristatellus* species group: electrophoretic analysis. *J. Herpetol.* 11:337–340.

Gorman, G. C., Buth, D. G., and Wyles, J. S. (1980) *Anolis* lizards of the eastern Caribbean: a case study in evolution. III. A cladistic analysis of albumin immunological data, and the definition of species groups. *Syst. Zool.* 29:143–158.

Greene, H. W. (1986) Diet and arboreality in the emerald monitor, *Varanus prasinus,* with comments on the study of adaptation. *Fieldiana Zool.,* no. 1370, pp. 1–12.

Greene, H. W., and Burghardt, G. M. (1978) Behavior and phylogeny: constriction in ancient and modern snakes. *Science* 200:74–77.

Greer, A. E. (1979) A phylogenetic subdivision of Australian skinks. *Rec. Aust. Mus.* 32:339–371.

Greer, A. E. (1980) Critical thermal maximum temperatures in Australian scincid lizards: their ecological and evolutionary significance. *Aust. J. Zool.* 28:91–102.

Haldane, J. B. S. (1949) Suggestions as to the quantitative measurement of rates of evolution. *Evolution* 3:51–56.

Hennig, W. (1966) *Phylogenetic Systematics.* Urbana, Ill.: University of Illinois Press.

Huey, R. B. (1982) Temperature, physiology, and the ecology of reptiles. In *Biology of the Reptilia,* vol. 12, ed. C. Gans and F. H. Pough, pp. 25–41. New York: Academic Press.

Huey, R. B., and Bennett, A. F. (1986) A comparative approach to field and laboratory studies in evolutionary biology. In *Predator-Prey Relationships: Perspectives and Approaches from the Study of Lower Vertebrates,* ed. M. E. Feder and G. V. Lauder, pp. 82–98. Chicago: University of Chicago Press.

Huey, R. B., and Bennett, A. F. (1987) Phylogenetic studies of coadaptation: preferred temperatures versus optimal performance temperatures of lizards. *Evolution* 41:1098–1115.

Huey, R. B., and Slatkin, M. (1976) Costs and benefits of lizard thermoregulation. *Q. Rev. Biol.* 51:363–384.

Huey, R. B., and Stevenson, R. D. (1979) Integrating thermal physiology and ecology of ectotherms: a discussion of approaches. *Am. Zool.* 19:357–366.

Huey, R. B., and Webster, T. P. (1976) Thermal biology of *Anolis* lizards in a complex fauna: the *cristatellus* group on Puerto Rico. *Ecology* 57:985–994.

Janzen, D. H., and Martin, P. S. (1982) Neotropical anachronisms: the fruits the gomphotheres ate. *Science* 215:19–27.

King, M.-C., and Wilson, A. C. (1975) Evolution at two levels: molecular similarities and biological differences between humans and chimpanzees. *Science* 188:107–116.

Kluge, A. G., and Farris, J. S. (1969) Quantitative phyletics and the evolution of anurans. *Syst. Zool.* 24:244–256.

Krebs, H. A. (1975) The August Krogh Principle: "For many problems there is an animal on which it can be most conveniently studied." *J. Exp. Zool.* 194:309–344.

Larson, A. (1984) Neontological inferences of evolutionary pattern and process in the salamander family Plethodontidae. *Evol. Biol.* 17:119–217.

Larson, A., Wake, D. B., Maxson, L. R., and Highton, R. (1981) A molecular phylogenetic perspective on the origins of morphological novelties in the salamanders of the tribe Plethodontini (Amphibia, Plethodontidae). *Evolution* 35:405–422.

Lauder, G. V. (1981) Form and function: structural analysis in evolutionary morphology. *Paleobiology* 7:430–442.

Lauder, G. V. (1982) Historical biology and the problem of design. *J. Theor. Biol.* 97:57–67.

Liem, K. F., and Wake, D. B. (1985) Morphology: current approaches and concepts. In *Functional Vertebrate Morphology,* ed. M. H. Hildebrand, D. M. Bramble, K. F. Liem, and D. B. Wake, pp. 366–377. Cambridge, Mass.: Belknap.

Mace, G. M., Harvey, P. H., and Clutton-Brock, T. H. (1981) Brain size and ecology in small mammals. *J. Zool. (Lond.)* 193:333–354.

Maddison, W. P., Donoghue, M. J., and Maddison, D. R. (1984) Outgroup analysis and parsimony. *Syst. Zool.* 33:83–103.

Maslin, T. P. (1952) Morphological criteria of phyletic relationships. *Syst. Zool.* 1:49–70.

Maxson, L. R., and Maxson, R. D. (1979) Comparative albumin and biochemical evolution in plethodontid salamanders. *Evolution* 33:1057–1062.

Mayr, E. (1981) Biological classification: toward a synthesis of opposing methodologies. *Science* 241:510–516.

Moynihan, M. (1985) Why are cephalopods deaf? *Amer. Natur.* 323:298–299.

Norris, K. S., and Møhl, B. (1983) Can odontocetes debilitate prey with sound? *Am. Natur.* 122:85–104.

Reid, M. L., Eckert, C. G., and Muma, K. E. (1986) Booming odontocetes and deaf cephalopods: putting the cart before the horse. *Am. Natur.* 128:438–439.

Ridley, M. (1983) *The Explanation of Organic Diversity: The Comparative Method and Adaptations for Mating.* Oxford: Oxford University Press (Clarendon Press).

Sibley, C. G., and Ahlquist, J. (1984) The phylogeny of the hominoid primates, as indicated by DNA-DNA hybridization. *J. Mol. Evol.* 20:2–15.

Simpson, G. G. (1944) *Tempo and Mode in Evolution.* New York: Columbia University Press.

Sokal, R. R. (1985) The continuing search for order. *Am. Natur.* 126:729–749.

Stanley, S. M. (1979) *Macroevolution: Pattern and Process.* San Francisco: Freeman.

Stearns, S. C. (1983). The influence of size and phylogeny on patterns of covariation among life-history traits in the mammals. *Oikos* 41:173–187.

Stearns, S. C. (1984) The effects of size and phylogeny on patterns of covariation in the life history traits of lizards and snakes. *Am. Natur.* 123:56–72.

Stevenson, R. D., Peterson, C. R., and Tsuji, J. S. (1985) The thermal dependence of locomotion, tongue flicking, digestion, and oxygen consumption in the wandering garter snake. *Physiol. Zool.* 58:46–57.

Straney, D. O., and Patton, J. L. (1980) Phylogenetic and environmental determinants of geographic variation of the pocket mouse *Perognathus goldmani* Osgood. *Evolution* 34:888–903.

Taylor, M. A. (1986) Stunning whales and deaf squid. *Nature* 323:298–299.

Thomas, R. D. K., and Olson, E. C. (1980) *A Cold Look at the Warm-Blooded Dinosaurs*, A.A.A.S. Selected Symp. no. 28. Boulder, Colo.: Westview Press.

Throckmorton, L. H. (1975) The phylogeny, ecology and geography of *Drosophila*. In *Handbook of Genetics*, vol. 3, ed. R. C. King, pp. 421–469. New York: Plenum.

van Berkum, F. H. (1985) The thermal sensitivity of sprint speeds in lizards: the effects of latitude and altitude. Ph.D. dissertation, University of Washington, Seattle.

van Berkum, F. H. (1986) Evolutionary patterns of the thermal sensitivity of sprint speed in *Anolis* lizards. *Evolution* 40:594–604.

Van Valen, L. (1973) Are categories in different phyla comparable? *Taxon* 22:333–373.

Vawter, L. T., and Brown, W. W. (1986) Nuclear and mitochondrial DNA comparisons reveal extreme rate variation in the molecular clock. *Science* 234:194–196.

Wanntorp, H.-E. (1983) Historical constraints in adaptation theory: traits and non-traits. *Oikos* 41:157–159.

Waterman, T. H. (1975) Expectation and achievement in comparative physiology. *J. Exp. Zool.* 194:309–344.

Watrous, L. E., and Wheeler, Q. D. (1981) The out-group comparison method of character analysis. *Syst. Zool.* 30:1–11.

Webb, P. W. (1986) Locomotion and predator-prey relationships. In *Predator-Prey Relationships: Perspectives and Approaches from the Study of Lower Vertebrates*, ed. M. E. Feder and G. V. Lauder, pp. 21–41. Chicago: University of Chicago Press.

Wiley, E. O. (1981) *Phylogenetics: The Theory and Practice of Phylogenetic Systematics*. New York: Wiley.

Wilson, A. C., Carlson, S., and White, T. J. (1977) Biochemical evolution. *Annu. Rev. Biochem.* 46:573–639.

Discussion

KOEHN: Why did you think it reasonable that the relationship between optimal temperature for sprint and temperature selection in skinks should be correlated one to one? Surely sprint speed is not the only activity of skinks relating to fitness. Other things must be affected by the level at which an organism behaviorally maintains its temperature. You should feel quite good about your results; you have a significant relationship explaining about five percent of the variance, and that's not bad in this business. You should not expect a stronger relationship than that, because there are other physiological processes to be considered. If you had chosen to measure digestive efficiency, I assume that you might have gotten an equally positive result.

HUEY: I would certainly encourage work on digestive efficiency. In studies of other reptiles, however, the thermal dependence of speed is rather similar to that of other traits, which also have thermal optima generally correlated with thermal preference. In any case, one might well expect a close correspondence between thermal preferences and the optimal temperature for sprinting, simply because of the probable influence of sprint speed on fitness.

KOEHN: I objected because you raised a straw man, in a sense, when you said, "Here's the expected relationship, and this is what we saw – animals doing something counterintuitive." I don't think it is counterintuitive at all. I believe you just don't have enough information.

BENNETT: Remember, that is the optimal temperature for running, it is not optimal for the entire suite of organismal traits.

HUEY: The hypothesis we were testing, that of a close correspondence between thermal optima for traits and thermal preferences, is over forty-years old and well established in the literature – it is hardly a novel straw man. In any case, we can draw a lesson from Arnold's approach here: linking physiology or morphology to performance may be easier than linking performance to fitness. Running faster may sometimes promote fitness, but this is not necessarily always true. In certain circumstances, being active at low body temperatures, even if this reduces sprint speed and other performance traits, may increase overall fitness. Bennett and I suspect in fact that this may be going on here. The evolution of a low thermal preference is clearly associated with the evolution of nocturnality in these skinks. Nocturnality results in activity of these skinks at low body temperatures and hence slow speeds; but perhaps it is not all that important to be very fast at night. Prey may be sluggish, and predators few. So despite having the disadvantage of being slow, these skinks may also have compensatory advantages. This general issue relates to an interesting point that Dawson raised. In looking at performance traits, we need to be very careful to interpret these traits in an appropriate ecological context.

KOEHN: Well, absolutely. Clearly the answer you are going to get when you ask that question is very much dependent upon the conditions for which you ask the question.

ARNOLD: The relationship between measurements of performance and fitness might be directional; that is, there might be directional selection on that trait, in which case you detect it with simple correlation. On the other hand, the relationship might be like this: I am drawing a line concave downward, concave upward – so that you could have absolutely no correlation between the two variables, yet there is a curvilinear relationship. That would describe a situation of stabilizing selection. Physiological performance may be under stabilizing selection, and at equilibrium you would expect a population to be under that peak. Detection of such a circumstance will require large samples, if the curvature is slight. Thus the jury is still out. As Huey points out, it is not an issue of merely validating measures of physiological performance that have endured a long history of laboratory work. We need to find out what the forms of selection are.

FUTUYMA: You have now explained why these skinks have a difference between temperature preferendum and the optimum temperature. It still remains interesting that their physiological optimum for running is at a higher temperature than their preferred temperature. It is a statement that organisms are not completely coadapted – in this case – that it probably is physiologically anachronistic, and that is interesting to know.

BURGGREN: Huey mentioned that species are not independent sampling units. How do you determine when species are sufficiently far apart to be regarded as independent?

HUEY: Unfortunately, there is no hard and fast answer to that. One recent technique, that by Ridley for discrete characters, looks for independent evolutionary transitions, not independent species per se. If, for example, you are examining whether or not toe fringes in lizards are associated with sandy environments, you do not really count as evidence each species of fringe-toed lizard that lives in deserts. You look within a phylogenetic assemblage and count the number of times that fringes have evolved and see if those correlate with invasions of sandy habitats. However, the problem is more complicated if one is dealing with continuous characters, as most physiological characters are. Felsenstein's [1985] paper in *American Naturalist* and Cheverud and colleagues' [1985] paper in *Evolution* deal with this. The Cheverud et al. paper basically develops an analysis of variance akin to that employed in quantitative genetics, where both genetic and environmental effects must be taken into account. Here they allocate the variance in some trait into phylogenetic and environmental components. However, they are not really trying to determine when two species are different enough to be treated as independent. Felsenstein develops other approaches to this issue. [References to the papers mentioned are given in Raymond Huey's paper.]

DAWSON: How does one cope with the matter of variation among populations within a species? You are probably dealing with quantitative characters there.

HUEY: One can use a phylogenetic approach to study variation among populations also. This has some advantages over working among species. For example, by working within species, one obviously reduces the phylogenetic artifacts that can confound interspecific studies. Even so, such interspecific studies do not solve the problem of independence among populations in this case. In a cline, for example, adjacent populations are likely to have similar genetic backgrounds and thus are likely to respond similarly to an environmental factor – so the problem of lack of independence is not avoided. Some approaches indirectly try to control for those factors. Straney and Patton [see Huey's chapter for reference] looked at what factor (ecogeographic vs. phylogenetic) best accounts for geographic variation in morphological characters within a species. Similar approaches could easily be used to study variation in physiological characters.

5 A multidisciplinary approach to the study of genetic variation within species

DENNIS A. POWERS

Introduction

Darwin's *On the Origin of Species* irrevocably affected the direction of biological investigation from the time of its publication until the present. Clearly, the theory of evolution has since become the greatest unifying principle in biology.

While everyone agrees that the general concept of Darwin's theory is correct, it has been modified as each new scientific discipline has made its contribution. For example, since Darwin's time, the field of genetics has greatly clarified our understanding of inheritance and has provided a theoretical framework for phenomena that natural historians had previously described but were unable to interpret. Clearly, the fields of genetics, systematics, biogeography, paleontology, physiology, embryology, ecology, and more recently biochemistry and molecular biology have each contributed significantly in their own specialized way to our modern understanding of evolution.

Although each specialized field has shed new light on the details of evolution, a unique perspective has come from those able to bridge two or more disciplines. Ecological physiology, for example, weds biochemistry, physiology, behavior, and ecology as it bridges the gap between molecules, cells, and organisms, and combines the reductionistic and holistic philosophies of its constituent subdisciplines. Disciplines that combine the resolving power of several fields make unique contributions because they are able to capitalize on the knowledge and methodologies of each of their constituent subdisciplines and are able to synthesize and theorize from a multidisciplinary vantage point. As new techniques expand a discipline's resolving power, the interdisciplinary scientist is in a unique position to transfer this new technology to related fields and thereby invigorate them. For example, the explosion of molecular biology since the mid-1970's is beginning to find its way into the fields of physiology, ecology, and evolutionary biology. This new wave of technology has been championed by interdisciplinary scientists interested in molecular evolution and molecular aspects of ecological physiology.

Similarly, population biologists bring a unique perspective to the field of ecological physiology. In turn, they acquire a physiological point of view that can be applied to problems in population biology. In their attempts to unravel the mysteries of evolution, ecological physiologists have traditionally employed species as their primary unit of comparison. They often assume that variation within a species is unfortunate experimental noise. Population biologists, by contrast, stress the population or the individual as an equally appropriate, if not a more appropriate, unit of comparison. Their rationale is that variation within species provides the genetic material for speciation and the potential for contributing to our understanding of the mechanisms of evolution. Ecological physiologists have focused on species comparisons because broad comparisons between species groups are more readily drawn and because the transmission of genetic information between species is minimal. However, such comparisons are so broad that they often have little predictive value and thus conclusions are inevitably correlative and often speculative (Chapter 4). In addition, the study of individuals from a single population as representative of an entire species can lead to erroneous conclusions concerning species. While the same potential problem exists for comparisons within and among populations, there also exists the potential for the generation of testable hypotheses that can provide information about the detailed mechanisms of evolution. The power of such an approach lies in one's ability to predict experimental results and the potential for uncovering causal relationships.

The study of genetic variation within and among species has been a major research focus of evolutionary biologists for more than a century. The analysis of morphological variation during the first half of this century gave way in the 1960's to the study of genetic variation at the molecular level (reviewed by Lewontin, 1974). The application of electrophoresis of proteins to the study of populations exposed a tremendous amount of genetic variation at the molecular level, which was totally unexpected. The extent of this variation was so great that questions arose concerning the significance, or lack thereof, of these polymorphic loci. In fact, few subjects in biology have been more debated than the evolutionary significance of protein polymorphisms. Most of the debate has centered around two contrasting views: the "selectionist" and the "neutralist." Proponents of the "selectionist" school assert that natural selection maintains protein polymorphisms, whereas those of the "neutralist" persuasion argue that the vast majority of such variation is selectively neutral. Other biologists favor intermediate positions.

Some biologists have addressed the conflict between these two extreme views by developing theoretical mathematical models, in hopes that such models might be falsified by available estimates of evolutionary rates, mutation rates, genetic loads, effective population sizes, genome sizes, and other parameters. Stebbins and Lewontin (1972), however, have shown that a purely mathematical treatment cannot resolve the conflict. Moreover, Clarke (1973) has pointed out that the estimates of evolutionary rates and other parameters are so variable that values can be found to favor any of the available models. In other words, the same biogeographic data can be used to support diametrically opposed theories. Lewontin (1974) has summarized the failure of evolutionary biologists to resolve this important issue.

The neutralist hypothesis implies that most genetic variation is functionally equivalent (Harris, 1976); that is, different allelic variants behave in an identical fashion in terms of cellular and organismal function. Therefore, this hypothesis can be rejected for specific loci whenever functional nonequivalence can be established between the genetic alternatives. These genetically based traits could be morphological, physiological, ecological, behavioral, biochemical, and so on. However, a problem with testing this hypothesis for many nonbiochemical traits (e.g., eye color, bristle number, etc.) is that their function is not often well defined. On the other hand, the application of biochemical techniques and the study of allelic isozyme traits has contributed to the solution of this problem because the specific functions of these enzymes were previously established.

Of course, establishment of in vitro functional nonequivalence of genetic alternatives is just the first step in the long process of determining if an enzyme polymorphism is subject to natural selection. Clarke (1975) has established a four-step strategy toward addressing the problem of genetic variation at enzyme-synthesizing loci. A similar approach could be used to study other types of genetic variation both within and between species. As outlined in Clarke's (1975) paper, the four steps are:

1. We must make a detailed biochemical and physiological study of enzymatic products of the alleles, noting any differences between them.
2. Knowing the nature of the differences, the function of the enzyme, and the ecology of the organism, we must postulate one or more selective factors and suggest a mechanistic connection between the selective factor and the gene product.
3. We must test our postulated mechanism by experimentally manipulating the environment to produce predictable responses.
4. In the light of the experiments, we must reexamine the natural populations, and seek a comprehensive explanation for the observed patterns of gene frequencies.

It goes without saying that Clarke's (1975) four-point strategy assumes, of course, that one has already established the existence of allelic variation and that patterns of gene frequencies have been observed for these loci. A major advantag^ of Clarke's (1975) strategy over classical field studies is one's ability to test the first part of step (1) above, that is, to make a detailed biochemical study of functional differences between allelic isozymes.

Some scientists are using an interdisciplinary approach (i.e., population genetics, physiology, and biochemistry) to address various aspects of Clarke's strategy. Examples of these types of studies are:

1. Alcohol dehydrogenase from *Drosophila* (Clarke, 1975; Daly and Clarke, 1981; Vigue, Weisgram, and Rosenthal, 1982; Dorado and Barbancho, 1984). Kinetic differences between allelic isozymes were associated with differences in survivorship and developmental time.
2. Aminopeptidase from mussels, *Mytilus edulis* (Koehn and Immerman, 1981; Koehn and Siebenaller, 1981). Allozyme polymorphism was associated with improved osmoregulation.
3. Glucose phosphate isomerase in *Colias* butterflies (Watt, 1977, 1983; Watt, Cassin, and Swan, 1983; Watt, Carter, and Blower, 1985). Allelic variation was correlated with survivorship and mating success.
4. α-Glycerophosphate dehydrogenase allozymes from *Drosophila* (O'Brien and MacIntyre, 1972; Miller, Pearcy, and Berger, 1975; Sacktor, 1975; Curtsinger and Laurie-Ahlberg, 1981). Kinetic differences were correlated with environmental temperature, flight metabolism, and power output.
5. Esterase-6 in *Drosophila*. Allelic isozyme activity (Richmond et al., 1980) and other kinetic parameters were associated with reproduction.
6. Glucose-6-phosphate dehydrogenase and 6-phosphogluconate dehydrogenase from *Drosophila*. Metabolic flux and partitioning differences (Bijlsma, 1978; Cavener and Clegg, 1981) were reflected in differences in fitness (Hughes and Lucchesi, 1977).
7. Lactate dehydrogenase from the "heart" or B locus of the fish *Fundulus heteroclitus* (reviewed by Powers, DiMichele, and Place, 1983; Powers et al., 1986b). Differences in the kinetic properties of the purified allelic isozymes were used to predict and subsequently establish significant differences in metabolism, oxygen transport, swimming performance, developmental time, and relative fitness.

Not all of these elegant studies take full advantage of the power of this approach – namely one's ability to generate testable hypotheses that may be validated or rejected by appropriate experimental design. For example, if biochemical analysis of allelic isozymes leads to predictable differences in cell physiology, organism response, etc. that can be substantiated by experimentation, then the "neutralist" hypothesis can be rejected for that locus. Experiments designed to test these cellular predictions should yield results that allow the scientist to make other testable predictions at higher levels of biological organization. As each new cycle of predictions is followed by experimental validation, one can ultimately be led to accepting either the "selectionist" or the "neutralist" paradigm. If predictions can be followed by experimental validation, then the "selectionist" viewpoint would be supported; otherwise the "neutralist" position would be favored. It is this cycle of a priori predictions that provides the power of such a multidisciplinary approach. Moreover, such an approach is useful for genetic studies between species as well as for an analysis of variation within a species.

Fundulus heteroclitus: a model for intraspecific studies

Although many of the studies cited above could illustrate the strength of the intraspecies approach, I shall call upon my own work because it emphasizes an ecological physiology approach (see Chapter 8 for another example). My colleagues and I are addressing the potential role of natural selection upon allelic variants of several representative enzyme-synthesizing loci in the fish *Fundulus heteroclitus.*

Fundulus heteroclitus, commonly known as killifish or mummichog, is an abundant Atlantic Coast fish ranging from the Mantanzas River in Florida to Port au Port Bay in Newfoundland, Canada. They have continued to be a popular experimental organism for over 100 years. A recent symposium on this biological model (DiMichelle, Powers, and Taylor, 1986) amply illustrates both the historic and current role of this important organism for a wide array of scientific studies. Although space is not adequate to document all aspects of our research, I shall attempt to provide a flavor for the intraspecies approach by appropriate illustrations of the *Ldh-B* locus. As mentioned above, before Clarke's four-point strategy can be utilized, one must first establish the existence of allelic isozyme variation and then establish the pattern of genetic variation over the species's geographical range.

Zoogeographical genetic variation

Just as biogeographers have examined the diversity of species in different parts of the world, population biologists have focused considerable attention on the distribution of gene diversity over a species's range. Sometimes directional changes in gene frequency (i.e., clines) are detected. Such directional changes, when correlated with an important environmental parameter like temperature, pH, salinity, etc., are often touted as evidence for natural selection. However, such clines can also be explained by a number of mechanisms, including some that do not invoke natural selection as a driving force. In fact, the major problem with such studies is that the data can often be explained by two or more different but equally valid hypotheses. Recent technical developments may allow biologists to distinguish unequivocally between some of these alternatives (see example of mitochondrial DNA below).

Fundulus populations maintain a high level of protein polymorphism (reviewed by Powers et al., 1986b). Examination of the geographic distribution of sixteen of these polymorphic loci has uncovered significant directional changes with latitude in gene frequency (i.e., clines) and in degree of

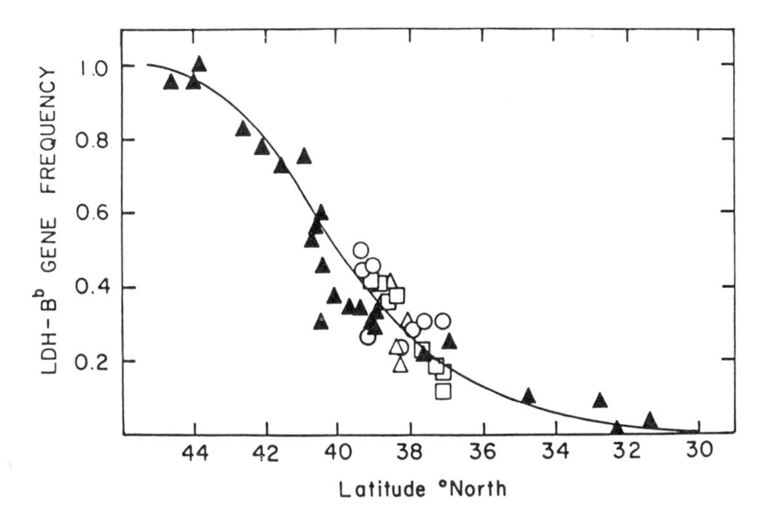

FIGURE 5.1 Frequency of *Ldh-B*^{b} gene frequency in relation to latitude (°N). Solid triangles designate populations along the East Coast of North America; open symbols designate populations along various tributaries to Chesapeake Bay.

genetic diversity. The distributions of these gene frequencies have been divided arbitrarily into four classes. Class I loci are clinal, having two predominant alleles, one fixed in northern populations, the other in southern populations. Figure 5.1 illustrates *Ldh-B* as a typical class I locus. Class II loci are fixed for a single allele at the northern extreme of a species's range but have substantial genetic variation at other latitudes. Class III loci are fixed for an allele having the same electrophoretic mobility at both the northern and southern extremes, but show variability at middle latitudes. Class IV loci are not clines, as defined by Huxley (1938). Rather, they show no significant change in gene frequency with latitude.

Directional changes in genetic characters with geography (i.e., clines) have classically been described by two general models: primary and secondary intergradation (reviewed by Endler, 1977). In the primary intergradation model (i.e., the Ongoing Model, Figure 5.2), adaptation to local conditions along an environmental gradient or genetic drift may lead to genetic differences along the gradient. Gene flow may not eliminate these differences either because it is too small or because of nonrandom dispersal along the gradient. In the secondary intergradation model (i.e., the Historical Model, Figure 5.2), populations are first separated by some barrier that prevents gene flow. Next, either adaptation to local conditions or genetic drift produces genetic differences between these disjunct populations. Finally, when the barrier is removed, the formerly disjunct populations interbreed, producing a

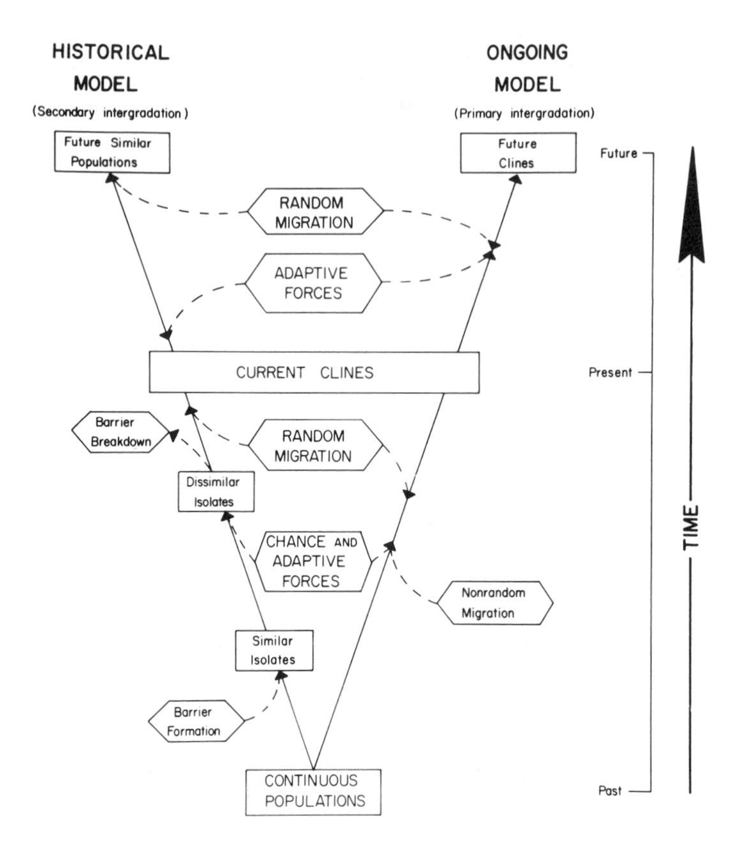

FIGURE 5.2 Model of the various forces and routes to cline formation. According to Powers and Place (1978), primary intergradation is described as an "ongoing" process whereas secondary intergradation is identified as an "historical" process because the latter requires previous genetic isolation. The various driving forces are identified in the hexagons with arrows indicating the direction of force relative to the directionality of the models.

cline in gene frequencies among them. Therefore, the main difference between these two models is the need for the previous existence of isolating barriers to gene flow in the latter. Figure 5.2 diagrammatically illustrates these two models and some of the driving forces that may affect them.

The present-day spatial patterns of *Fundulus heteroclitus* gene frequencies could have arisen by either type of intergradation (reviewed by Powers et al., 1986b). One cannot distinguish between these on the basis of classical zoogeographic data unless these data are available within a few hundred generations of an alleged secondary contact (Endler, 1977). However, direct analysis of mitochondrial DNA (mtDNA) can allow the distinction between the primary and secondary models at much greater intervals after an alleged secondary contact. For example, *Fundulus heteroclitus* populations were analyzed by studying mtDNA fragments obtained by digestion with each of eighteen restriction endonucleases (Powers et al., 1986b; Gonzalez-Villasenor and Powers, unpublished data). The mtDNA from individuals representing four localities were analyzed by the general procedure of Gonzalez-Villasenor, Burkhoff, Corces, and Powers (1986) using a radiolabeled clone of *Fundulus* mtDNA as a probe. Analysis of the mtDNA restriction fragment data indicated that a previous barrier to gene flow existed at or near 41 °N latitude (Gonzalez-Villasenor and Powers, unpublished data). This conclusion was based on the fact that the mtDNA restriction patterns of fish at specific localities could be interrelated by a network of single nucleotide base changes. However, populations on each side of 41 °N latitude required many nucleotide changes. The extent of those differences indicated that the populations had to have been separated for several thousand years. The presence of a migrant-derived genotype in one of the populations suggested that gene flow was reestablished probably sometime during the past 10,000 years. Those data clearly and unequivocally support the secondary intergradation model of cline formation.

Natural selection as a potential driving force

While this finding is very important in that it allows better understanding of the general clinal nature of gene diversity in this species, it does not provide insight concerning the relative contributions of chance (e.g., genetic drift) and adaptive (e.g., natural selection) forces (see Figure 5.2) prior to, during, and/or after the genetic isolation event. Yet it is the relative roles of chance and adaptive forces that strike at the very heart of the "neutralist"/"selectionist" controversy. Thus, additional approaches must be undertaken to determine the role of natural selection, if any, as a major driving force for generating and/or maintaining gene diversity in this species.

Fundulus heteroclitus is found in one of the steepest thermal gradients in the world and, being a poikilotherm, must be influenced by this environmental parameter. Figure 5.3a illustrates that the fish's range is dominated by a mean water temperature change of approximately 1 °C per degree latitude. In the southern marshes, summer water temperatures in excess of 40 °C are commonly recorded whereas summer temperatures of the northernmost marshes are relatively cool. Although northern marshes are often covered by ice in the winter, southern marshes are generally free of such severe low water temperatures (see Figure 5.3b). South of 41 °N latitude the minimum water temperature increases at 1 °C per degree latitude (Figure 5.3b). The gene diversity of this species decreases at latitudes greater than 41 °N, but is unchanged at southern latitudes (Figure 5.3c). Powers et al. (1986b) have

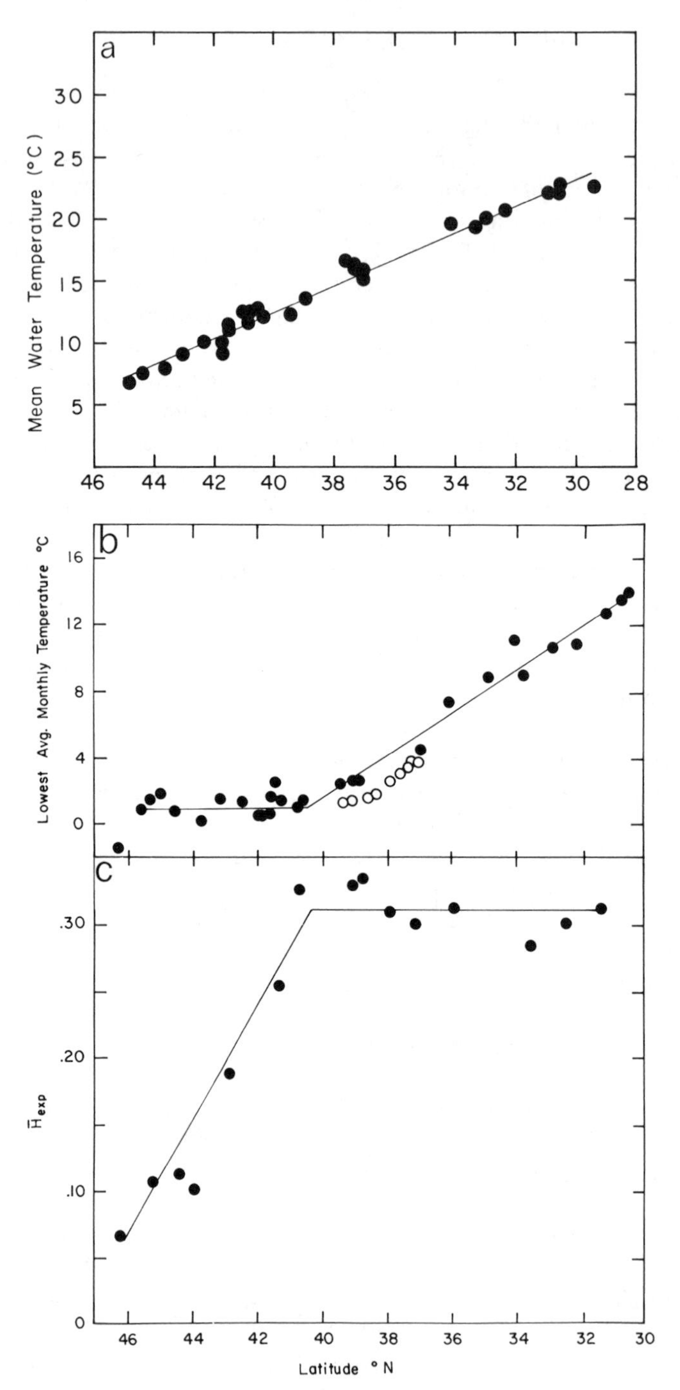

FIGURE 5.3 (a) Mean surface water temperature (°C) vs. latitude (°N), (b) lowest average monthly surface water temperature (°C) vs. latitude (°N), and (c) the degree of gene diversity ($\overline{H}_{exp}$; see Nei, 1975) for polymorphic enzyme loci vs. latitude (°N).

pointed out that this phenomenon is correlated with the amount of time that each population spends at or below freezing temperatures.

Thus, if temperature differentially affects the survival of *Fundulus* with specific allelic isozymes, then natural selection could be, or may have been, acting to change the gene frequency of populations that experience different thermal regimes along the East Coast. This possibility is supported by the finding of Mitton and Koehn (1975) that some loci are shifted toward "southern" phenotypes in fish inhabiting thermal effluents of power plants. It is further supported by shifts in gene frequency in several loci at localities where temperature anomalies are found (see Powers et al., 1986b). While these are suggestive correlations, they are not definitive and could be explained by stochastic arguments that would be impossible to test. As mentioned previously (e.g., Powers et al., 1986b), the problem with the classical natural historical and biogeographic analysis is that it often leads to two or more diametrically opposed hypotheses that can explain the same data. For that reason, such studies often raise more questions than they answer.

I have previously discussed (Powers et al., 1986b) how alternative driving forces, that is, chance and adaptive forces, could not be resolved by the classical approaches alluded to above. Thus, we shall attempt to falsify the hypothesis that natural selection was (or is) a major driving force responsible for the observed gene diversity. As stated earlier, the strategy is to begin with molecules and progress through higher levels of biological organization by a linked series of predictions followed by experimental validation. So far, we have found evidence consistent with selective differences between allelic isozymes of five loci. This includes six kinds of evidence: (1) kinetic analysis of purified allelic isozymes, (2) predictable differences in cell metabolism, (3) differential developmental rates, (4) genotype-specific swimming ability, (5) differential hatching survivorship, and (6) field selection experiments. We shall summarize some of the representative data in relation to the general strategy (Clarke, 1975) outlined earlier in this paper. Since work on *Ldh-B* has been most extensive to date, we shall present those findings as illustrative of our general approach and afterward show how in vitro kinetic studies have been used to predict physiological, developmental, and whole-organism performance responses. If needed, the fundamentals of enzyme kinetics, including lactate dehydrogenase, are reviewed in the appendix to this chapter.

Functional differences between the allelic isozymes are usually characterized by classical "steady-state" kinetic parameters (see Appendix). In brief, the velocity (v) at which an enzymatic reaction occurs is proportional to the concentration of the substrate(s) and the concentration of the enzyme. In the absence of products, when substrate(s) is (are) in excess (i.e., not limiting), a reaction will proceed at its maximum velocity (V_{max}). The V_{max} of an enzyme reflects its catalytic efficiency (k_{cat}) at saturating substrate. If, however, substrate concentration(s) are limited, the reaction will be less than V_{max}. Since

$V_{max} = k_{cat}[E_0]$ (see Appendix), differences in this parameter may be attributed to either different first-order rate constants (k_{cat}) or different enzyme concentrations, $[E_0]$. The substrate concentration at which the velocity (v) is 50% of V_{max} is called the K_m (Michaelis-Menten constant) or the apparent dissociation constant for the enzyme-substrate complex (see Appendix for qualification). While a single substrate reaction has one K_m, multisubstrate reactions have a K_m for each substrate. In addition to V_{max} and K_m, another useful parameter for describing the kinetics of enzymes is V_{max}/K_m which reflects the catalytic efficiency at low substrate concentrations (see Appendix, for physiological significances of this parameter). Since living cells contain substrates, reaction products, and other metabolites that can inhibit enzyme reactions, it is also useful to determine the inhibition constants (K_I) for products, substrates, and/or other metabolic inhibitors.

Lactate dehydrogenase

The enzyme lactate dehydrogenase (LDH) catalyzes the interconversion of pyruvate and lactate, and is thus involved in both the catabolism and anabolism of carbohydrates. During anaerobic glycolysis, the conversion of pyruvate to lactate by LDH is essential for continued ATP production. LDH also may convert lactate to pyruvate, which in turn may be used in gluconeogenesis or in the generation of ATP by aerobic metabolism.

LDH is a tetrameric enzyme. In somatic cells of vertebrates, at least two loci (*Ldh-A* and *Ldh-B*) code for subunits of LDH. A third locus (*Ldh-C*) is expressed in eye and nervous tissue of bony fish. The products of these genes are designated LDH-A_4, LDH-B_4, and LDH-C_4, respectively. These three genes appear to be independently regulated and show tissue specificity. Although *Ldh-A* and *Ldh-B* are simultaneously expressed in the same tissues of many vertebrates, there is a remarkable tissue specificity and exclusivity of *Ldh* expression in the tissues of many marine fishes, including *Fundulus heteroclitus.* White skeletal muscle, whose metabolism is predominantly anaerobic, expresses the *Ldh-A* locus. Red muscle and liver, which have significant aerobic metabolism, express almost exclusively *Ldh-B. Fundulus* erythrocytes, which have some aerobic capability (about 5 to 10%), also express *Ldh-B* exclusively. The suggested functional significance of the difference in LDH isozymes is that LDH-A_4 is principally involved in the conversion of pyruvate to lactate (i.e., anaerobic glycolysis), while LDH-B_4 is principally involved in the conversion of lactate to pyruvate (i.e., gluconeogenesis and aerobic metabolism) (Everse and Kaplan, 1973).

In *Fundulus heteroclitus*, the only locus that will be examined is *Ldh-B*, which has two codominant alleles: *Ldh-B^a* and *Ldh-B^b*. As illustrated in Figure 5.1, the relative proportions of these alleles vary with latitude. *Ldh-B^b* is predominant in the northern (i.e., colder) portions of the range, and *Ldh-B^a* is predominant in the southern (i.e., warmer) portion of the range. According to Clarke's scheme, the first step in the analysis of this pattern should be to establish any functional differences between the products of these two alleles. We have used various biochemical and physiological techniques to accomplish this step. Our conclusions are that the LDH-B_4 allelic isozymes differ in (1) catalytic efficiency at low substrate levels, (2) degree of inhibition by lactate, and (3) steady-state in vivo enzyme concentration. We shall elaborate on these points below.

At low temperatures the LDH-B_4^b allelic isozyme, whose gene frequency is greatest in the northern colder waters (Figure 5.1), has a greater catalytic efficiency (V_{max}/K_m) at low substrate than does LDH-B_4^a (Place and Powers, 1979). This is true in either catalytic direction (Place and Powers, 1984b). At higher temperatures, the situation is reversed with LDH-B_4^a, whose gene frequency is greatest in the southern warm waters, having a greater catalytic efficiency (i.e., V_{max}/K_m). A possible confounding factor is that the pH of cells decreases as their temperature increases. However, even after pH changes are taken into account, these differences between LDH-B_4^b and LDH-B_4^a are still evident (Figure 5.4). Thus, the catalytic functional assays clearly demonstrate that the LDH-B_4 allelic isozymes are functionally nonequivalent at the molecular level. Moreover, the nonequivalence can be correlated with the temperatures at which each allelic isozyme is most common.

Another difference between the allelic isozymes is in terms of product and substrate inhibition. Very high concentrations of reaction product(s) or substrate(s) may inhibit an enzyme's function. For example, during the conversion of pyruvate to lactate, the LDH-B_4^b isozyme is much less susceptible to product (i.e., lactate) inhibition than is the LDH-B_4^a isozyme. For the conversion of lactate to pyruvate, the LDH-B_4^b isozyme is more susceptible to substrate (i.e., lactate) inhibition than is the LDH-B_4^a isozyme. Moreover, the magnitude of inhibition is greater at cool temperatures than at warm ones. The putative selective significance of this difference may relate to the accumulation of lactate during extreme activity in fish, which can exceed 20 mM (see DiMichele and Powers, 1982b; Place and Powers, 1984a, 1984b).

Place and Powers (unpublished data) observed that the specific activity of LDH-B_4 in the livers of fish collected from different localities varied with latitude. For example, the LDH-B_4 of fish livers collected from Maine had a specific activity (μmoles$\cdot$min$^{-1}\cdot$mg protein^{-1}) twice that of the livers collected from Georgia. Maine fish are essentially only *Ldh-B*^{b} genotype and Georgia fish *Ldh-B*^{a} genotype. These differences can be equated with differences in LDH-B_4 concentration (see Appendix). As stated earlier, the water temperatures in Maine and Georgia are very different. Is this difference a function of different genes or is it induced by different environments? To answer this question, fish from Maine and Georgia were acclimated at 20 °C for thirty days and their LDH-B_4 specific activities determined. Crawford, Place, Cashon, and Powers (1985) found a consistent twofold difference in

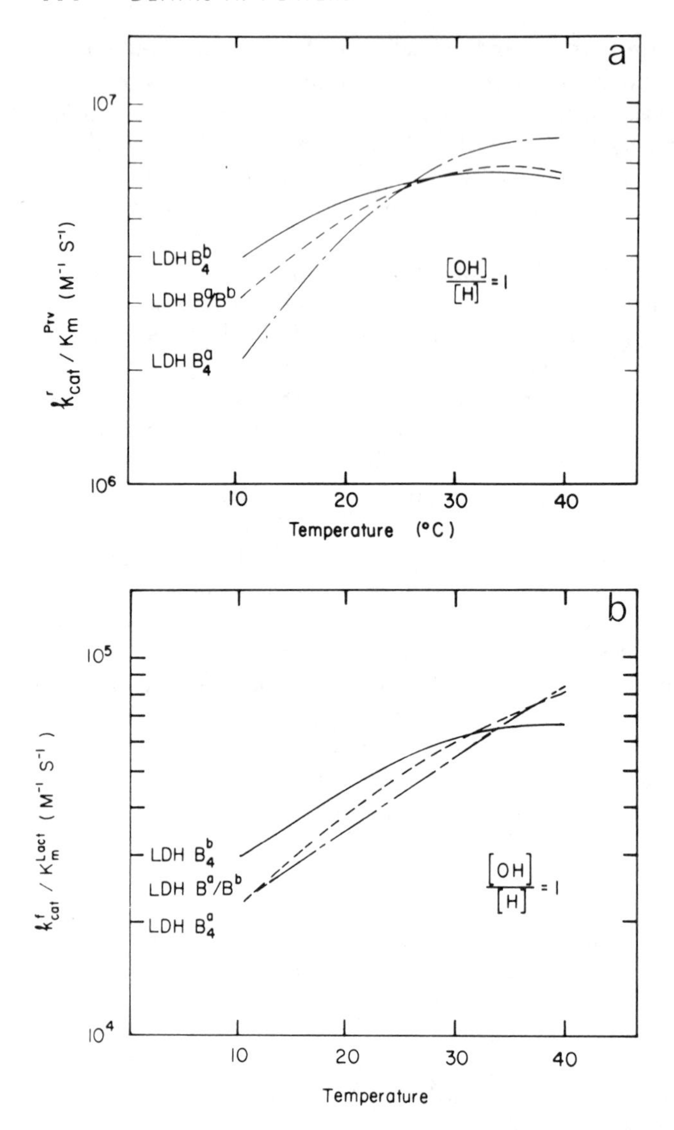

FIGURE 5.4 Effects of temperature (°C) on (a) pyruvate reduction and (b) lactate oxidation at a constant $[OH^-]/[H^+]$ ratio of unity. The values used were supplied by interpolation of the pH and temperature dependence of V_{max}/K_m for the reverse (a) and forward (b) catalytic directions. The pH at each temperature with a constant $[OH^-]/[H^+]$ ratio is described by: pH = ½pKw_i + C. In the above case, C = 0, and $V_{max} = k_{cat} [E_0]$.

specific activity and no apparent effect of thermal acclimation. However, similar analysis of populations from an intermediate latitude indicated LDH-B_4 activity intermediate to Maine and Georgia fish. Interestingly, when fish were scored for *Ldh-B* genotype, there were consistent differences between livers of different genotypes; *Ldh-B*^{b} genotypes had greater specific activity than did *Ldh-B*^{a}. While Crawford et al. (1985) have provided convincing evidence for differences in specific activity, long-term acclimation (perhaps years) has not been totally ruled out. Moreover, the mechanism for maintenance of these differences needs to be elucidated. Such differences in in vivo enzyme concentration could be due to any number of causes including a different number of gene copies, transcriptional factors, rates of mRNA degradation, rates of protein synthesis and/or degradation, and so on. These possibilities are under study.

Functional differences in erythrocytes

Ideally, one would like to predict the functional physiological differences at the cellular and organismal level that should result from catalytic differences. However, such predictions assume that in vitro kinetic differences are large enough to produce significant variation in vivo. Thus, any physiological differences predicted on the basis of in vitro kinetic studies must be rigorously tested experimentally at the cellular, organismal, and population level.

Because the LDH-B_4^b enzyme had a greater catalytic efficiency and a higher in vivo specific activity than the LDH-B_4^a enzyme, we reasoned that the metabolic flux or initial metabolic rates of cells that rely heavily on glycolysis should be greater for *Ldh-B*^{b} genotypes at temperatures less than 20 °C. These differences should result in higher steady-state ATP concentrations. That prediction was confirmed when we demonstrated that the red cells with LDH-B_4^b had 2.11 ± 0.22 ATPs/Hb whereas cells with LDH-B_4^a had only 1.65 ± 0.12 ATPs/Hb (Powers, Greaney, and Place, 1979).

The finding of differences between *Ldh-B* genotypes in intraerythrocyte ATP concentrations was particularly important because it allowed the prediction of hemoglobin-oxygen (Hb-O_2) affinity differences between adult fish (Powers et al., 1979), differences in swimming performance (DiMichele and Powers, 1982b), and differential hatching of embryos with different *Ldh-B* genotypes (DiMichele and Powers, 1982a). All of these predictions revolve around the fact that ATP is an allosteric effector of fish hemoglobin; that is, it affects Hb-O_2 affinity.

The preferential allosteric binding of organic phosphates to deoxyhemoglobin results in a decrease in hemoglobin-oxygen affinity (Sugita and Chanutin, 1963; Benesch and Benesch, 1967; Chanutin and Curnish, 1967). The major organic phosphate in fish erythrocytes is either adenosine triphosphate (ATP; e.g., Gillen and Riggs, 1971) or guanosine triphosphate (GTP; e.g.,

Geohegan and Polkuhowich, 1974), both of which decrease the affinity of hemoglobin for oxygen.

Fundulus with the *Ldh-B[a]* genotype have erythrocyte ATP levels that are significantly less than those with the *Ldh-B[b]* genotype, while heterozygotes have intermediate concentrations. We predicted that fish with higher levels of intraerythrocyte ATP (i.e., *Ldh-B[b]*) should have blood with a lower oxygen affinity at pH values favoring ionic interactions between ATP and hemoglobin (reviewed by Powers, 1980; Powers, Dalessio, Lee, and DiMichele, 1986). As predicted, *Ldh-B[a]* erythrocytes had a higher oxygen affinity than *Ldh-B[b]* genotypes, which can unload oxygen more readily. From these differences in oxygen loading and unloading, we can make testable predictions about whole-organism responses. I shall illustrate this point with two examples: (1) the prediction of differential hatching rates, and (2) differential swimming ability at low temperature.

Developmental rate and hatching time

DiMichele and Taylor (1980, 1981) have shown that respiratory stress triggers the hatching mechanism in *F. heteroclitus*. In view of this, we (DiMichele and Powers, 1982a) reasoned that the hatching rates of *Ldh-B* genotypes should differ because of differences in hemoglobin-oxygen affinity. We predicted that the *Ldh-B[a]* genotype embryos should feel oxygen stress before *Ldh-B[b]* genotypes and thus should hatch first. Consistent with that expectation we found that *F. heteroclitus* embryos hatched at rates that were highly correlated with *Ldh-B* genotype. *Ldh-B[a]* genotypes hatched before *Ldh-B[b]* genotypes, and the heterozygotes had an intermediate hatching distribution (Figure 5.5).

Mostly *Ldh-B[a]* genotype eggs hatched in the first three days of the hatching period, and mostly *Ldh-B[b]* eggs hatched in the last half. The heterozygote eggs hatched over the entire time span. The overall mean hatching times for offspring were 11.9 days for the *Ldh-B[a]* genotype, 12.4 days for the heterozygotes, and 12.8 days for the *Ldh-B[b]* homozygotes (DiMichele and Powers, 1982a). In a much larger experiment that included three loci (*Ldh-B*, *Mdh-A*, and *Gpi-A*) and slightly different physical conditions, the hatching order remained the same and the differences were even more pronounced (DiMichele, Powers, and DiMichele, 1986).

The time of hatching is very important to *Fundulus* populations because of its unique reproductive strategy. The eggs of *F. heteroclitus* are laid in empty mussel shells or between the leaves of the marsh grass, *Spartina alterniflora* (Taylor, DiMichele, and Leach, 1977). Under these conditions, the eggs incubate in air for most of their developmental period. Hatching occurs when eggs laid on one spring tide are immersed in water by the following spring tide. As water covers the eggs, environmental oxygen decreases at the egg surface, which is the hatching cue for the embryo (DiMichele and Tay-

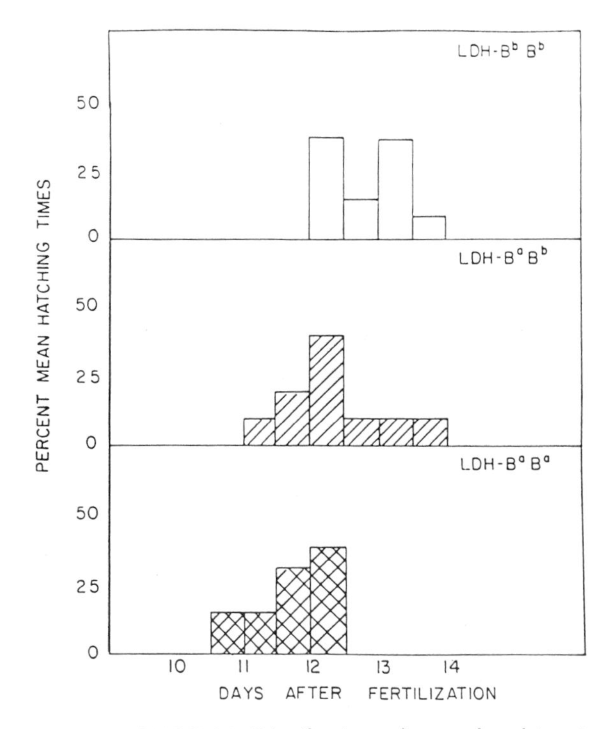

FIGURE 5.5 Distribution of mean hatching times among three *Ldh-B* genotypes from twenty random crosses. Analysis of variance indicated that there were differences between the crosses ($p < .05$). Duncan's Multiple Range Test showed that some of this variation was due to *Ldh-B* genotype. (From DiMichele and Powers, 1982a.)

lor, 1980). Hatching at the correct time would seem to be important for survival of the fry. Therefore, overall plasticity in hatching times may be important in protecting *F. heteroclitus* populations that live under variable environmental conditions. Our data suggest that premature hatching cues (e.g., rainstorms) would select mostly against *Ldh-B^a* homozygotes, whereas late hatching (i.e., after the tide has retreated) would primarily select against the *Ldh-B^b* homozygous genotypes. This has been recently substantiated by field selection experiments (see DiMichele et al., 1986a). This is particularly interesting in light of the finding of Meredith and Lotrich (1979) that the mortality of *F. heteroclitus* in age class zero (eggs to fry of 59 mm) is greater than 99.5%.

If extremes in hatching time are selected against, then there would be a net heterozygote advantage in a variable or uncertain environment. Such an advantage would result in the maintenance of genetic variability at the *Ldh-B* locus as well as a stability in gene frequency at those localities where such selection operates. This is consistent with the temporal stability in *Ldh-B* gene frequencies at several localities (Powers and Place, 1978; Cashon, Van Beneden, and Powers, 1981). However, spatial changes in the gene frequencies for a number of loci are probably due to other physical, biological, and stochastic factors.

Since erythrocyte ATP concentrations are correlated with *Ldh-B* genotype and ATP regulates hemoglobin-oxygen affinity, the simplest interpretation is that hypoxia-induced hatching of *Ldh-B* variants results from functional differences between LDH-B_4 allelic isozymes that affect ATP levels. DiMichele and Powers (1984) have presented evidence that there are also differences in oxygen consumption and developmental rates such that oxygen stress levels are achieved earlier in the *Ldh-B^a* homozygous embryos than in those with the *Ldh-B^b* genotype.

Physiological basis for swimming endurance differences between LDH-B genotypes

Our analysis of purified LDH-B_4 allelic isozymes indicates that the greatest catalytic differences between LDH-B_4^a and LDH-B_4^b exist at low temperature (10 °C), while no significant difference exists at 25 °C (see Figure 5.4). We reasoned that if the LDH-B_4 enzyme has a direct influence on erythrocyte ATP concentration, then differences in ATP and blood oxygen affinity should exist only at body temperatures below 25 °C. Furthermore, since organic phosphate amplifies the Bohr effect of *F. heteroclitus* hemoglobins (Mied and Powers, 1978), these phenomena should be exaggerated at low pH values, like those produced during swimming performance. To test these predictions after an acclimation period, fish of each of the two homozygous LDH-B_4 phenotypes were swum to exhaustion in a closed water tunnel. As predicted, swimming performance was highly correlated with genetic variation at the *Ldh-B* locus for *Fundulus* acclimated to and swum at 10 °C, while no such difference existed for the 25 °C treatment (DiMichele and Powers, 1982b).

Among resting fish acclimated to 10 °C, hematocrit, blood pH, blood oxygen affinity, serum lactate, liver lactate, and muscle lactate did not differ significantly between the two *Ldh-B* homozygous genotypes. Fish exercised to fatigue at 10 °C showed a significant change in all of these parameters. The LDH-B_4^b phenotype fish were able to sustain a swimming speed 20% higher than that of LDH-B_4^a fish. Blood oxygen affinity, serum lactate, and muscle lactate also differed between the phenotypes. Since the rate of lactate accumulation was the same for the LDH-B_4 phenotypes, fish with LDH-B_4^b accumulated more lactate in the blood and muscle simply because they swam longer.

In an extensive analysis of the binding of ATP to carp deoxyhemoglobin, Greaney, Hobish, and Powers (1980) showed that the organophosphate-hemoglobin affinity constants change by two orders of magnitude between pH 8 and pH 7. The same general phenomenon appears to be true for *F. heteroclitus* hemoglobins (Powers, 1980). In resting *Fundulus* at 10 °C, the blood pH was about 7.9. At this pH, the difference in erythrocyte ATP between LDH-B_4 phenotypes (ATP/Hb were 1.65 ± 0.12 and 2.11 ± 0.22 for LDH-B_4^a and LDH-B_4^b, respectively) is not reflected as a significant difference in blood oxygen affinity. However, as blood pH falls with increasing exercise, the organophosphate-hemoglobin affinity constant increases, and differences in blood oxygen affinity between homozygous *Ldh-B* genotypes become apparent (Figure 5.6). As blood pH is lowered, ATP amplifies the dissociation of oxygen from *F. heteroclitus* hemoglobin; the more ATP, the greater the effect. This difference is translated into a differential ability to deliver oxygen to muscle tissue, which in turn affects swimming performance (DiMichele and Powers, 1982b).

Fish acclimated to 25 °C did not differ significantly in erythrocyte ATP concentrations. The ATP/Hb ratios were 1.45 ± 0.24 and 1.65 ± 0.31 for the *Ldh-B^a* and *Ldh-B^b* genotypes, respectively. In addition, there were no significant differences between LDH-B_4 phenotypes in any of the other parameters. Since there were swimming differences at 10 °C but none at 25 °C, the data validate our predictions based on kinetic analyses, metabolic studies, selection experiments, and so on. These collective data strongly suggest that the *Ldh-B* genotypes in *F. heteroclitus* are differentially affected by natural selection. On the other hand, we have consistently pointed out that such data do not absolutely rule out the possibility that other genes, very tightly linked to the *Ldh-B* locus, may be responsible for the observed cellular and whole-organism responses alluded to above. In fact, one could invoke a linked "mystery" locus for most published genetic and physiological studies. While it is difficult to rule out this possibility, recent studies by DiMichele and his colleagues (unpublished) clearly indicate that developmental rate can be altered by microinjection of additional LDH enzyme or by the introduction of an inhibitor of LDH. These studies show a specific effect of LDH activity on developmental rate that is independent of genetic background.

Finally, the approach illustrated by the *Ldh-B* locus of *Fundulus heteroclitus* may lead one to conclude that natural selection is acting on that locus or a tightly linked locus. However, demonstrating functional nonequivalence of allelic products, or even selection for a single locus, will not resolve the "neutralist"/"selectionist" controversy. The major question is not, Does selection operate at the molecular level? Rather it is, What fraction of the observed polymorphic loci are a function of natural selection? Therefore, a sufficient number of enzyme loci must be evaluated in order to determine what portion of the genome can actually be affected by natural selection and how specific gene combinations interact at the functional or regulatory levels to increase or reduce relative fitness.

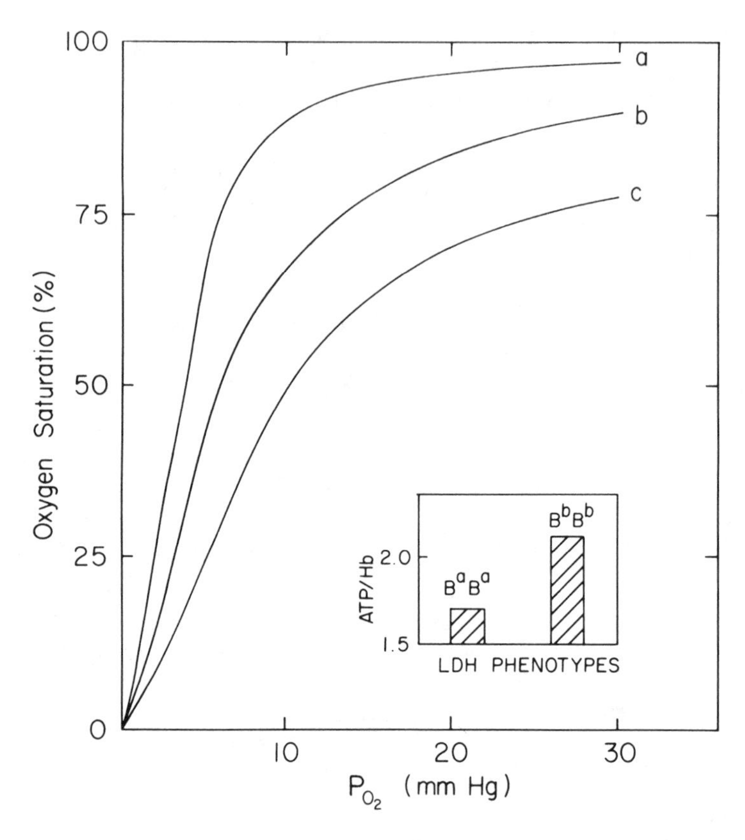

FIGURE 5.6 Oxygen equilibrium curves for whole blood of *Fundulus heteroclitus* acclimated to 10 °C, as determined with an oxygen dissociation analyzer (Aminco). The ordinate is the percent saturation of hemoglobin by oxygen, and the abscissa is the partial pressure of oxygen (P_{O_2}). Indicated in the figure are oxygen equilibrium curves of blood from (a) resting fish of both *Ldh-B* genotypes, (b) LDH-B_4^a fish swum to exhaustion, and (c) LDH-B_4^b fish swum to exhaustion. The intraerythrocyte ratio of ATP to hemoglobin (ATP/Hb) in resting fish is 1.65 ± 0.12 for LDH-B_4^a and 2.11 ± 0.22 for LDH-B_4^b. (From DiMichele and Powers, 1982b.)

Conclusions

The example of *Fundulus heteroclitus* demonstrates a novel approach that is being used by some scientists interested in intraspecies biochemical adaptation of specific genotypes. This example illustrates the utility of Clarke's (1975) approach and should act as a cautionary note for physiological ecologists who are interested in interspecific comparisons. For example, oxygen consumption rates, swimming ability, hatching rate, etc. are variable within some species, and thus values collected from a few individuals from a particular locality may not be representative of the species as a whole. In fact, variation within a species may be as great or greater than variation between species.

Our work on *Fundulus heteroclitus* indicates that a multidisciplinary approach to problems of intraspecies variation provides answers to complex problems that cannot be addressed by a more monolithic approach. We have emphasized the importance of starting with simple molecular systems and making predictions that can be tested by experimentation at a higher level of biological complexity – leading from molecules to cells to organ systems to organisms and eventually to ecological interactions.

While this approach has proven fruitful, the use of modern molecular techniques may help answer yet other questions that previously could not be resolved. For example, as one clones and sequences the DNAs of various allelic alternatives, it will be possible to determine (l) the presence of silent mutations, (2) the number of genomic copies, (3) the quantity of specific mRNA, (4) the mechanisms of gene expression, etc. for organisms from different populations and species. The cloning of specific genomic DNAs will open the door to investigation of evolutionary aspects of tissue specificity, evolutionary rate, gene expression, and perhaps the creation of new "species." Moreover, the use of noninvasive methods like nuclear magnetic resonance (NMR) will allow one to study the dynamic change of critical metabolites (e.g., ATP, phospholipids, glycolytic intermediates, etc.) during the development of living embryos. Such an approach, which can also be applied to other living tissues, allows the determination of free as well as bound cofactors (e.g., $NADP^+$, NADPH, NAD^+, NADH) and high-energy compounds (e.g., ATP and GTP). These and other molecular techniques will allow scientists to unravel questions that are not approachable by classical methods. However, they will not replace the more classical physiological and ecological approach. Rather, the molecular analysis will be complementary and expand the depth in which scientists can address fundamental questions in ecological physiology. This new exciting frontier is limited only by our imagination and willingness to apply modern molecular technology.

Appendix: Enzyme kinetics

Single-substrate reaction

Steady-state kinetic analysis of an enzyme is important as a basis for understanding metabolic functions. For a single substrate reaction like that in scheme 1 (Figure 5.7), the reaction velocity (v) is described by equation 5.1.

$$v = \frac{k_{cat}\,[E_0]\,[S]}{K_m + [S]} \tag{5.1}$$

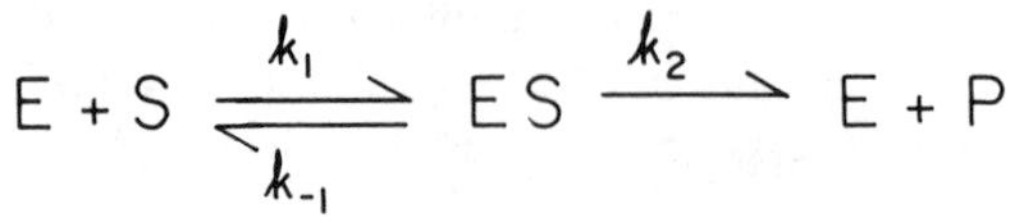

$$E + S \underset{k_{-1}}{\overset{k_1}{\rightleftharpoons}} ES \xrightarrow{k_2} E + P$$

FIGURE 5.7 Scheme 1: a simple reaction of an enzyme (E) with a single substrate (S), where k_1 and k_2 are forward rate constants and k_{-1} is the rate of enzyme-substrate complex breaking down to form E and S.

where k_{cat} is the catalytic rate constant or turnover number (units, s^{-1}); $[E_0]$ is the enzyme concentration at time zero; $[S]$ is the substrate concentration; and K_m is the classical Michaelis-Menten constant for substrate, S. The Michaelis-Menten mechanism illustrated in equation 5.1 assumes that the ES complex is in thermodynamic equilibrium with free enzyme (E) and substrate (S). This is true only when $k_2 << k_{-1}$. Under such conditions, the K_m is equal to the dissociation constant for the ES complex. While K_m is equal to the dissociation constant for many reactions, this is not always the case (see Fersht, 1977, regarding deviations from Michaelis-Menten conditions).

At very high substrate levels (e.g., $[S] >>> K_m$) equation 5.1 approaches:

$$v = k_{cat} [E_0] \tag{5.2}$$

Because the enzyme is saturated with substrate, further increases in $[S]$ have no effect on the reaction rate. Therefore, the reaction is said to be zero order with respect to substrate, but first order with respect to enzyme concentration, $[E_0]$. Thus, k_{cat} is the first-order rate constant with respect to enzyme at saturating substrate and is denoted by $V_{max} = k_{cat} [E_0]$. When the reaction velocity is a one-half of it maximum (i.e., $v = ½V_{max}$), the substrate concentration $[S]$, will be equal to the Michaelis-Menten constant, K_m (Note: substitute K_m for $[S]$ in equation 5.1 and simplify.)

At very low $[S]$, the velocity increases linearly with substrate. When $[S] <<< K_m$, equation 5.1 simplifies to:

$$v = \frac{k_{cat} [E_0] [S]}{K_m} \tag{5.3}$$

Equation 5.3 illustrates a two-substrate reaction (i.e, E_0 and S) with a second order rate constant of k_{cat}/K_m (units: $M^{-1} \cdot s^{-1}$). If the enzyme is held constant and not considered to change significantly during the reaction, then equation 5.3 can be written as:

$$v = \frac{V_{max} [S]}{K_m} \tag{5.4}$$

such that V_{max}/K_m becomes a pseudo-first-order rate constant (units, s^{-1}) with respect to substrate, but only when $[S] <<< K_m$.

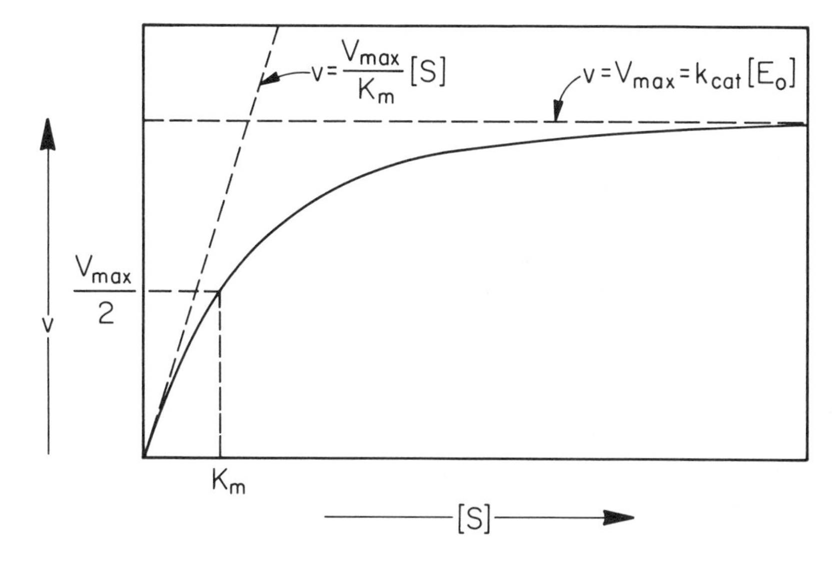

FIGURE 5.8 Reaction rate (v) plotted against substrate concentration, $[S]$, for a reaction obeying a single-substrate Michaelis-Menten kinetic mechanism.

The simple, single-substrate Michaelis-Menten equation is graphically represented by a rectangular hyperbola with asymptotes of V_{max} in the first quadrant and $-K_m$ in the third quadrant with a tangent of V_{max}/K_m at $v=0$ and $[S]=0$. Figure 5.8 illustrates the first quadrant.

Two-substrate reaction

Most enzyme reactions involve multiple substrates and products. Dehydrogenases, including LDH, generally have two substrates and two products. Of these, there are two major types of mechanisms: (1) the ternary-complex mechanisms, wherein the reaction proceeds through an enzyme-substrate complex that contains both substrates; and (2) the substituted enzyme mechanism, wherein the reaction proceeds through an enzyme-transfer group stage without either substrate bound. Lactate dehydrogenase follows the ternary complex mechanism. Moreover, for LDH the addition of substrates must be done in a specific order (i.e., in either catalytic direction, cofactor must be added first). Therefore, LDHs obey a compulsory ordered bisubstrate ternary-complex mechanism.

In the absence of product inhibition, the reaction for LDH can be most simply illustrated by the diagrammatic notation of Cleland (1970) (scheme 2, Figure 5.9). The reaction is described as proceeding from left to right. The rate constants with positive subscripts indicate rates for the addition of substrate whereas negative values are "off" constants with respect to the forward

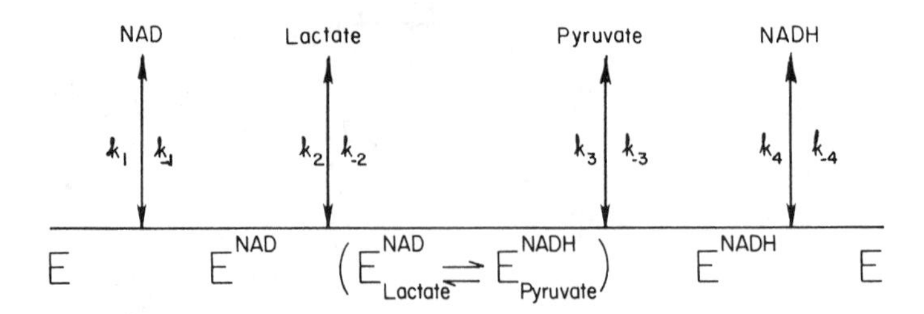

FIGURE 5.9 Scheme 2: the LDH reaction in the absence of product inhibition, illustrated by the diagrammatic notation of Cleland (1970). See text for further information.

direction. The large E represents the enzyme, and the enzyme-substrate complexes are indicated with the appropriate substrate(s) or product(s) bound. In the absence of product inhibition, scheme 2 can be described by equation 5.5 for the forward reaction (lactate to pyruvate) and equation 5.6 for the reverse direction (i.e., pyruvate to lactate). In the absence of products, the velocity for the forward direction (v^f) is:

$$v^f = \frac{V^f_{max}[\mathrm{NAD}][\mathrm{Lactate}]}{K_a^{\mathrm{NAD}}\, K_m^{\mathrm{Lact}} + K_m^{\mathrm{Lact}}\,[\mathrm{NAD}] + K_m^{\mathrm{NAD}}\,[\mathrm{Lactate}] + [\mathrm{NAD}]\,[\mathrm{Lactate}]} \tag{5.5}$$

and for the reverse direction (v^r) is:

$$v^r = \frac{V^r_{max}[\mathrm{NADH}][\mathrm{Pyruvate}]}{K_a^{\mathrm{NADH}} K_m^{\mathrm{Prv}} + K_m^{\mathrm{Prv}}[\mathrm{NADH}] + K_m^{\mathrm{NADH}}[\mathrm{Pyruvate}] + [\mathrm{NADH}]\,[\mathrm{Pyruvate}]} \tag{5.6}$$

where K_m^{NAD} is the Michaelis-Menten constant for NAD^+; K_a^{NAD} is the estimated dissociation constant for the NAD^+-LDH complex; K_a^{Lact} is the Michaelis-Menten constant for lactate; and V^f_{max} is the maximum rate of lactate oxidation. For the reverse reaction (equation 5.6), K_m^{NADH} is the Michaelis-Menten constant for NADH; K_a^{NADH} is the estimated dissociation constant for the NADH-LDH complex; K_m^{Prv} is the Michaelis-Menten constant for pyruvate; and V^r_{max} is the maximum rate of pyruvate reduction. Each of these parameters is defined by two or more of the individual rate constants depicted in scheme 2. These parameters are defined in Table 5.1; see also Figure 5.8.

If cofactor binding does not vary between allelic isozymes, kinetic studies can be done at saturating cofactor. Under those conditions, equations 5.5 and 5.6 reduce to simple Michaelis-Menten equations 5.7 and 5.8, respectively.

$$v^f = \frac{V^f_{max}\,[\mathrm{Lactate}]}{K_m^{\mathrm{Lact}} + [\mathrm{Lactate}]} \tag{5.7}$$

$$v^r = \frac{V^r_{max}\,[\mathrm{Pyruvate}]}{K_m^{\mathrm{Prv}} + [\mathrm{Pyruvate}]} \tag{5.8}$$

TABLE 5.1 LDH kinetic parameters and equivalent rate constant expressions for equations 5.5 and 5.6 as illustrated[a]

Forward reaction		Reverse reaction	
Constant	Rate expression	Constant	Rate expression
K_m^{NAD}	$\frac{k_3 k_4}{k_1(k_3 + k_4)}$	K_m^{NADH}	$\frac{k_{-1} k_{-2}}{k_{-4}(k_{-1} + k_{-2})}$
K_a^{NAD}	$\frac{k_{-1}}{k_1}$	K_a^{NADH}	$\frac{k^4}{k^{-4}}$
K_m^{Lact}	$\frac{k_4(k_{-2} + k_3)}{k_2(k_3 + k_4)}$	K_m^{Prv}	$\frac{k_{-1}(k_{-2} + k_3)}{k_3(k_{-1} + k_{-2})}$
V_{max}^f	$\frac{k_3 k_4 [E_0]}{k_3 + k_4}$	V_{max}^r	$\frac{k_{-1} k_{-2} [E_0]}{(k_{-1} + k_{-2})}$

[a]See Figure 5.8.

Since $V_{max} = k_{cat} [E_0]$, equations 5.7 and 5.8 can be described by the two fundamental rate constants k_{cat} and k_{cat}/K_m as well as by the enzyme concentration $[E_0]$.

Inhibition

The LDH reaction is inhibited by the presence of high levels of substrate and products. The inhibition is observed in both directions for both LDH-A_4 and LDH-B_4, but it is most pronounced for the latter. These phenomena are related to the formation of abortive ternary complexes like LDH-NAD^+-pyruvate and LDH-NADH-lactate. Since cells have both substrates and products, a realistic analysis of LDH allelic isozyme reactions should include these complexes (e.g., see Place and Powers, 1984b). With the use of Cleland's (1970) notation the LDH reaction can be depicted as in scheme 3 (Figure 5.10). The equations analogous to equations 5.5 and 5.6 that involve the inhibition constants for pyruvate and lactate (K_I^{Prv} and K_I^{Lact}, respectively) are very large and complicated and their significance cannot be explained here because of space limitations. A detailed account of simple and multiple inhibition can be found in Segel (1975).

The physiological and evolutionary significance of V_{max}/K_m

Substrate concentrations in vivo are normally too low to saturate enzymes. For example, the measured intracellular pyruvate and lactate concentrations are generally lower than their respective K_m values (Fersht, 1977). When the substrate concentrations are one-tenth the K_m or less, the substrate terms in

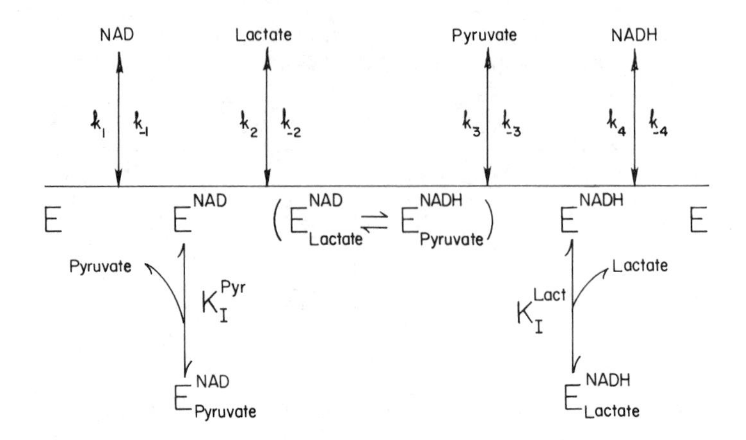

FIGURE 5.10 Scheme 3: the reaction illustrated in scheme 2, but including product inhibition. See text for further information.

the denominator of the Michaelis-Menten equation can be neglected and V_{max}/K_m becomes a pseudo-first-order rate constant (equation 5.4). However, at very low substrate levels where the enzyme must be considered a substrate, then k_{cat}/K_m is a much more appropriate parameter. Moreover, sometimes in vivo enzyme concentration varies and thus, while k_{cat}/K_m is independent of $[E_0]$, V_{max}/K_m is not.

Considerable discussion has centered on the evolutionary importance of V_{max}/K_m (e.g., Crowley, 1975; Cornish-Bowden, 1976; Brocklehurst and Cornish-Bowden, 1976; Brocklehurst, 1977; Fersht, 1977). Because k_{cat}/K_m cannot be larger than any of the second-order rate constants of the appropriate reaction pathway, its value sets a lower limit on the rate of enzyme-substrate association. When k_{cat}/K_m is large (e.g., $10^{-8}\ s^{-1} \cdot M^{-1}$), the reaction is maximized and becomes diffusion limited. Since $V_{max}/K_m = k_{cat}[E_0]/K_m$, comparisons of V_{max}/K_m reflect the catalytic efficiency at low substrate concentration. If the prime evolutionary objective is to attain the highest possible rates, then both V_{max} and K_m should increase within the constraints imposed by the diffusion-limited value of V_{max}/K_m. According to Crowley (1975) and Fersht (1977), V_{max} should be as large as possible and K_m should be large compared with the physiological substrate levels. Cornish-Bowden (1976) suggests that there is little to be gained in increasing the K_m much more than an order of magnitude over the substrate concentration.

Finally, if there are differences between the two allelic isozymes in V_{max}/K_m for one catalytic direction (e.g., pyruvate to lactate), there must be differences in V_{max}/K_m for the other catalytic direction (e.g., lactate to pyruvate) because the ratio of the V_{max}/K_m values for the two directions is equivalent to the thermodynamic equilibrium constant for the chemical reaction, which is independent of the enzyme (Haldane, 1930). In other words, although kinetic differences might be detected between the allelic isozymes, the changes must occur within the thermodynamic constraints imposed by the thermodynamic equilibrium constant. A change in one or more rate constants in one direction must be accompanied by an equivalent compensation in one or more rate constants in the opposite catalytic direction.

References

Benesch, R., and Benesch, R. E. (1967) The effect of organic phosphates from the human erythrocyte on the allosteric properties of hemoglobin. *Biochem. Biophys. Res. Commun.* 26:162–167.

Bijlsma, R. (1978) Polymorphism at the glucose-6-phosphate dehydrogenase and 6-phosphogluconate dehydrogenase loci in *Drosophila melanogaster*. Part 2. Evidence for interaction in fitness. *Genet. Res.* 31:227–238.

Brocklehurst, K. (1977) Evolution of enzyme catalytic power characteristic of optimal catalysis evaluated for the simplest plausible kinetic model. *Biochem. J.* 163:111–116.

Brocklehurst, K., and Cornish-Bowden, A. (1976) The pre-eminence of k_{cat} in the manifestation of optimal enzymatic activity delineated by using the Briggs-Haldane two-step irreversible kinetic model. *Biochem. J.* 159:165–166.

Cashon, R. E., Van Beneden, R. J., and Powers, D. A. (1981) Biochemical genetics of *Fundulus heteroclitus* (L.). IV. Spatial variation in gene frequencies of *Idh-A*, *Idh-B*, *6-Pgdh-A*, and *Est-S*. *Biochem. Genet.* 19:715–728.

Cavener, D. R., and Clegg, M. T. (1981) Evidence for biochemical and physiological differences between genotypes in *Drosophila melanogaster*. *Proc. Natl. Acad. Sci. U.S.A* 78:4444–4447.

Chanutin A., and Curnish, R. R. (1967) Effect of organic phosphates on the oxygen equilibrium of human erythrocytes. *Arch. Biochem. Biophys.* 121:96–102.

Clarke, B. (1973) Neutralists vs. selectionists. *Science* 180:600–601.

Clarke, B. (1975) The contribution of ecological genetics to evolutionary theory: detecting the direct effects of natural selection on particular polymorphic loci. *Genetics* 79:101–108.

Cleland, W. W. (1970) Steady-state kinetics. In *The Enzymes*, 3rd ed., vol. 2, ed. P. Boyer, pp. 1–65. New York: Academic Press.

Cornish-Bowden, A. (1976) The effect of natural selection on enzymic catalysis. *J. Mol. Biol.* 101:1–9.

Crawford, D. L., Place, A., Cashon, R., and Powers. D. A. (1985) Thermal acclimation or differential regulation. *Am. Zool.* 25:122.

Crowley, P. H. (1975) Natural selection and K_m. *J. Theor. Biol.* 50:461–475.

Curtsinger, J. W., and Laurie-Ahlberg, C. C. (1981) Genetic variability of flight metabolism in *Drosophila melanogaster*. I. Characterization of power output during tethered flight. *Genetics* 98:549–564.

Daly, K., and Clarke, B. (1981) Selection associated with alcohol dehydrogenase locus in *Drosophila melanogaster*: differential survival of adults maintained on low concentration of ethanol. *Heredity* 46:219–226.

DiMichele, L., and Powers, D. A. (1982a) LDH-B genotype specific hatching times of *Fundulus heteroclitus* embryos. *Nature* 296:563–564.

DiMichele, L., and Powers, D. A. (1982b) Physiological basis for swimming endurance differences between LDH-B genotypes of *Fundulus heteroclitus. Science* 216:1014–1016.

DiMichele, L., and Powers, D. A. (1984) Developmental and oxygen consumption rate differences between lactate dehydrogenase genotypes of *Fundulus heteroclitus* and their effect on hatching time. *Physiol. Zool.* 57:52–56.

DiMichele, L., and Taylor, M. H. (1980) The environmental control of hatching in *Fundulus heteroclitus. J. Exp. Zool.* 214:181–187.

DiMichele, L., and Taylor, M. H. (1981) The mechanism of hatching in *Fundulus heteroclitus*: developmental physiology. *J. Exp. Zool.* 217:73–79.

DiMichele, L., Powers, D. A., and DiMichele, J. A. (1986a) Developmental and physiological consequences of genetic variation of enzyme synthesizing loci in *Fundulus heteroclitus. Am. Zool.* 26:201–208.

DiMichele, L., Powers, D. A., and Taylor, M. H., eds. (1986b) The Biology of *Fundulus heteroclitus. Am. Zool.* 26:107–288.

Dorado, G., and Barbancho, M. (1984) Differential responses in *Drosophila melanogaster* to environmental ethanol: modification of fitness components at the *Adh* locus. *Heredity* 53:309–320.

Endler, J. A. (1977) *Geographic Variation, Speciation and Clines.* Princeton, N.J.: Princeton University Press.

Everse, J., and Kaplan, N. O. (1973) Lactate dehydrogenases: structure and function. *Adv. Enzymol.* 37:61–133.

Fersht, A. (1977) *Enzyme Structure and Mechanism.* San Francisco: W. H. Freeman.

Geohegan, W. H., and Polkuhowich, J. J. (1974) The major erythrocyte organic phosphates of the American eel *Anguilla rostrata. Comp. Biochem. Physiol.* 49B:281–290.

Gillen, R. G., and Riggs, A. (1971) The hemoglobins of a freshwater teleost *Cichlasoma cyanoguttatum*: the effects of phosphorylated organic compounds upon the oxygen equilibria. *Comp. Biochem. Physiol.* 38B:585.

Gonzalez-Villasenor, L. I., Burkhoff, A. M., Corces, V., and Powers, D. A. (1986) Characterization of cloned mtDNA from the teleost *Fundulus heteroclitus* and its usefulness as an interspecies hybridization probe. *Can. J. Fish. Aquat. Sci.* 43:1866–1872.

Greaney, G. S., Hobish, M. K., and Powers, D. A. (1980) The effects of temperature and pH on the binding of ATP to carp (*Cyprinus carpio*) deoxyhemoglobin. *J. Biol. Chem.* 255:445–453.

Haldane, J. B. S. (1930) *Enzymes.* London: Longmans.

Harris, H. (1976) The neutralist vs. selectionist controversy. *Proc. Fed. Am. Soc. Exp. Biol.* 35:2079–2082.

Hughes, M. B., and Lucchesi, J. C. (1977) Genetic rescue of a lethal "null" activity allele of 6-phosphogluconate dehydrogenase in *Drosophila melanogaster. Science* 196:1114–1115.

Huxley, J. S. (1938) Clines: an auxiliary taxonomic principle. *Nature* 142:219–220.

Koehn, R. K., and Immerman, F. W. (1981) Biochemical studies of aminopeptidase polymorphism in *Mytilus edulis.* I. Dependence of enzyme activity on season, tissue and genotype. *Biochem. Genet.* 19:1115–1142.

Koehn, R. K., and Siebenaller, J. F. (1981) Biochemical studies of aminopeptidase polymorphism in *Mytilus edulis.* II. Dependence of reaction rate on physical factors and enzyme concentration. *Biochem. Genet.* 19:1143–1162.

Lewontin, R. C. (1974) *The Genetic Basis of Evolutionary Change.* New York: Columbia University Press.

Meredith, W. H., and Lotrich, V. A. (1979) Production dynamics of a tidal creek population of *Fundulus heteroclitus* (L). *Estuar. Coast. Mar. Sci.* 8:88–118.

Mied, P. A., and Powers, D. A. (1978) Hemoglobins of the killifish *Fundulus heteroclitus*: separation, characterization, and a model for the subunit composition. *J. Biol. Chem.* 253:3521–3528.

Miller, S., Pearcy, R. W., and Berger, E. (1975) Polymorphism at the α-glycerophosphate dehydrogenase locus in *Drosophila melanogaster.* I. Properties of adult allozymes. *Biochem. Genet.* 13:175–188.

Mitton, J. B., and Koehn, R. K. (1975) Genetic organization and adaptive response of allozymes to ecological variables in *Fundulus heteroclitus. Genetics* 79:97–111.

Nei, M. (1975) *Molecular Population Genetics and Evolution.* New York: Elsevier Science.

O'Brien, S. J., and MacIntyre, R. J. (1972) The α-glycerophosphate cycle in *Drosophila melanogaster.* I. Biochemical and developmental aspects. *Biochem. Genet.* 7:141–161.

Place, A. R., and Powers, D. A. (1979) Genetic variation and relative catalytic efficiencies: lactate dehydrogenase B allozymes of *Fundulus heteroclitus. Proc. Natl. Acad. Sci. U.S.A.* 76:2354–2358.

Place, A. R., and Powers, D. A. (1984a) Purification and characterization of the lactate dehydrogenase (LDH-B_4) allozymes of *Fundulus heteroclitus. J. Biol. Chem.* 259:1299–1308.

Place, A. R., and Powers, D. A. (1984b) The lactate dehydrogenase (LDH-B) allozymes of *Fundulus heteroclitus* (L.): II. Kinetic analyses. *J. Biol. Chem.* 259:1309–1318.

Powers, D. A. (1980) Molecular ecology of teleost fish hemoglobins: strategies for adapting to a changing environment. *Am. Zool.* 20:139–162.

Powers, D. A., and Place, A. R. (1978) Biochemical genetics of *Fundulus heteroclitus* (L.). II. Temporal and spatial variation in gene frequencies of *Ldh-B*, *Mdh-A*, *Gpi-B*, and *Pgm-A. Biochem. Genet.* 16:593–607.

Powers, D. A., Greaney, G. S., and Place, A. R. (1979) Physiological correlation between lactate dehydrogenase genotype and haemoglobin function in killifish. *Nature* 277:240–241.

Powers, D. A., DiMichele, L., and Place, A. R. (1983) The use of enzyme kinetics to predict differences in cellular metabolism, developmental rate, and swimming performance between *Ldh-B* genotypes of the fish, *Fundulus heteroclitus.* In *Isozymes: Current Topics in Biological and Medical Research*, vol. 10. *Genetics and Evolution*, ed. M. C. Rattazzi, J. G. Scandalios, and G. S. Whitt, pp. 147–170. New York: Alan R. Liss.

Powers, D. A., Dalessio, P. M., Lee, E., and DiMichele, L. (1986a) The molecular ecology of *Fundulus heteroclitus* hemoglobin-oxygen affinity. *Am. Zool.* 26:235–248.

Powers, D. A., Ropson, I., Brown, D. C., Van Beneden, R., Cashon, R., Gonzalez-Villasenor, L. I., and DiMichele, J. A. (1986b) Genetic variation in *Fundulus heteroclitus*: geographical distribution. *Am. Zool.* 26:131–144.

Richmond, R. C., Gilbert, D. G., Sheehan, K. B., Gromko, M. H., and Butterworth, F. M. (1980) Esterase-6 and reproduction in *Drosophila melanogaster*. *Science* 297:1483–1485.

Sacktor, B. (1975) Biochemistry of insect flight. Part I. Utilization of fuels by muscle. In *Insect Biochemistry and Function.* ed. D. J. Candy and B. A. Kilby, pp. 1–88. London: Chapman and Hall.

Segel, I. H. (1975) *Enzyme Kinetics.* New York: Wiley.

Stebbins, G. L., and Lewontin, R. C. (1972) Comparative evolution at the levels of molecules, organisms, and populations. In *Proc. Sixth Berkeley Symp. on Mathematical Statistics and Probability*, vol. 5, ed. L. M. Le Cam, J. Neyman, and E. L. Scott, pp. 23–42. Berkeley and Los Angeles: University of California Press.

Sugita, Y., and Chanutin, A. (1963) Electrophoretic studies of red cell hemolysates supplemented with phosphorylated carbohydrate intermediates. *Proc. Soc. Biol. Med.* 112:72–75.

Taylor, M. H., DiMichele, L., and Leach, G. J. (1977) Egg stranding in the life cycle of the mummichog, *Fundulus heteroclitus. Copeia* 1977:397–399.

Vigue, C. L., Weisgram, P. A., and Rosenthal, E. (1982) Selection at the alcohol dehydrogenase locus of *Drosophila melanogaster*: effects of ethanol and temperature. *Biochem. Genet.* 20:681–688.

Watt, W. B. (1977) Adaptation at specific loci. I. Natural selection on phosphoglucose isomerase of *Colias* butterflies: biochemical and population aspects. *Genetics* 87:177–194.

Watt, W. B. (1983) Adaptation at specific loci. II. Demographic and biochemical elements in the maintenance of *Colias* PGI polymorphism. *Genetics* 103:691–724.

Watt, W. B., Cassin, R. C., and Swan, M. S. (1983) Adaptation at specific loci. III. Field behavior and survivorship differences among *Colias* PGI genotypes are predictable from in vitro biochemistry. *Genetics* 103:725–739.

Watt, W. B., Carter, P. A., and Blower, S. M. (1985) Adaptation at specific loci. IV. Differential mating success among glycolytic allozyme genotypes of *Colias* butterflies. *Genetics* 109:157–175.

Discussion

FEDER: Could you describe how your mitochondrial DNA analysis could be used to indicate the nature of gene flow within a population or relationships among several populations?

POWERS: Well, there are a number of ways. Basically you want a means to determine if one genotype is going from one place to another and at what frequency. Classically, this was done by mark-and-recapture studies, which have obvious limitations. We are considering the possibility of releasing animals possessing a unique genotype in a given area, and then determining how far they move by sampling different populations. One then assesses the migration rate of that mitochondrial kind of genotype. There are lots of different approaches to the problem. The difficulties that I have had to deal with include not knowing the effective population size [N_e], whether certain genotypes were randomly distributed across the environment, whether males and females migrate equally, and what their real migration rate was.

DAWSON: Does the fact that you are really tracking maternal inheritance with mitochondrial DNA involve any special considerations?

POWERS: Yes, that is really the advantage; it makes it simpler.

ARNOLD: You discussed the effect of LDH variation on variation in hatching date. That is looking from a gene's-eye view at hatching date. Have you thought about a phenotypic view of LDH from the standpoint of hatching date? What proportion of genetic variation in hatching date would be linked to a variation in LDH?

POWERS: I do not understand the question.

ARNOLD: Well, suppose you look for the correspondence in hatching date between offspring and parents. Let us say that three-quarters of the phenotypic variation in hatching date is due to additive genetic effects and could be accounted for by the resemblance between parents and offspring. We could actually put a number on the genetic variation in hatching date. I am asking what proportion of that variation is due to LDH variation?

POWERS: I do not know exactly how to quantify the proportion. What we can do, and have done at different field localities, is to introduce eggs of the known genotype and see what fraction of each of the genotypes hatch and try to sort that out relative to the gene-frequency fitness of the individuals in the population. It seems to track, but I think you are stuck. I do not know how to test that statistically. Maybe Koehn has a better idea.

KOEHN: That is a really interesting question. What you are asking is how much of the total variation in hatching time could you partition among LDH genotypes? I do not think that has ever been done. I wish somebody would do it.

ARNOLD: The procedure required would probably involve determining the hatching dates of fish, raising them to maturity, and then breeding them, so that you know the hatching dates of parents and offspring and also the LDH genotype. Then one could ask what proportion of the variation in the offspring is explained by parents, holding the LDH constant.

POWERS: Should the genetic background be variable or constant?

ARNOLD: One wants the genetic background to be variable so that one can distinguish between *Ldh* genotype and other unmeasured things as accounting for the genetic variation in hatching date.

KOEHN: Powers probably has the answer Arnold would like. Is LDH the only gene for which you can detect an effect on hatching date?

POWERS: We already know that there are other factors. In fact, one of the things that led us to do those experiments was that a series of individual crosses indicated significant variation. The only way that you can get all of the crosses to hatch in a uniform distribution is to mix the eggs and sperm randomly, so that you are randomizing the genetic background. Otherwise you get a discrete series, all of which are consistent, but that cannot be compared to one another. So you already know that there are other components in the genetic background.

FUTUYMA: You ended your talk today with a highly molecular slant that some people might call reductionist in nature, just to use an inflammatory word. I can imagine reasons why people would be interested in knowing whether the variation in physiological function was due to the differences in translation rate, or the number of gene copies, or other possible molecular mechanisms. However, I was uncertain why most of the people in this room, who are primarily interested in the whole-organism level of physiological functions and how those relate to the environment, would want to know the precise molecular details of the kind that you were urging them to pursue.

POWERS: I was not urging everyone to pursue it. I think that it is important in these areas for ecologists, physiologists, biochemists, geneticists, and so forth, who are interested in bridging gaps, to look at the whole profile of physiological and biological parameters.

DAWSON: I have been troubled by the seeming hostility of some ecologists and evolutionary biologists to reductionism. It does not seem appropriate to say that the principal analytical thrust of physiological ecology should be confined to the organismal level. I believe this amounts to imposing constraints on inquiry. You do not have to do that: your modal activity can certainly be with the organism, and occasionally it becomes highly interesting to know just where a change in enzymatic activity, for example, is mediated and what the change involves. Please, let's not impose intellectual constraints on ourselves, particularly where these may discourage interactions of the type Powers was describing.

FEDER: There is a pragmatic issue of some importance here. Powers thinks that a program in which graduate students and faculty could learn specific techniques might be useful. Suppose Powers takes on the burden of providing such training, whether in molecular biology or some other advanced technique. When the trainee returns to his or her home institution, it will be necessary to have a fairly large equipment setup to apply the techniques. How transmissible is this knowledge, given the large equipment base needed to carry out the techniques that one has learned?

POWERS: It depends on the technique. If it involves Southern blot analysis, it doesn't take much. If you need an ultracentrifuge and there isn't one in the department, that may be a problem. Most biology departments, maybe not the field-oriented departments, own an ultracentrifuge or at least have access to one. It is not that bad. If you need high-resolution NMR, that could be a problem. Clearly, people are going to different labs all the time; in fact, one of the best ways to pick up a technique is to spend a sabbatical leave in somebody's lab, or to go there and learn how to do peptide mapping or how to use cladistic analysis. For a person switching fields, or going into an area very strongly, it would be appropriate to spend a year's sabbatical or more, publish a paper or two, and then return to the home institution to set up a lab of his or her own. He or she then could contribute to the dissemination of a certain set of techniques. It depends upon the problem involved.

BURGGREN: I would emphasize the point that Dawson brought up about the dangers of serial amateurism. For example, I could spend a semester in Powers's laboratory learning molecular techniques and take them back to my own lab. Nonetheless, I am concerned that I would never be able to apply these techniques as efficiently or accurately as someone who does them routinely. Is it really profitable for me to retool mentally and physically to learn new techniques, when in fact, through collaboration I could really do first-class work? I could bring the expertise that I have to bear on one aspect of the problem, while the person who really has the basic training and setup for these specialized techniques, which are getting increasingly more complex, could bring his or her expertise to bear, and we could mutually solve the research problem. I am all for collaboration. My chapter makes a strong pitch for that.

POWERS: It depends on the extent. If you are talking about a really long term sort of thing, where you have an expertise and someone else has an expertise, that is ideal. You can look around for collaborators and you can find experts, but nobody is going to care about your problem as much as you do. They've got problems in which they are interested, and it is very hard to find someone who is even interested in your question. So sometimes you have to learn a particular technique yourself.

BARTHOLOMEW: One of the reasons for frustration among physiological ecologists is that many work at a couple of steps of reduction from the ecological level. They find it hard to get ecologists to listen. One of the bridges that has to be made is to change the mindset of many ecologists so that they come to regard reductionism as not being inherently bad. This is the first step in one kind of understanding of phenomena. It is the reductionist way. The other is the synthetic way. I concur absolutely that the ideal, but only infrequently practical, solution would take the analysis of phenomena all the way from the ecosystem to the molecule.

6 Comparisons of species and populations: a discussion

WILLIAM R. DAWSON

DAWSON: Physiological ecology depends heavily on the comparative method in attempts to interpret within an environmental context the diversity of functional capacities evident in animals. Comparisons frequently are made at the interspecific level, with species employed as a variable in attempts to discern patterns of adaptation to the environment. In view of their centrality to physiological ecology, it is useful in this general consideration of the future of the field to define on what basis and for what important purposes such comparisons properly should be made.

At the risk of stating a truism, proper application of the comparative method in physiological ecology requires identification of the critical things to compare. In dealing with the question of adaptation, it is crucial to heed Feder's remarks concerning the importance of null hypotheses. Moreover, it is important to realize that comparisons of species within a clade may not meet statistical requirements for independence of sampling units. The burden of proof concerning the adaptive significance of differences detected in comparisons must lie with the investigator. Identification of the critical things to compare requires sufficient information on the natural history of the species under consideration to define their ecological niche and details of their behavior in nature. Moreover, it is no longer adequate to deal exclusively with the functional capacities of adult animals. Quite possibly, definition of developmental trajectories for these capacities will be at least as critical as more traditional analyses of adults.

Huey's presentation appropriately emphasizes the contribution that modern systematics can make to the design and interpretation of studies in physiological ecology. Knowledge of the phylogenetic relations of the species assemblages with which we deal, and inclusion of outgroups in our comparisons, are essential. In this connection it is important to avoid gratuitous inferences concerning phylogeny based on the assumption that the respective criteria used to define genera and other higher categories are consistent from group to group.

Physiological ecologists necessarily must study living forms, and this requirement hinders interpretation of the evolutionary history of functional patterns that appear adaptive within the current environmental context.

Only with adequate information on the systematics of the animals under study can physiological ecologists hope to formulate valid historical interpretations. Central to these interpretations is the capacity to distinguish between ancestral and derived capacities, where the latter may represent special adjustments to particular environmental situations.

Inevitably, physiological ecology will require more information on the genetics of species under investigation. Estimates of levels of heterozygosity, and, where feasible, association of variation in physiological performance with particular polymorphic loci add a whole new dimension to analyses in physiological ecology, as Powers's presentation nicely documents.

The considerations discussed thus far are rivaled by the need to design comparisons that are as free as possible from anthropomorphic perceptions. Bartholomew notes in his presentation that adequacy rather than perfection is probably a general theme in the evolution of capacities that enhance the ability of animals to deal with particularly demanding features of the biotic and abiotic segments of their environment. Investigations of species in extreme environments indicate that their survival may depend heavily on some combination of behavior and of tolerance of transiently severe conditions or on reconciling essentially antagonistic requirements (e.g., supporting extensive evaporative cooling and managing water balance in hot arid situations), rather than on enhancement of regulatory capacities. Competition as well as physiological requirements may affect the temporal and spatial aspects of habitat utilization of animals such as reptiles [Tracy and Christian, *Ecology* 67:609–615, 1986].

It is also important in interspecific comparisons to determine what constitutes adequate sampling of the species under study. Physiological ecologists generally seem to appreciate that acclimatization state can affect the performance of their experimental subjects. In widely distributed forms there is the additional complication of geographic variation in response.

If physiological ecology is to continue to prosper, implementation of effectively designed analyses must be accompanied by expansion of present goals. Clearly, a primary objective will continue to be the analysis of pattern and process in functional and behavioral adjustments to the environment. If analysis of pattern and process is to consist of something more than affirming that animals can occur where they do, attention must be given to how the physiological and behavioral armamentaria of these organisms influence the manner and cost, in terms of energy and materials, of their environmental usage. Moreover, examination must be made of the question of limits upon distribution and, to the extent that functional analyses are of primary relevance, to that of why certain species are stenotopic and other closely related ones eurytopic. Bartholomew's comments indicate that this latter question can be expanded to inquiry concerning the basis of the broad ecological range characterizing some groups of species such as ground squirrels. A related question concerns identification of constraints that may have served to restrict the distribution of stenotopic forms.

Feder's presentation cogently reminds us that the analyses of concern to physiological ecologists should and can have relevance to understanding the consequences of phenotypic variation for neo-Darwinian fitness. He and others who will speak later identify the opportunity for physiological ecology to operate in a new and broader context than has been the norm.

The comments made thus far relate exclusively to interspecific comparisons involving more or less closely related species. However, another type of interspecific comparison in which animals are grouped by habit rather than by taxonomic affinity can be made. Here the goal should be to identify patterns of divergence and convergence among the animals sharing a habit such as burrowing. Doubtless, interpretation of these patterns will be facilitated by judicious investigation of close relatives of the individual species comprising the study set.

It is appropriate now to discuss the various presentations on interspecific studies. It should be noted that this discussion has been edited in the interests of space limitations, clarity, and preserving flow.

EDWARDS: I was very pleased with the way that Huey explicated the need for and the uses of phylogenetic analyses in defining how various organisms should be regarded, how experimental subjects should be chosen, and how various questions should be asked, using those phylogenetic analyses as a basis. My comment is that there are many systematists who I think would also enjoy working with physiological ecologists in some mutually profitable collaborations.

Both Dawson and I were struck during the talks with the suggestions that physiological ecologists ought to collaborate with systematists, molecular biologists, population geneticists, and others. I was also impressed by Bartholomew's remarks: first, that it may be more difficult to find new generalities in physiological ecology at this relatively mature stage in its development than it was earlier; and second, that the paradigms appropriate for one era of a scientific field may not be the most useful in another era. In some ways the development of a scientific field resembles a speciation event. Formation of your own scientific journal is the reproductive barrier separating your field from others. However, given the remarks made in the presentations today concerning the need for collaboration with workers in other fields, it seems appropriate to ask whether physiological ecology is still a viable paradigm. Should physiological ecologists disperse and identify themselves instead as molecular or evolutionary biologists or as population geneticists? Is something remaining in the field that is best served by keeping it a separate entity, or should one attempt to set up an integrative biology to yield an organismally based science that is greater than the sum of its parts?

FEDER: I think your questions are important. Physiological ecology in many ways is currently a field unto itself. In preparing my talk, I looked in vain at Doug Futuyma's book, which is one of the major textbooks on evolutionary biology, to find mention of physiological ecology.

FUTUYMA: There is only one reference, on marine iguanas.

FEDER: My perception seems accurate. By and large, what is going on in ecological physiology does not impact the field of evolutionary biology. I would also hold that, with some exceptions, ecological physiology also does not impact the field of molecular biology. An apparent explanation for this is that physiological ecology generally has failed to articulate questions that people outside the field perceive as important ones. Evolutionary biology, for example, has assimilated our observations on equilibrium or adaptation to the environment and quite reasonably has moved on to other issues. If ecological physiology is not interested in generating questions of general interest and importance, what then should its mission be? One possibility that does not fill me with enthusiasm is that it could be a "service discipline" for other fields. Ladd Prosser recently made a case in *The Physiologist* [see Chapter 3 for reference] for the mission of "adaptational physiology" as providing evolutionary biologists with functional significance for the characters they work on, helping molecular biologists integrate their work with an organismal level of organization, aiding systematists to build better phylogenies; that is, not a leadership role for the field. The field can do more. However, it is incumbent upon us, if we want to maintain a viable field, to come up with compelling questions that draw in people from other disciplines.

FUTUYMA: Since you have mentioned that particular textbook on evolution, let me also point out that some recent textbooks of ecology also include very little in the way of physiological ecology. It is all population dynamics, interspecific interactions, community structure, and so on. Here is something which was once at the heart of ecology, which is being left out of some ecology books as well. Edwards just made a very important point: that the importance of ecological physiology as perceived by other disciplines may be largely a function of where its findings are published. The one reference in my textbook that deals with physiological ecology is to a paper by three of you in this room, which I was aware of only because it appeared in the journal *Evolution*, which is one that I read.

DAWSON: Bennett, Bartholomew, and I published on the marine iguana in *Evolution* because we wanted to be explicit about the evolutionary implications of our study. Physiological ecologists should not operate as an isolated priesthood, and we should try to address our more general findings to other audiences.

BENNETT: I would really like to echo that and say that I think that some of the difficulty stems from the absolute size of animal physiological ecology. Our field is so big that evolutionists and ecologists may be forced to ignore most of it to maintain a focus on their primary interest.

POWERS: With respect to interaction with other disciplines, my department emphasizes molecular biology; otherwise it would be very difficult for me to use its techniques. It would be very nice for more people to bridge such gaps, either through training programs or workshops. Some people are starting to make advances in these areas.

DAWSON: I would like to see the best of both worlds, where we keep the allegiance to questions and concern with them, but are flexible about incorporating techniques.

BARTHOLOMEW: I concur completely. It is up to us who identify ourselves as physiological ecologists to make the overtures. *We* must form the bridges, not only toward the molecularly inclined, biochemically inclined people, but also the other way. We have a great deal of relatively hard data; we can see more or less where these data should fit. But we need support, at the level of population biologists, population geneticists, and the theoretical ecologists. In fact, because we have this wealth of material, I think it behooves us, in the name of science, to go out and search for these collaborations.

KOEHN: In Britain, a significant need has been perceived for creating a marriage of some kind among some of those disciplines that have traditionally been isolated from one another. The British have recently established a fund to support individual exchanges between research programs. For example, people working in genetics, whether it be molecular genetics or population genetics, and in physiology are joined within the context of a particular problem. Basically, the problem must be sufficiently crystallized so that it is clear that both elements – that is, genetics and physiology – are to be involved. The funding is for a period of up to about two years, and it is actually a respectable amount. It should be highly beneficial, because it has been clear to me that many of us in different fields are actually working on very similar problems. You may not solve the problem, but you are going to get a good hold on it.

BARTHOLOMEW: I need not tell you that we are all drowning in a flood of data. My recent rough calculation indicates that the mass of published data has grown twenty-five times what it was when I graduated from college. I believe the only feasible way that we can survive this deluge of information is by thoughtful collaboration. We must thoughtfully seek out colleagues whose information, philosophical perspective, and techniques complement our own.

EDWARDS: Concerning what Bartholomew just said, the operative word, I think, is *thoughtful*. He pointed out that some physiological ecologists feel that they are being assigned a service function; that is, that other folks want to use their knowledge without giving much in return. Systematists feel similarly; they are afraid that ecologists or physiological ecologists will simply use their expertise and not really involve them intellectually in the research. Apparently, even molecular biologists are confronted by a similar circumstance. Certain laboratories known for teaching molecular techniques are filling up, and now are refusing to take anyone else. The way that these kinds of cooperation can work is through assurance that both sides are going to gain something, intellectually or monetarily, and certainly in terms of publications.

DAWSON: We have unexploited areas within physiological ecology. A little attention to those would facilitate some of the interactions that we seek. On the ecological side, one that Bartholomew alluded to is nutritional physiology. Also, the potential is great for study at several levels of how animals deal with ingested secondary plant compounds. Problems of this type afford rich opportunities for meaningful collaboration among physiologists, biochemists, ecologists, and evolutionary biologists.

FUTUYMA: I was just going to raise that very point. One of the notions advanced this morning was that physiological ecology should stop the cataloging of adaptations, and move on to new kinds of theoretical constructs and questions. I am all in favor of conceptual approaches, but it should be recognized that part of the field's value is in providing a wealth of information important in and of itself. You know much about such things as adaptation to desert conditions, but less about other areas, specifically toxicological aspects of secondary compounds. How herbivores deal with them is something that I confront in my own work. But there is essentially no information on them.

ARNOLD: Physiological ecologists have a lot to offer evolutionary biologists, as collaborators and as representatives of an adjoining discipline. I will give one example with which I am familiar. Currently there is interest among the evolutionary biologists in trying to measure fitness of variants in natural populations. Often this kind of work is undertaken by people who have established mark-and-recapture programs for these populations. They have measured the fecundities and survivorships of individuals in the population, making it a relatively simple step to ask, What are important phenotypic traits within the population, and how do they relate to fitness? At this point, collaborations with physiological ecologists are particularly valuable. People who have the capacity to measure selection in natural populations often need to be directed to characters that have a particular physiological, phylogenetic, or ecological significance. It is not so obvious from the standpoint of the demographer what those characters are and how they can be measured.

SCHEID: In my experience, there are some pragmatic difficulties connected with establishing associations between physiology and molecular biology. We have undertaken a project involving such an association. Funding initially proved difficult because of the General Research Council's reservations about our expertise in molecular biology and about their funding still another molecular biology laboratory. We ultimately were funded and obtained some interesting results. Then came the problem of publishing the work. Most journals sent it back with very harsh critiques from reviewers, who were identified as experts in the field. These reviewers attacked the account of the work, but they did not really understand our approach of trying to marry physiology and molecular biology. The problem is that we are faced with very strong intellectual families among which it is difficult to achieve intermarriage.

DAWSON: I can see that as a problem; it requires a special form of pioneering. What we are talking about is probably a more mutually congenial liaison than a marriage. Perhaps the promise of utilizing in a new context techniques of a field such as molecular biology should be emphasized. Our questions define this context and must remain the primary focal point.

ARNOLD: The problems of both funding and journals are real ones, when you are bridging between fields. If you can identify the two fields that you are bridging, you have to take the trouble to explain the rationale in the language of both.

BENNETT: This particular problem, in having difficulty in getting articles accepted to journals because of the interdisciplinary nature of our research, is not just a problem for the future. This is something that we presently face all the time in physiological ecology. Arnold and I just had a paper dealing with the morphological basis of locomotion rejected from a journal. The subject matter did not differ substantially from that of other papers published there; however, it was evidently regarded as too evolutionary in emphasis. Criticism by editors concerning emphasis seems quite common.

POWERS: Much of the work that I do is very biochemical, especially the kinetics and enzymology. I try to make it of the highest possible quality, and therefore few journals are appropriate for it. A problem is that journals like the *Journal of Biological Chemistry* now will not accept comparative biochemistry. So we have to change the title, adjust the abstract, and so forth, to get it accepted there.

SCHEID: A simple question: if you feel that you have to create a marriage, why not create a new journal for the marriages?

ARNOLD: One problem is that some of the synergistic interactions between the fields initially are so small that you have insufficient papers to justify a new journal. One solution that I know that Bennett has tried with some success is specifically to approach editors of journals and say, "Look, you're going to think that this is physiological, but couldn't it be viewed as part of animal behavior?"

DAWSON: Rather than increasing the number of journals, I would make the best of the existing situation but modify it in a manner more receptive to interdisciplinary efforts.

BURGGREN: One additional point concerning formation of a new journal is, Who is going to read it besides the "converted" following? It is important for physiological ecology to be represented in ecological journals, so that the ecologists see that information. Additionally, it is important for physiological ecology to appear in physiological journals, so that the physiologists get some exposure to the ecological aspects of the subject. This requires persuasion of editors to be a little more receptive concerning interdisciplinary material.

BENNETT: We are not talking here about establishing only one new journal. That would not do any good, because it might be a journal of molecular physiology, or it might be genetic physiology, or it might be any number of different things.

BARTHOLOMEW: I support what Burggren has said, but there is an additional consideration. Physiological ecology or ecological physiology represents integrative disciplines between a series of points of view. If we discipline ourselves to publish in the leading journals for the component fields, we are meeting the technical criteria for specialists. Publishing in journals emphasizing different fields is hard, because we have to use different vocabularies, push different buttons; but for the dignity of the field and its standards, we should do this.

METCALFE: One thing confusing me – and I am neither a physiologist nor an ecologist but a cardiologist – is that at one time we defined our specialties (and our professional affiliations) pretty much by the technology that we used. Anatomists used quite a different technology than did physiologists and clinicians. That is one of the things that I think helped in the definition of our journals. Appropriately, that is disappearing as people strive to establish the cross-disciplinary linkages needed in their research.

DAWSON: Given that one wanders on from technique to technique in establishing these linkages, I worry a bit about serial amateurism. I also worry about the appropriateness for physiological ecology of having its practitioners confine their allegiance to a single scientific society.

JACKSON: I feel that very acutely. I came out of the more physiological side. Knut Schmidt-Nielsen (Duke University) and Ted Hammel (University of California, San Diego) and people in their labs strictly went to American Physiological Society meetings, and I never went to an American Society of Zoologists meeting until about two or three years ago. Things have improved in the last few years with joint meetings of ASZ and APS, which seem highly successful. Yet many people still never cross from one side to the other. It would be an excellent thing if an umbrella organization existed that could bring all of these people together, at least once in a while.

DAWSON: We have tended to profit from the idea that there is not one single pathway that you have to follow in entering physiological ecology. I think that that has been very positive; I would like to see it continue.

DAWSON and EDWARDS: Let us make some concluding remarks. The range of matters dealt with in the discussion just presented is impressive. Topics considered extended from increasing the range of questions that should be pursued by physiological ecologists, through technical points relating to the papers presented, to logistical and procedural matters concerning how to implement our research more effectively. Increasing interaction with persons representing disciplines extending from population ecology to molecular biology was viewed as a vital need. However, such interaction and resultant collaborative efforts must proceed on the basis of mutual benefit. Molding physiological ecology into a service discipline for other fields will in the long term serve neither its needs nor those of biology generally.

There was cogent criticism of encyclopedism as a primary goal of physiological ecology. Nevertheless, appreciation developed for the fact that increasingly comprehensive physiological data bases for various groups of animals can be a useful by-product of problem-oriented endeavors. These endeavors will require a higher level of sophistication than has sometimes been evident in the past. In dealing with questions of adaptation, appropriate null hypotheses are necessary. Rigorous information on phylogenetic relations must be taken into account in designing studies in physiological ecology and in interpreting the results. It is also important that these studies be drawn in terms that are ecologically relevant to the species under consideration and that they deal with fully representative samples of each of these taxa. Hopefully, it will be increasingly feasible to supplement these basic steps with genetic information. We should move vigorously to exploit the potential our field has for analyzing the influence of various aspects of physiological performance on neo-Darwinian fitness. Clearly, interactions among systematists, population biologists, behaviorists, physiological ecologists, and population geneticists in this effort can place functional studies in a firmer ecological and evolutionary context, to the benefit of all the fields involved.

There was general concern in our discussions about the difficulties of publishing work that cuts across traditional intellectual boundaries.

In its role as an important part of organismal biology, physiological ecology should not avoid reductionism, where this approach can lead to a fuller understanding of pattern and process in the adjustment of animals to their respective environments. However, a reductionist approach will be most beneficial in tandem with a continuing commitment to furthering analysis at the organismal level. Implementation of the approach will involve the introduction of new techniques into physiological ecology. This surely has implications for patterns of collaboration and for the actual support base of our field. Such a circumstance cannot help but create a need for planning at the levels of individual laboratories, the universities, and above.

Dissatisfaction exists regarding the status quo in physiological ecology. Happily, the remedy for this dissatisfaction principally involves a reaffirmation of the basic goals of the field, which have always been the rigorous exploration of the ecological and evolutionary implications of physiological variation.

PART TWO

Interindividual comparisons

Individual animals live, reproduce, and die. Individual animals move, eat, thermoregulate, and excrete. Each is a unique combination of genes, physiological capacities, and morphological structures that permit them to do these things more or less well. It is the unique individual animal that is either more or less fit to contribute to the next generation. Although emphasizing the theoretical importance of the individual, organismal biology has most often studied the population or species level of biological organization, not the individual level. Part Two examines the contributions that studies of traits of individual animals can make to ecological physiology.

The first chapter of this section (Chapter 7 by Albert F. Bennett) discusses the emphasis in physiological studies on central tendency rather than on individual differences. It demonstrates that differences in physiological functions among individuals can be repeatably measured and argues that this repeatability has analytical utility. Different applications of this variability are proposed, including its uses in functional analysis and in determining the selective value of the trait under natural conditions.

The next two chapters discuss analyses of the genetic determinants of physiological and ecological performance. The authors present two complementary approaches. These are designated by Stevan J. Arnold as the gene-to-physiology and physiology-to-genetics perspectives. In the former, discussed by Richard K. Koehn (Chapter 8), genetic variation is initially measured at a single gene locus or at multiple loci. Correlates of this variation are then sought at the organismal level and their selective advantage is examined. The strengths of this approach are discussed and several examples of its successful application are presented. In contrast, the physiology-to-genetics approach (Stevan J. Arnold, Chapter 9) begins with observations on the variability of characters within a population. It then seeks to determine whether this variation has a genetic component and to analyze selection and genetic constraints on the evolution of the trait in question. This chapter discusses the problem of genetic correlation among traits, leading to the possibility of antagonistic adaptation and counterintuitive evolutionary consequences. Both of these genetic approaches offer insight into the coupling between genetic structure and phenotype, and both require an analysis of interindividual variability within a population.

The last chapter (Gary C. Packard and Thomas J. Boardman, Chapter 10) discusses the proper analysis of comparative data, particularly when differences in body size are a confounding variable. Ratio-based analyses are systematically shown to be inferior to analysis of covariance in discriminating real differences among experimental groups. These comments, which have applicability in both interindividual and interspecific studies, again emphasize the importance of analyzing differences among individual animals in comparative studies.

7 Interindividual variability: an underutilized resource

ALBERT F. BENNETT

Introduction

Two principal analytical approaches have been used in studies of organismal physiology. These are represented by the terms "comparative physiology" and "physiological ecology." The former compares functional characters in two or more populations, species, or higher taxa in an attempt to understand mechanism. Biological diversity is used to help understand principles of physiological design. Often the experimental species are chosen specifically because their systems demonstrate an extreme phenomenon or because the experimental preparation is technically accessible. The selection of a species on these grounds is known as the Krogh Principle (Krogh, 1929; Krebs, 1975), which has been very influential and successful in guiding studies in comparative physiology for more than fifty years.

The second approach, physiological ecology or ecological physiology, examines the physiological attributes of a species and interprets them in the context of the natural environment or ecological niche of an animal. These studies concentrate on analysis of adaptive pattern, of how physiology, morphology, and behavior interact to permit survival and reproduction in a given environment. In this approach, emphasis is placed on ecological and evolutionary aspects of physiological function. Monitoring the organism in its natural environment and speculation on selective factors that influenced the evolution of characters are the principal interpretive contexts of these studies.

These two approaches are by no means exclusive and have often proved complementary. They have yielded a substantial understanding of how animals work and function in the natural world. However, my thesis here is that both approaches have overlooked a valuable source of information. In their concentration on population-, species-, or higher-level phenomena, they have failed to analyze and take advantage of biological differences among individuals. As traditionally practiced, physiological studies neglect differences among individual animals and attempt to describe the functional response in the average animal of the group. I believe that this approach has been very short-sighted and that the study of interindividual differences has much to contribute to both comparative physiology and physiological ecology. I will argue that the analysis of the bases and consequences of interindividual variability can provide new tools for both types of physiological analysis. I believe that it is also capable of building new and important bridges to other allied fields of biology, especially ecology, ethology, evolution, and genetics.

The tyranny of the Golden Mean

The framework of physiological studies implicitly emphasizes the description and analysis of central tendency. Depending on the data, this involves the calculation of mean values or the development of least-squares regression equations. After these values are determined, they take on a life of their own and become the only point of analysis and comparison. The complete breadth of biological variation determined in the investigation then is forgotten. Measures of variability (e.g., variance, standard deviation) are calculated and reported only to stipulate confidence limits about the mean or slope of the regression line. Groups are then compared to determine whether they are different from one another or from hypothesized values. The variability inherent in the original data is seen only as "noise," through which the "true" value of the central tendency can be glimpsed with appropriate statistical techniques.

This assumption of a "true" or "real" central tendency, which biological reality only approximates, stems from Platonic philosophical traditions. These maintain that ideal archetypes exist that can be perceived only imperfectly through perceptual sensation. The concept of an ideal form of a structure or process was central to the thinking of medical physiologists of post-Renaissance Europe and heavily influenced the functional biologists of the nineteenth century. These physiologists and morphologists, in their search for proximate causation, maintained a typological approach to experimentation and analysis and were largely unaffected by contemporaneous developments in evolutionary biology and genetics (cf. Mayr, 1982, for a more detailed discussion). Analysis of variability played an important role in these latter fields, but it was ignored by functional biologists at the time and remains largely unexplored by them even today.

To dispel any doubt that analysis of central tendency and neglect of variability is the dominant or exclusive analytical mode in organismal physiology, I reviewed all papers published during 1985 in the *Journal of Comparative Physiology*, the *Journal of Experimental Biology*, and *Physiological Zoology*. These are some of the best and most forward-looking journals in the field. Nearly all the articles reported mean values or regression equations and did statistical analyses. However, less than 5% of the articles even reported the range of values of the data obtained, and out of more than 250

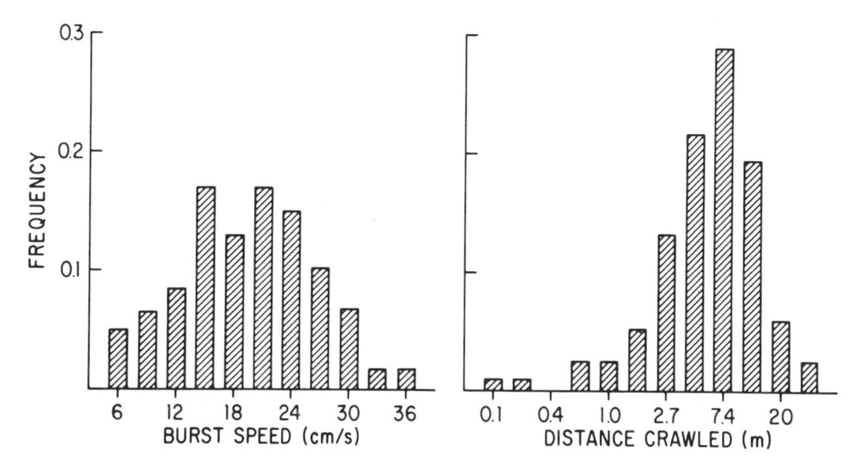

FIGURE 7.1 Frequency distributions of burst speed and total distance crawled under pursuit by individual newborn garter snakes (*Thamnophis radix*). Each individual observation is the mean of two trials conducted on two successive days; individual repeatability is highly significant ($r = 0.60$ for burst speed and 0.55 for distance; $p < .001$). Distance crawled is reported on a logarithmic axis. (Data from Arnold and Bennett, in press.)

articles, only one (Taigen and Wells, 1985) analytically examined the variability in the observations.

The concentration on central tendency has been and will continue to be very useful in testing certain hypotheses, but it has distracted us from an examination of the causes and consequences of biological variability. An example of this variability is given in Figure 7.1, in this case variability in locomotor performance capacity of newborn garter snakes. Maximal burst speed and the total distance crawled under pursuit were measured in nearly 150 laboratory-born animals shortly after birth (Arnold and Bennett, in press). These behaviors are individually repeatable (see below) and represent the breadth of response of the population at birth, before natural selection by the external environment has had the opportunity to act. Both these performance measures show strong central tendencies, but they also show enormous interindividual variability. The fastest snake has a burst speed ten times that of the slowest; the endurance of some individuals is more than twenty times that of others. Assuming for a moment that these individual differences are real (see below), these observations immediately suggest two sorts of questions. First, what is the functional basis of these individual performance differences? Which physiological or morphological factors make a fast snake fast and which account for the relatively low stamina of some other animals?

Second, what are the ecological and evolutionary consequences of these differences? Is there differential survivorship or growth under natural conditions based on locomotor performance capacities? These questions reflect the somewhat artificial dichotomy raised earlier between comparative physiology and physiological ecology, but both of them reflect compelling questions of general biological interest. They are obscured, however, if one concentrates only on central tendency. This is the tyranny of the Golden Mean: it restricts our vision of the data and narrows our conceptual framework so that we cannot take advantage of all the analytical possibilities of biological variability.

The failure to consider interindividual variability is not that of ecological or comparative physiology alone. Almost identical comments and comparisons could be made about any other field of organismal biology.

In our concentration on central tendency, we have failed in several respects:

1. We have ignored interesting biological problems and questions.
2. We have not been particularly interested in the consequences of the data we have gathered for survivorship or fitness.
3. We have failed to utilize the breadth of our data in assessment of physiological hypotheses.
4. We have failed to provide sufficient information in our research reports that would permit others to analyze biological variability.

The reality of interindividual variability

I believe that part of the difficulty that most ecological and comparative physiologists have in reporting and utilizing variability is a suspicion of its reality and information content. Biological measurements are inherently highly variable as compared to those made by physicists or chemists. Coefficients of variation of 20 to 30%, values that would cause a physical scientist to blanch, are routine in most physiological measurements. To what extent, however, is this variability real and useful? It seems to me that there are three potential objections to its use:

1. *Extreme values are atypical or abnormal and do not reflect the true response of most individuals.*

This view is essentially a restatement of the typological concept: the average is the real. Extreme performance certainly is "atypical" and "abnormal" in the strict sense of the words, but that does not mean that it is not real. A physiologist must be sure that experimental animals are in good condition, but it should go without saying that one must have external cause to doubt any data point. It cannot be questioned only because it happens to lie on the extreme of the range.

This view suggests that the experimenter has more confidence in values that lie closer to the mean than those at the extremes. If this is the case, then not all points should receive equal weighting: those closer to the mean should be weighted more highly. The circularity of this logic is apparent. Further, normal parametric statistics are inappropriate in such a circumstance. Either all data points receive equal confidence and equal weight, or the analytical methods we normally use are inapplicable; one cannot have it both ways.

2. *Observed variability is due to instrumentation or procedural error; the observed range does not result from real biological differences but from inaccuracies in experimental setups or procedures.*

According to the type of measurement, this objection may have some validity. However, the precision of modern physiological equipment is typically less than 1% and is consequently a doubtful explanation of much higher apparent biological variability. Further, if such errors are felt to be important, their magnitude must be quantified and analyzed (although they almost never are) even in studies that are interested only in central tendency. If the errors are random, then the mean values will be correct, but the measurements of variance and standard deviation of the means will be inflated. As statistical comparisons between groups are dependent on the extent of intragroup variability, incorrect judgments may be made if experimental or instrumentation error is not analyzed and removed. Consequently, if this type of error is a problem, it is not a special problem in the analysis of variability alone. It also affects any kind of analysis, including that of central tendency.

3. *The variation measured is real but reflects random and unrepeatable responses of individuals; that is, intraindividual variability is so high that there is no significant interindividual component to total variance.*

This is by far the most serious potential objection to the analysis of variability: if the responses are random with respect to individuals, then analyzing the differences among individuals is futile. The measurements required to demonstrate whether this is an important problem are a series of repeated observations on the same individuals and analysis of the significance of the individual component. For instance, if one is interested in oxygen transport capacity, one might measure maximal oxygen consumption in each of several individuals on sequential days to determine whether some individuals have consistently high or low capacities.

Given the general lack of interest in interindividual variability, analyses of intra- versus interindividual variability are relatively few in ecological or comparative physiological studies. Most of these relate to data on locomotor performance capacity, and many of the examples in this discussion will be drawn from this area. Individual locomotor performance ability has a significant repeatable interindividual component in every study in which it has

TABLE 7.1 Studies demonstrating significant interindividual variability in locomotor performance

Group	Performance	No. of species
Lizards		
	burst speed	6[a]
		1[b]
		2[c]
		2[d]
		1[e]
	stamina	6[a]
		1[f]
		1[d]
	defensive behavior	1[b]
Snakes		
	burst speed	1[g]
		1[h]
	stamina	1[h]
	defensive behavior	1[i]
Anurans		
	stamina	2[j]

[a]Bennett (1980). [b]Crowley and Pietruszka (1983).
[c]Huey and Hertz (1984). [d]Garland (1984, 1985).
[e]Crowley (1985). [f]John-Alder (1984).
[g]Garland and Arnold (1983). [h]Arnold and Bennett (in press).
[i]Arnold and Bennett (1984). [j]Putnam and Bennett (1981).

been examined (Table 7.1). An example of individual constancy of day-to-day differences in locomotor performance is given in Figure 7.2 (Bennett, 1980). Maximal burst speed was measured in fifteen adult fence lizards on five sequential days. Rank order of performance was conserved through the repetitive trials ($p < .001$). These individual differences in burst speed capacity were independent of both sex and body mass. Similarly, individual performance rank is stable even when the internal environment of the animals is grossly altered, as during changes in body temperature. Individual rankings of burst speed performance of alligator lizards at different body temperatures are given in Table 7.2. Again, individual differences are highly significant ($p < .001$): some animals are fast and some are slow at all body temperatures (see also Huey and Hertz, 1984).

I believe that locomotor measurements would a priori be among the *least* repeatable of any of the potential spectrum of "physiological" measurements. They may be influenced by a great many motivational and psychological factors, as well as differences in underlying physiological or morphological

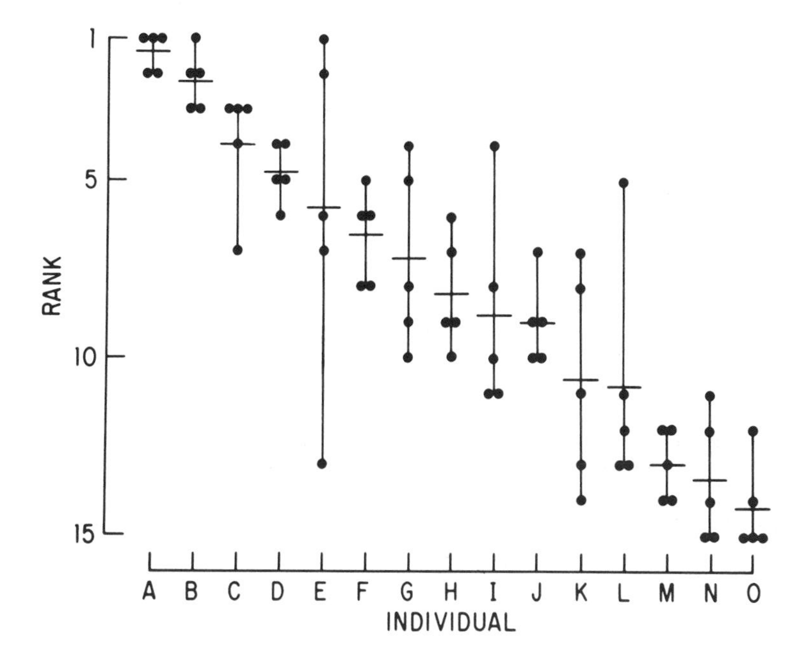

FIGURE 7.2 Rank order performance of burst speed in fifteen adult fence lizards (*Sceloporus occidentalis*) measured on five successive days. Rank 1 is the fastest animal, rank 15 is the slowest. Dots indicate rank performance on each day; vertical bars, range; horizontal bars, mean rank. Individual ranking effects are highly significant ($p < .001$ by Kendall's coefficient of concordance). (Data from Bennett, 1980, and unpublished observations.)

capacity. From that viewpoint, a significant interindividual component in measurements of locomotor capacity might suggest that many other physiological variables would also have individual fidelity.

Significant interindividual differences have been demonstrated in such diverse systems and measures as maximal oxygen consumption in amphibians and lizards (Pough and Andrews, 1984; Wells and Taigen, 1984; Sullivan and Walsberg, 1985), enzymatic activities in fruit flies (Laurie-Ahlberg et al., 1980), cuticular water loss in cicadas (Toolson, 1984), muscular morphology of birds (Berman, Cibischino, Dellaripa, and Montren, 1985), skeletal morphology of salamanders (Hanken, 1983), kinematics and muscle activity patterns during feeding in salamanders (Shaffer and Lauder, 1985a, 1985b), foraging tactics in fish (Ringler, 1983), food preferences in snakes (Arnold, 1981), and regulated body temperatures of lizards (Christian, Tracy, and Porter, 1985).

TABLE 7.2 Rank order of burst speed at different body temperatures in twelve individual alligator lizards (*Gerrhonotus multicarinatus*)

Temperature (°C)	Individual											
	A	B	C	D	E	F	G	H	I	J	K	L
10	8	7	3	11	4	9.5	1	9.5	2	5	12	6
15	8	5	1	11	3	6	2	10	7	9	12	4
20	7	6	2	11	3	8.5	5	8.5	1	4	12	10
25	9	8	2	11	3	4	6	7	1	12	10	5
30	6	9	2	12	7.5	5	4	10	1	7.5	11	3
35	8	10	7	11	5.5	2	4	9	1	5.5	12	3
37.5	9	8	7	11	6	3.5	1	10	2	3.5	12	5

Note: Rank 1 = fastest; $p < .001$ by Kendall's coefficient of concordance.
Source: Bennett (1980).

In my opinion, the large majority of physiological variables that can be sampled repeatedly will show real and significant interindividual variation. The question then becomes how we can utilize this variability to our benefit in asking analytical questions.

The analytical utility of interindividual variation

I suggest four different types of studies in which the exploration of interindividual variability might play a crucial role. Some represent new sorts of investigations for ecological or comparative physiology. Others permit a new approach to both current and classical questions in the fields.

The testing of correlative hypotheses

A common analytical approach in comparative physiology is to measure the correlation between two or more variables in two or more groups (e.g., populations, species) and to infer mechanistic relationships if significant correlations exist. For example, if positive associations are found between the length of the loops of Henle in kidneys of various mammals and their ability to concentrate urine, one might conclude that these may be functionally linked. These correlational examinations have been central in building the field of comparative physiology. They have, however, been criticized for their failure to take into account the phylogenetic history of the experimental animals involved (Gould and Lewontin, 1979; Felsenstein, 1985; Chapter 4).

A companion approach to interspecific analyses is the examination of interindividual correlations of variables within a species. This approach maintains the benefits of comparative analysis without some of the objections associated with using organisms that are distantly related phylogenetically (see Chapter 4). If two factors are functionally related, they should be significantly correlated among individuals within a species. In fact, if evolutionary or functional trends are proposed, the argument is strengthened if intraspecific associations can be demonstrated, because selection on traits within populations must be the ultimate source of adaptation.

The experimental protocol required to investigate interindividual variability is similar to that of interspecific comparative studies, except that observations on functional traits of interest must be made on the same individuals and analyzed on that basis. The researcher then correlates one trait with the other to determine whether they are positively or negatively associated. If so, the hypothesis of functional relationships among the traits is supported, and further experimentation can be planned to explore the nature of the relationship (see Huey and Bennett, 1986). If no significant association is found, then the traits are not functionally linked and the hypothesis is rejected.

One important step in this analysis is the determination of the dependence of the traits in question on body size (mass) and the elimination of such a dependence in the analysis. So many morphological, physiological, and behavioral traits are dependent on body size (see Calder, 1984; Schmidt-Nielsen, 1984) that it is very easy to obtain positive correlations among otherwise unrelated traits because of their mutual dependence on mass (see Appendix for a further discussion and example). Allometric analyses should be performed (see Chapter 10) and, if mass effects are significant, the mass-corrected residuals should be analyzed for correlation.

An illustrative example of the use of interindividual variability in testing correlative hypotheses may be beneficial here. These data are drawn from some observations on the skeletal muscle physiology and locomotor performance of tiger salamanders (Else and Bennett, 1987, and unpublished observations). Close (1964, 1965) proposed a correlation between the speed of isometric and isotonic contractions of skeletal muscle: the maximal velocity of shortening (isotonic) is supposed to be positively related to the rate of tension development in an isometric twitch or tetanus. This proposal is a straightforward mechanistic linkage that is supported by interspecific comparative studies. We can test this hypothetical connection by making observations of all these factors on individual animals and determining whether they are associated within individuals. A further correlation that might also be investigated is the association between muscle contractile speed and locomotor speed: are the animals that have the greatest intrinsic speed of muscle contraction also the fastest? First, all variables are mass-corrected and the residuals are then correlated with each other in Table 7.3. Correlations are significant among isometric variables and between isotonic variables, but no associations are significant between any isotonic and isometric variable nor between burst speed and any measure of muscle contractile performance.

TABLE 7.3 Correlation coefficients (r) among mass-corrected residuals of locomotor performance and muscle contractile factors in the salamander *Ambystoma tigrinum nebulosum* at 20 °C ($n = 20$)

		Isometric muscle factors				Isotonic muscle factors	
	Locomotion (burst swim speed)	Tetanic force	Twitch force	Tetanic contraction rate	Twitch contraction rate	Maximal rate of shortening	Maximal power output
Burst run speed	.13	−.27	−.38	−.52	−.42	−.05	.31
Burst swim speed		.14	−.24	−.11	−.09	−.41	−.29
Tetanic force			.79*	.67*	.72*	−.67*	−.47
Twitch force				.85*	.86*	−.43	−.31
Tetanic contraction rate					.97*	−.27	−.27
Twitch contraction rate						−.38	−.24
Maximal rate of shortening							.71*

Note: Asterisks indicate significant correlations ($r > 0.56$, $p < 0.01$).
Source: Unpublished data of A. F. Bennett, P. L. Else, and T. Garland.

These results argue against any necessary mechanistic association among these factors.

This is only one example of an approach that can be utilized in many different physiological or functional studies. For instance, the role of maximal heart rate in limiting maximal oxygen consumption or that of a particular muscle in generating force during feeding or locomotion could be investigated using an appropriate analysis of interindividual variability.

Examining the functional bases of organismal or physiological variables

Another use that can be made of interindividual variability is the determination of which of a potential suite of characters might influence performance at a higher level of biological organization. This is a multivariate statistical approach based on an array of characters measured in identified individuals of a species. The researcher measures a performance variable, such as burst speed or lower critical temperature, and a number of morphological and/or physiological predictor variables that might reasonably be associated with it (e.g., limb length and maximal velocity of muscle shortening, or fur density and body temperature, respectively). All these measurements are made on the same series of individuals. Mass dependence of any of the factors is analyzed and removed, as discussed previously. Then stepwise multiple regression analysis (or another appropriate technique, such as canonical correlation) is used to determine which, if any, of the predictor variables are associated with the performance variables.

An example of this approach is provided by the study of Garland (1984) on locomotor performance by a lizard, *Ctenosaura similis*. Endurance, burst speed, and maximal distance run under pursuit were measured in a series of individuals, along with a variety of physiological and morphological variables, including body mass and length; standard and maximal rates of oxygen consumption and carbon dioxide production; mass of thigh muscle, heart, and liver; hematocrit and hemoglobin concentration of the blood; myofibrillar ATPase activity of thigh muscle; and activities of three selected metabolic enzymes in heart, liver, and skeletal muscle tissue. Body mass effects were removed by regressing all variables on mass and analyzing only mass-corrected residuals. Each measure of locomotor performance was then regressed as a dependent variable on the suite of morphological and physiological characters as independent variables. The results of these analyses are given in Table 7.4. Nearly 90% of the mass-corrected interindividual variation in endurance could be attributed to four predictive factors, including maximal oxygen consumption, skeletal muscle and heart mass, and hepatic aerobic enzyme activity. This is a remarkable amount of predictive power. More than half the variation in maximal distance run is correlated with maximal carbon dioxide production and anaerobic enzyme activity of the skeletal muscle. None of the variables measured in this study was significantly associated with

TABLE 7.4 Stepwise multiple regression analysis of locomotor performance of the lizard *Ctenosaura similis*

Performance	Variable	Partial R^2	
Endurance	Thigh muscle mass	0.540	
	Maximal oxygen consumption	0.187	
	Heart mass	0.086	
	Liver aerobic enzyme activity	0.080	
	Total	0.893	($p < .0001$)
Distance run	Maximal carbon dioxide production	0.405	
	Thigh anaerobic enzyme activity	0.177	
	Total	0.582	($p = .0022$)
Burst speed	None	N.S.	

Source: Garland (1984).

burst speed. Thus, a multivariate statistical approach does not necessarily find a significant association among any set of variables. It may uncover strong correlations (as in the case of endurance) or no correlation (as with burst speed). A subsequent investigation on another species of lizard found significant interindividual correlations between burst speed and glycolytic enzymatic activity of skeletal muscle and an inverse relationship between burst speed and muscle fiber diameter (T. Gleeson, unpublished data).

A multivariate statistical approach can be particularly powerful when numerous underlying variables might be expected to influence higher-level performance. It can help to single out the most significant factors from an entire array and allow a researcher to concentrate further on those. The result of the analysis may serve to confirm a priori associations or may suggest entirely unexpected linkages that can be explored further. This multivariate analysis should be regarded as a first-stage approach, to be followed by more detailed comparative and experimental research on the factors identified with this technique. These further studies may also take advantage of interindividual variability.

Measurement of selective importance of traits under field or experimental conditions

Physiological ecologists and comparative physiologists usually assume that the traits that they study are of adaptive significance, that is, that they enhance survivorship and reproductive potential. This assumption is, however, almost never tested directly (Arnold, 1983; Endler, 1986). Using interindividual variability, one can evaluate whether performance of any given physiological or organismal trait is in fact correlated with differential survivorship under natural conditions. The observations required involve scoring a trait on a large number of individual animals, releasing them into their natural environment, and recapturing the survivors after exposure to this environment. The survivors are then examined to determine whether they are drawn from any subset of the original distribution.

Selection might operate in a number of ways to favor different portions of the original distribution of the trait (Simpson, 1953; Lande and Arnold, 1983; Endler, 1986). It might be *directional* and favor individuals at one end of the range of variability. For example, do birds with greater insulation survive better during the winter or do caterpillars that eat more metamorphose more rapidly and successfully? Do newborn snakes that are very fast or have a high endurance (see Figure 7.1) accrue an advantage under natural conditions such that they are more likely to survive to reproductive age? Selection may also be *stabilizing*, favoring animals with modal values for a given trait, thereby reducing variability and reinforcing central tendency in the population. In these cases, both very well and poorly insulated birds, caterpillars with both large and small appetites, and very fast and very slow snakes would be selected against. Selection might also be *disruptive*, favoring animals at both extremes of the distribution and tending to increase overall variability. The null hypothesis against which the presence of selection must be tested is the absence of any detectable effect of the variable on such indices of fitness as survivorship, growth, or reproduction. In the examples above, variability in plumage quality, feeding capacity, or speed would have no detectable influence on fitness under field conditions.

This correspondence between physiological or performance characters and survivorship or fitness under field conditions is termed the "fitness gradient" (Arnold, 1983). Its determination is judged to be essential for the characterization of the ecological and evolutionary implications of any physiological variable. However, comprehensive studies of the fitness gradient have rarely been attempted for any variable. The effects of natural selection on physiological variation generally are unknown (Endler, 1986). A lack of correspondence between maximal oxygen consumption and some measures of reproductive performance has been reported in adult male toads (Wells and Taigen, 1984; Sullivan and Walsberg, 1985), but its effect on differential survivorship up to adulthood has not been measured. Studies on the effects of locomotor performance on postnatal survivorship are currently underway on fence lizards (R. Huey, University of Washington) and garter snakes (my laboratory). The direct measurement of the impact of a character on performance under natural conditions, in spite of its obvious importance to field ecology and evolutionary biology, is almost unexplored. It may be operationally difficult or even impossible on some types of organisms, but I believe it is in fact feasible for many different types of animals in many different environments.

This approach has great potential to measure the importance of selection on traits in natural populations in natural environments. It also can be used in situations in which the environment has been experimentally altered. In this case, the response of the trait in the population can be compared to a priori expectations about the effect of such alteration. For instance, one might remove predators and determine whether burst speed or endurance declines in a population in the absence of this particular selective agent. One might supplement animals living in saline ponds or deserts with fresh water and investigate whether osmotic tolerance or fluid-concentrating capacity changes as a result of altered environmental circumstances. An excellent example of this experimental approach is provided by the study of Ferguson and Fox (1984). A combination of studies, examining responses of populations in both natural and experimentally manipulated environments, has a great deal of potential to help us understand the importance of various physiological processes to total fitness of organisms. This approach presents a protocol for testing assumptions about adaptation, not simply asserting them axiomatically. I believe this is one of the most exciting new developments and directions for physiological ecology as a field.

Determination of heritabilities of organismal or physiological characters

For adaptation and evolution of a trait to occur, it must have a genetic basis. Without a heritable basis, selection on a trait within each generation will not influence the variability or distribution of the trait in ensuing generations. It is necessary, for example, for fast parents to have fast offspring if the population is to respond to a new agent that selects against slower individuals. Studies of the heritability of physiological traits are a valuable supplement to ecological studies because they permit the determination of both the potential of the trait to evolve and the rapidity with which the response can occur.

Some progress has been made in particular systems in identifying effects of individual loci on organismal physiology and performance (e.g., Watt, 1977, 1983; DiMichele and Powers, 1982; Chappell and Snyder, 1984; Barnes and Laurie-Ahlberg, 1986; Chapters 5 and 8). While individual loci may have identifiable effects, many of the traits of interest to a physiological ecologist will be under multilocus control. Consequently, the techniques of quantitative genetics will be the most appropriate for examining the inheritance of these characters (see Falconer, 1981, and Chapter 9 for a general discussion of the field and appropriate methodology). Techniques involve examining the similarities of traits in parents and offspring and/or among the offspring of given parents. They require that the organisms in question can be bred successfully in the laboratory or that gravid females can be obtained that will deliver offspring in the laboratory.

Few studies have examined the heritability of physiological or performance characters. Most of them have dealt with the inheritance of locomotor performance. Significant heritabilities have been found for speed in race horses (Langlois, 1980; Tolley, Notter, and Marlowe, 1983), speed in humans (Bouchard and Malina, 1983a, 1983b), burst speed and stamina in lizards (van Berkum and Tsuji, in press; R. B. Huey, unpublished data) and snakes (S. J. Arnold and A. F. Bennett, unpublished data; T. Garland, unpublished data). Defensive behaviors in snakes are also heritable (Arnold and Bennett, 1984). In these locomotor studies, a minimum of 30 to 50% of the variability among individuals is genetic. Other types of physiological traits have also been found to be heritable: for example, growth rate and efficiency in pigs (Smith, King, and Gilbert, 1962), reproductive output of chickens (Emsley, Dickerson, and Kashyap, 1977), and thermoregulatory behaviors of mice (Lacy and Lynch, 1979). Observations are so few at this point that a case may be made for a general investigation of the topic of heritability per se of physiological systems in different types of animals. Future studies may concentrate on more specific genetic issues concerning this inheritance, but, given the multilocus nature of these traits, these are bound to be more difficult.

Conclusions

Interindividual physiological variability is rarely studied. However, this variation is real and repeatable in many physiological traits. I believe that the analysis of the causes and consequences of interindividual variability has major promise as an analytical tool in physiological studies. Ecological and comparative physiology have often been characterized as major branches of organismal biology, but their view of the organism has been ideal or typological. It has been that of the nonexistent animal that possesses the average value of all physiological, morphological, and behavioral attributes of the population. Such animals do not exist. Real individuals are unique combinations of traits, some above and some below average. It is time to recognize the uniqueness of the individual and to turn it to our advantage as biologists.

The analysis of variation can be useful in studies on physiological correlation and mechanism, on the importance of the variable to fitness under natural conditions, and on the potential for inheritance of the trait, with the consequent possibility of its adaptation and evolution. I do not suggest that the study of variation should supplant other approaches nor that it is even feasible for all physiological variables. But where such study is applicable, it can be a powerful analytical tool, for analysis of both mechanism and adaptation. Its particular advantage is that it can pull together so many different aspects of biology, not only physiology and ecology, but also behavior, morphology, population biology, and evolution. Biologists often treat these as different areas, but of course individual organisms do not make these arbitrary divisions and distinctions. They react to problems and opportunities as integrated organisms. Appreciating and studying individual differences can be a synthetic approach that puts the individual organism back into organismal biology and gives us a much broader understanding of animals and their evolution.

Appendix

The problem of mass-dependent correlation is so ubiquitous in correlation analysis that I will provide an illustrative example of the utility of the analysis of residuals. In Figure 7.3a, two physiological traits are found to be positively associated when one is plotted as a function of the other. These might, for example, be length of a Malphigian tubule and secretion rate in an insect, or tidal volume and anatomical dead space during ventilation in a mammal. Such a result might lead one to conclude that the traits are positively linked functionally. If, however, both traits are plotted as a function of body mass (Figures 7.3b and 7.3c), each is also found to be strongly and positively mass dependent. Are the traits truly linked to each other or is their apparent association due to their mutual but independent relationship to body mass? This size influence may be removed by examining the deviation of each data point from the mass regression line (i.e., the residuals of the regression). If these mass-corrected residuals are then plotted against each other (Figure 7.3d), their relationship can be examined without the interfering effects of body size. In the case illustrated, the traits are found to be negatively related to each other, which is exactly the opposite of the original conclusion based on Figure 7.3a. Their apparent positive association was due only to their mutual correlation with body mass. This was, of course, a contrived example: the residuals might also have been positively associated or not significantly correlated with each other. The point is that an examination of the original, uncorrected data in Figure 7.3a would not have permitted this assessment.

Acknowledgments

I thank S. J. Arnold, M. E. Feder, T. Garland, R. B. Huey, G. V. Lauder, H. B. Shaffer, and C. R. Taylor for helpful discussions and/or comments on the manuscript. Support for the workshop was provided by NSF Grant BSR 86–07794. Support for the author's research cited herein is from NSF Grants BSR 86–00066, DCB 85–02218, DEB 81–14656, and PCM 81–02331.

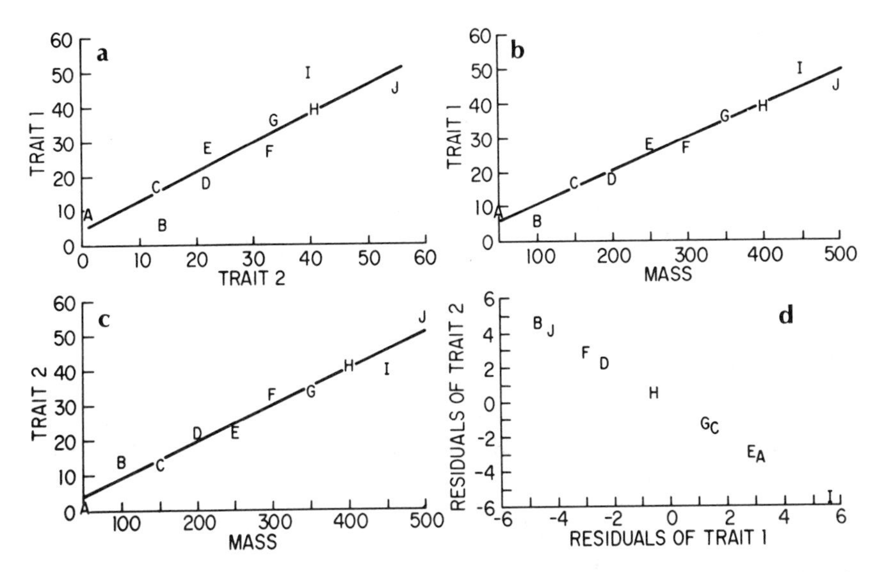

FIGURE 7.3 A hypothetical example of the effect of mass-correlated effects on the apparent association between two traits. Data are reported for two traits and body mass in arbitrary units for ten individuals (A through J) along with least-squares regression lines. (a) The two traits are positively and significantly correlated when they are related to each other directly. (b) and (c) Each trait is positively mass-correlated. (d) Mass effects are removed by calculating residuals, that is, the difference between the observed value for each individual and the value predicted by the mass regression. These are plotted against each other, demonstrating a significant negative association between the traits after the confounding effects of mass are eliminated. Note that opposite conclusions about the relationship between the traits would be derived from (a) and (d).

References

Arnold, S. J. (1981) Behavioral variation in natural populations. I. Phenotypic, genetic and environmental correlations between chemoreceptive responses to prey in the garter snake, *Thamnophis elegans*. *Evolution* 35:489–509.

Arnold, S. J. (1983) Morphology, performance and fitness. *Am. Zool.* 23:347–361.

Arnold, S. J., and Bennett, A. F. (1984) Behavioural variation in natural populations. III. Antipredator displays in the garter snake *Thamnophis radix*. *Anim. Behav.* 32:1108–1118.

Arnold, S. J., and Bennett, A. F. (in press) Behavioural variation in natural populations. V. Morphological correlates of locomotion in the garter snake *Thamnophis radix*. *Biol. J. Linn. Soc.*

Barnes, P. T., and Laurie-Ahlberg, C. C. (1986) Genetic variability of flight metabolism in *Drosophila melanogaster*. III. Effects of *Gpdh* allozymes and environmental temperature on power output. *Genetics* 112:267–294.

Bennett, A. F. (1980) The thermal dependence of lizard behaviour. *Anim. Behav.* 28:752–762.

Berman, S. L., Cibischino, M., Dellaripa, P., and Montren, L. (1985) Intraspecific variation in the hindlimb musculature of the house sparrow. *Am. Zool.* 25:21A.

Bouchard, C., and Malina, R. M. (1983a) Genetics for the sport scientist: selected methodological considerations. *Exerc. Sport Sci. Rev.* 11:274–305.

Bouchard, C., and Malina, R. M. (1983b) Genetics of physiological fitness and motor performance. *Exerc. Sport Sci. Rev.* 11:306–339.

Calder, W. A., III. (1984) *Size, Function, and Life History.* Cambridge, Mass.: Harvard University Press.

Chappell, M. A., and Snyder, L. R. G. (1984) Biochemical and physiological correlates of deer mouse α-chain hemoglobin polymorphisms. *Proc. Natl. Acad. Sci. U.S.A.* 81:5484–5488.

Christian, K. A., Tracy, C. R., and Porter, W. P. (1985) Inter-individual and intra-individual variation in body temperatures of the Galapagos land iguana (*Conolophus pallidus*). *J. Therm. Biol.* 10:47–50.

Close, R. (1964) Dynamic properties of fast and slow skeletal muscles of the rat during development. *J. Physiol. (Lond.)* 173:74–95

Close, R. (1965) The relation between intrinsic speed of shortening and duration of the active state of muscle. *J. Physiol. (Lond.)* 180:542–559.

Crowley, S. R. (1985) Insensitivity to desiccation of sprint running performance in the lizard *Sceloporus undulatus*. *J. Herpetol.* 19:171–174.

Crowley, S. R., and Pietruszka, R. D. (1983) Aggressiveness and vocalization in the leopard lizard (*Gambelia wislizennii* [sic]): the influence of temperature. *Anim. Behav.* 31:1055–1060.

DiMichele, L., and Powers, D. A. (1982) Physiological basis for swimming endurance differences between LDH-B genotypes of *Fundulus heteroclitus*. *Science* 216:1014–1016.

Else, P. L., and Bennett, A. F. (1987) The thermal dependence of locomotor performance and muscle contractile function in the salamander *Ambystoma tigrinum nebulosum*. *J. Exp. Biol.*, 128:219–233.

Emsley, A., Dickerson, G. E., and Kashyap, T. S. (1977) Genetic parameters in progeny-test selection for field performance of strain-cross layers. *Poult. Sci.* 56:121–146.

Endler, J. (1986) *Natural Selection in the Wild.* Princeton, N.J.: Princeton University Press.

Falconer, D. S. (1981) *Introduction to Quantitative Genetics*, 2nd ed. New York: Longman.

Felsenstein, J. (1985) Phylogenies and the comparative method. *Am. Natur.* 125:1–15.

Ferguson, G. W., and Fox, S. F. (1984) Annual variation of survival advantage of large juvenile side-blotched lizards, *Uta stansburiana*: its causes and evolutionary significance. *Evolution* 38:342–349.

Garland, T., Jr. (1984) Physiological correlates of locomotory performance in a lizard: an allometric approach. *Am. J. Physiol.* 247:R806–R815.

Garland, T., Jr. (1985) Ontogenetic and individual variation in size, shape and speed in the Australian agamid lizard *Amphibolurus nuchalis*. *J. Zool. (Lond.)* 207:425–440.

Garland, T., Jr., and Arnold, S. J. (1983) Effects of a full stomach on locomotory performance of juvenile garter snakes (*Thamnophis elegans*). *Copeia* 1983:1092–1096.

Gould, S. J., and Lewontin, R. C. (1979) The spandrels of San Marco and the Panglossian paradigm: a critique of the adaptationist programme. *Proc. R. Soc. Lond.* [*B*] 205:581–598.

Hanken, J. (1983) High incidence of limb skeletal variants in a peripheral population of the red-backed salamander, *Plethodon cinereus* (Amphibia: Plethodontidae), from Nova Scotia. *Can. J. Zool.* 61:1925–1931.

Huey, R. B., and Bennett, A. F. (1986) A comparative approach to field and laboratory studies in evolutionary ecology. In *Predator-Prey Relationships: Perspectives and Approaches from the Study of Lower Vertebrates*, ed. M. E. Feder and G. V. Lauder, pp. 82–98. Chicago: University of Chicago Press.

Huey, R. B., and Hertz, P. E. (1984) Is a jack-of-all-temperatures a master of none? *Evolution* 38:441–444.

John-Alder, H. B. (1984) Seasonal variations in activity, aerobic energetic capacities, and plasma thyroid hormones (T_3 and T_4) in an iguanid lizard. *J. Comp. Physiol.* 154:409–419.

Krebs, H. A. (1975) The August Krogh Principle: "For many problems there is an animal on which it can be most conveniently studied." *J. Exp. Zool.* 194:221–226.

Krogh, A. (1929) Progress of physiology. *Am. J. Physiol.* 90:243–251.

Lacy, F. C., and Lynch, C. B. (1979) Quantitative genetic analysis of temperature regulation in *Mus musculus*. I. Partitioning of variance. *Genetics* 91:743–753.

Lande, R., and Arnold, S. J. (1983) The measurement of selection on correlated characters. *Evolution* 37:1210–1226.

Langlois, B. (1980) Heritability of racing ability in thoroughbreds–A review. *Livestock Prod. Sci.* 7:591– 605.

Laurie-Ahlberg, C. C., Maroni, G., Bewley, G. C., Lucchesi, J. C., and Weir, B. S. (1980) Quantitative genetic variation of enzyme activities in natural populations of *Drosophila melanogaster*. *Proc. Natl. Acad. Sci. U.S.A.* 77:1073–1077.

Mayr, E. (1982) *The Growth of Biological Thought: Diversity, Evolution, and Inheritance*. Cambridge, Mass.: Belknap.

Pough, F. H., and Andrews, R. M. (1984) Individual and sibling-group variation in metabolism of lizards: the aerobic capacity model for the evolution of endothermy. *Comp. Biochem. Physiol.* 79A:415–419.

Putnam, R. W., and Bennett, A. F. (1981) Thermal dependence of behavioural performance of anuran amphibians. *Anim. Behav.* 29:502–509.

Ringler, N. H. (1983) Variation in foraging tactics in fishes. In *Predators and Prey in Fishes*, ed. D. L. G. Noakes, D. G. Lindquist, G. S. Helfman, and J. A. Ward, pp. 159–171. The Hague: Dr. W. Junk.

Schmidt-Nielsen, K. (1984) *Scaling: Why Is Animal Size So Important?* Cambridge University Press.

Shaffer, H. B., and Lauder, G. V. (1985a) Patterns of variation in aquatic ambystomatid salamanders: kinematics of the feeding mechanism. *Evolution* 39:83–92.

Shaffer, H. B., and Lauder, G. V. (1985b) Aquatic prey capture in ambystomatid salamanders: patterns of variation in muscle activity. *J. Morphol.* 183:273–284.

Simpson, G. G. (1953) *The Major Features of Evolution.* New York: Columbia University Press.

Smith, D., King, J. W. B., and Gilbert, N. (1962) Genetic parameters of British Large White pigs. *Anim. Prod.* 4:128–143.

Sullivan, B. K., and Walsberg, G. E. (1985) Call rate and aerobic capacity in Woodhouse's toad (*Bufo woodhousei*). *Herpetologica* 41:404–407.

Taigen, T. L., and Wells, K. D. (1985) Energetics of vocalization by an anuran amphibian (*Hyla versicolor*). *J. Comp. Physiol.* 155:163–170.

Tolley, E. A., Notter, D. R., and Marlowe, T. J. (1983) Heritability and repeatability of speed for two-and three-year-old standard bred racehorses. *J. Anim. Sci.* 56:1294–1305.

Toolson, E. C. (1984) Interindividual variation in epicuticular hydrocarbon composition and water loss rates of the cicada, *Tibicen dealbatus* (Homoptera: Cicadidae). *Physiol. Zool.* 57:550–556.

van Berkum, F. H., and Tsuji, J. S. (in press) Among-family differences in sprint speed of hatchling *Sceloporus occidentalis. J. Zool. (Lond.)*

Watt, W. B. (1977) Adaptation at specific loci. I. Natural selection on phosphoglucose isomerase in *Colias* butterflies: biochemical and population aspects. *Genetics* 87:177–194.

Watt, W. B. (1983) Adaptation at specific loci. II. Demographic and biochemical elements in the maintenance of the *Colias* PGI polymorphism. *Genetics* 103:691–724.

Wells, D. K., and Taigen, T. L. (1984) Reproductive behavior and aerobic capacities of male American toads (*Bufo americanus*): is behavior constrained by physiology? *Herpetologica* 40:292–298.

Discussion

LINDSTEDT: I agree with Bennett that there are identifiable individuals which are low performers – they have a low maximum oxygen consumption, a low maximum running speed, etc. – and there are other individuals that are high performers. But on any given day a low-performing animal may outperform the high-performing animals: the ranges of their performances overlap, even if their means are repeatedly different. The likelihood of finding mechanistic differences to account for those mean differences may be rather low. We still need to have the broader overview between species, where we have a higher signal-to-noise ratio.

BENNETT: I am not advocating that we abandon all other approaches for the study of variation, nor that we should ignore the means. But we should take interindividual variations into account in addition. Sometimes the least and most able individuals overlap, but you can still find that the individual differences are statistically repeatable.

LINDSTEDT: Yes, we do try to do that. By the same token, I think we have to be cautious about throwing out simpler statistics because they are simple, especially if we risk losing some biological insight with greater statistical sophistication. For example, in using stepwise multiple regression, it can be hard to intuit what the result means. Also, we have found that as we increase the sample size, the total proportion of explained variation changes very little, yet the relative contributions of various independent variables to the total explained variation changes a great deal. Again, that leads me to gain less insight.

BENNETT: I agree that many times the simple statistics are adequate, but where they are inadequate, we should not continue using the old models and old ways of doing things. The stepwise multiple regression approach seems to me to be a tool to suggest further directions for study. If you find that no factors are correlated with performance (measured as burst speeds), that may tell you that you should be looking at other factors. When you do have significant correlations, then you have a basis for further experimental analysis. It is a first pass in looking for important variables.

SCHEID: I agree that interindividual variability is really important. Nature intends to tell us something that we have mostly neglected so far. Now, is it not true that if you want to address the interindividual variability, then you have to look at the intraindividual variability first? In fact, the only thing that remains beyond intraindividual variability is true interindividual variability.

BENNETT: That's right. You have to be able to make repeated measures on individuals. This is feasible for some factors and unfeasible for other factors. If you are looking at whole-body lactate content, for instance, you can do that only once. But there are a large number of physiological characters, such as blood flow parameters, that we can now sample nondestructively because of improved instrumentation.

SCHEID: We now have improved techniques to work on uninstrumented, nonanesthetized animals, which is mandatory if you want to ask questions about variability. I think that the techniques were not formerly available to address this variability in a meaningful way.

HUEY: The standard statistical method of measuring repeatability is the intraclass correlation coefficient, which measures the proportion of the variation that is due to difference among individuals versus within individuals. By that measure, sometimes the types of measurements that Bennett was referring to are highly repeatable: most of the variation is among individuals and not within individuals. For example, Art Dunham and I looked at sprint speeds of lizards in natural populations over a whole year, and the repeatabilities are on the order of 0.5 to 0.6, which is higher than in thoroughbred horses. That is probably high enough that we can begin to analyze the mechanistic basis of individual variation and also look at the adaptive significance of that variation.

POWERS: One problem of reproducing the same experiment on the same individual is that some organisms become trained. In addition, we found that we cannot put more than one individual in a track at a time because of behavioral interactions between them.

BENNETT: For running speed, in about half of the species that we observe, we see what we assume is a conditioning effect, from day one to day two, but after that the means stay exactly the same, the order of the individuals stays the same. Some species show this initial effect, others don't.

POWERS: One thing we have to do with fish is to acclimate them to water that is moving at a constant speed, for thirty to sixty days. Everything from then on is very reproducible. I am sure that a lot of the variation in the literature is a function of this training phenomenon and where the organisms came from.

FLORANT: I think that developmental effects can be extremely important, and I wondered whether you were rearing these animals in the lab or being careful about the developmental processes that were going on prior to, during, and after birth.

BENNETT: All the animals that we have been dealing with are adult animals, taken directly from the field and tested within a matter of days. In the breeding studies, gravid animals are collected and young animals are born under constant conditions in the laboratory. The whole issue of developmental effects and constancy of rank-ordered performance over time has not even begun to be explored.

FUTUYMA: Suppose you are interested in very short term acclimation effects, the capacity of the individual to change its phenotype from moment to moment, which is the opposite of repeatability. How do you deal with that? There are interesting questions there as well.

BENNETT: You begin by immediately asking questions about your equipment and techniques, and get that out of the way first. Then perhaps you can begin building correlations from moment to moment by measuring the variables sequentially, to see whether you are getting tracking of one variable by the other.

ARNOLD: We can examine the capacity to change performance over short or long periods of time as traits, and that's a virtually unexplored area. But it is not the opposite of repeatability. Suppose we define a new variable that represents the capacity to change performance as a function of an elevation in temperature from 10 °C to 20 °C. We can measure its repeatability, we can ask whether it is inherited, whether it is genetically correlated with other traits, and so forth. The statistical field for dealing with such traits is sometimes called profile analysis of variance.

FEDER: I want to shift the focus of this discussion to the point that Bennett made about the prospect for performing natural experiments using natural populations. I am very excited about this prospect. Using the variation in populations as a substrate, altering an environmental variable for individuals in a population or adding individuals to a population and looking at the effects could potentially be a very powerful technique. Dennis Powers said that it may soon be possible to take individual genes and move them into or out of individual organisms, which could offer us a lot of insight.

POWERS: It is already possible for some species. In lower vertebrates, it will probably take another year.

DAWSON: There are detraining or conditioning effects that go with captivity. When we studied cold resistance in small birds, we found that the animals maintained in outdoor flight cages, given the seeds of the type that they were using naturally, abandoned their winter fattening, perhaps because they had assured meals and more complicated cues. They also had much lower cold resistance than freshly captured animals. If one is dealing with badly distorted responses, which is sometimes a risk with wild animals, that ought to be determined. A good deal of what one may be dealing with in animals long standing in captivity may not be relevant to the natural situation.

FLORANT: In keeping hibernators for many years in the lab, the hibernators begin to free-run, and it is as if certain physiological responses occur at "nonadaptive" times of year. This obscures the optimal time that the animal performs a particular response under natural circumstances.

BENNETT: These are valid concerns. One way of keeping track of them is to run appropriate controls, so that we can place boundaries on the magnitude of the captivity responses.

DAWSON: By attempting to determine repeatability, if you start early enough, you can discern if there are any effects of that type. That is not done a lot. This is a caveat about use of material from animal dealers, which may have a very fuzzy history indeed.

8 The importance of genetics to physiological ecology

RICHARD K. KOEHN

Introduction

In making a contribution to this volume, one that attempts to project the study of physiological ecology into the future, I have been charged with addressing a number of questions that are both of relevance and of interest to physiological ecology. Why should physiologists care about genetics? Are there readily discernible links between physiological and genetic variations? If so, how might this relationship affect broader issues that are of concern to physiologists and geneticists? In general, physiologists, or specifically, physiological ecologists, have not been naive about the potential importance of genetics to physiology. Yet, this perception has not been widely applied in studies of physiological variation. This chapter seeks to stimulate a wider appreciation of the importance of genetics to physiological investigations. We will see that genetic variation can have very significant effects upon physiological performance. In some cases these effects can be ascribed to polymorphism at a single gene, while in other cases, variation at only a few genes can produce remarkably large differences in energy balance, and thereby in physiological performance among individuals. Results from the combined study of genetics and physiological ecology will bear uniquely upon the resolution of some fundamental and long-standing problems in both fields.

The study of genetics has not typically been held to have much relevance to the study of physiological ecology. Indeed, the influences of genetics and physiology, broadly conceived, upon each other have always been rather vague. Physiological genetics is neither a new concept nor a new discipline. Yet, the "genetic" part of physiological genetics has historically been little more than an assumption of a probable genetic basis for certain physiological traits. Physiological traits reflect the behavior of underlying metabolic machinery. These traits are known to result from various aspects of specific metabolic pathways; such pathways are structured interactions among the enzyme products of genes. From this view it follows quite obviously that virtually all physiological characteristics of an organism must be genetically determined. This view misses the point, since it would be difficult, if not impossible, to imagine a physiological trait that was not, in this sense, genet-

Genotype A (Species A) $\xrightarrow{\text{Env. 1}}$ Biochemical Phenotype 1 ⟶ Physiological Phenotype 1
Genotype B (Species B) $\xrightarrow{\text{Env. 2}}$ Biochemical Phenotype 2 ⟶ Physiological Phenotype 2

SYSTEMATICS or POPULATION GENETICS	COMPARATIVE BIOCHEMISTRY	COMPARATIVE PHYSIOLOGY or PHYSIOLOGICAL ECOLOGY

FIGURE 8.1 The traditional paradigm of the study of adaptation by the comparative method. Various disciplines tend to be isolated from one another. This isolation among disciplines militates against discovery of the mechanisms by which genetics can have an influence on physiological and biochemical performance.

ically determined. While physiologists have generally assumed physiological phenotypes to have a genetic basis, the study of physiological traits has not historically necessitated precise specification of a genetic system.

More importantly, the foregoing view does not consider how variation of physiological traits of an organism might result from genetic variability. Virtually all physiological characteristics are highly variable; physiological variation has largely been attributed to environmental influences that act upon complex metabolic systems (i.e., nongenetic), or at best, to polygenes. Since physiological traits result from a complex metabolism, it has been reasonably assumed that a significant proportion of the variation in any specific physiological trait could not be caused by variation of one, or a few, specific genes. However, while the metabolic machinery is complex, it is nevertheless highly structured, and this structure permits specific physiological variations to be attributed to a gene, or genes, of specific metabolic function. We would not, for example, attempt to demonstrate the importance of a glycolytic enzyme (and its genetic variants) to variations in amino acid catabolism, as glycolysis has no direct metabolic role in protein degradation. Variations in amino acid metabolism would more likely result from genetic variants of enzymes important to amino acid synthesis, amino acid transamination, and/or catabolism of proteins to amino acids. As we shall see later, this is not an arbitrary example.

The study of adaptation is the purview of several (perhaps all) traditional disciplines in biology, each separated from the others by a concern with different levels of biological organization and/or different traits of organisms (Figure 8.1). Each discipline in Figure 8.1 (and several others could be included) shares with the others a "comparative approach" to the study of adaptation.

Species differences, be they genetic, biochemical, and/or physiological, are "mapped" onto, or correlated with, environments. From this mapping, inferences about adaptation are made. Yet, this comparative approach precludes the discovery of any precise genetic basis for physiological variation in any but the most general sense. We cannot experimentally test the correlation of physiology to genes, since the ability to exchange genes between species is still rudimentary. In some cases, it is possible to "exchange environments," or species between environments, and comparatively measure species performance, but this too has many potential experimental pitfalls. A comparative approach to the study of adaptation, as it has been practiced in, for example, comparative biochemistry and physiological ecology, can make a compelling argument for why organisms exhibit specific biochemical and/or physiological traits, but in the final analysis the arguments are inferential since these cannot be rigorously tested.

What we seek in discovering the genetic basis of physiological performance is how variations at different levels of biological organization are mechanistically and causally related to each other. Whether we start with the genetics or the physiology, the question is precisely the same. For example, population genetics has been dominated for nearly two decades by attempts to assess how enzyme polymorphisms may or may not be important in the adaptation of organisms to ecological circumstances. The fitness differences among enzyme genotypes that geneticists have sought depend upon the physiological consequences of enzyme polymorphism. In this context, the goals of population genetics and physiological ecology are nearly identical.

Genetic studies that attempt to identify phenotypic (i.e., physiological) consequences of known genetic variations in single and relative simple multigene systems have begun to provide an understanding of the importance of genetic variation to physiological performance. These studies are of particular pertinence to the subject of this volume; each sends a clear signal to the field of traditional physiological ecology: polymorphism of enzymes of intermediary metabolism can produce physiological variation. Attempts by physiological ecologists to understand the causes of physiological differences must take account of genetic differences among individuals. Of course, the same message must be acknowledged by geneticists concerned with the potential adaptive importance of enzyme polymorphism, at least for those specific cases of polymorphisms that have been studied. Physiological variations, arising from the different biochemical properties of enzyme variants, can lead to differences in components of Darwinian fitness. Whether this is true for all gene products, or even all enzymes, will be addressed later.

The relationship(s) among genetic variation, physiological performance, and fitness (i.e., adaptation) can be pursued only *within* species (Figure 8.2). This is the appropriate (and I believe desirable) complement to the comparative approach discussed above. Natural polymorphisms amount to examples of gene substitution, and the biochemical, physiological, and fitness phenotypes of specific genotypes can be measured over a range of environments.

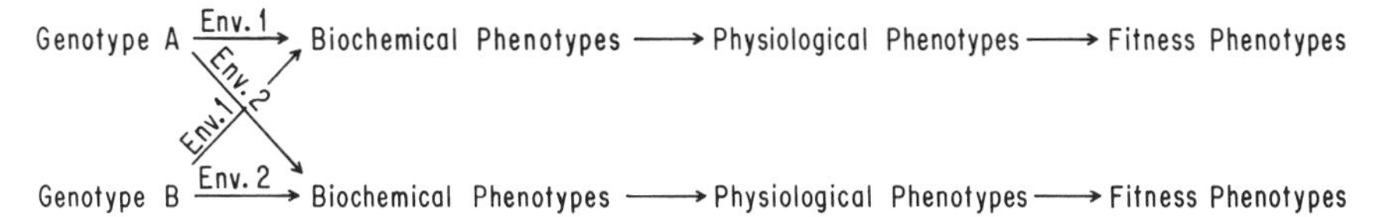

FIGURE 8.2 An appropriate paradigm for discovering the genetic basis of physiological variation. The biochemical and physiological phenotypes of genetic variants within a species can be studied, and the influence of different environments upon these can be measured. This permits an assessment of the relative physiological performance of different genotypes and the contribution of different performances to the evolutionary success (i.e., fitness) of individuals.

Because physiological ecologists engage this paradigm at the physiological level, and since physiological traits are nearly always continuously distributed, the genetic basis of such traits has been studied by quantitative genetic methods. Yet, this approach can be limited (but see Chapter 9). First, such an approach cannot be pursued unless a species is able to be bred under controlled laboratory conditions. This is impossible for the vast majority of species. In those species whose controlled breeding can be accomplished, any discovered genetic component will be genetically imprecise. It cannot tell us how many genes are involved in the determination of a trait, nor what the specific functions of such genes might be. Lastly, physiological ecology is the study of the *interaction* of physiology and the environment, and effects of the natural environment cannot be accurately duplicated in the laboratory.

If we cannot efficiently elucidate genetic mechanisms for physiological phenotypes through quantitative genetics, what approach then might we take? It might be possible to begin with a physiological phenotype and work downward to the biochemistry and then ultimately to those genes that may contribute to specific biochemical properties of an organism. In some instances, this may be a successful approach, particularly when a physiological characteristic has a relatively simple and well-understood biochemical basis. The remaining sections of this chapter will illustrate a different approach. Specific gene products can be identified and their variation (i.e., polymorphism) characterized. The metabolic function of such enzymes (when this is known) allows us to examine performance of those physiological traits that are potentially influenced by this metabolic function. We can then compare and contrast such metabolic (or physiologic) performances among individuals of different enzyme genotype.

In pursuing this research strategy, a number of important points will become obvious. First, polymorphism of a single gene can have measurable and biologically significant effects upon physiological performance, and thereby potentially upon adaptation. Second, polymorphism at only a few genes synthesizing enzymes of intermediary metabolism can have very dramatic effects upon the energy balance of an individual. This will result in significantly different levels of physiological performance among individuals of differing multilocus genotype. These results have two important major implications, one to genetics and the other to physiological ecology. Geneticists have great difficulty in ascribing the phenotypes associated with different genotypes to the action of a specific gene. Genes that are closely linked to a gene under study can be potentially involved in the determination of a phenotype since they cannot be easily "unlinked" from the gene of interest. However, when the gene products (i.e., enzyme variants) of an individual gene can be shown to have specific biochemical (and thereby physiological) properties, the causative action of that gene can be demonstrated without concern for the involvement of other linked genes. It will be apparent that future studies in physiological ecology should take into account the genetic characteristics of individuals and populations.

Genetic and physiological variations of single genes

Several authors have described the physiological consequences of single-gene enzyme polymorphism. These studies have been carried out by experimental population geneticists seeking an adaptive explanation for genetic polymorphism. To characterize the biochemical and physiological phenotypes associated with enzyme polymorphism is a tedious and difficult task, and for this reason, there are few examples where the physiological consequences of genetic variation are unequivocally known. These few "case studies" have been reviewed on several previous occasions (cf. Koehn, Zera, and Hall, 1983; Watt, 1985; Zera, Koehn, and Hall, 1985), and therefore I will present each in brief outline only. Interested readers should refer to the cited reviews.

Several studies have focused upon the potential importance of enzyme polymorphism in insect flight physiology and performance. Of such studies, those by Ward Watt and his collaborators (Watt, 1977, 1983; Watt, Cassin, and Swan, 1983; Watt, Carter, and Blower, 1985) are perhaps the most extensive. Phosphoglucose isomerase (PGI) is central to glycolysis, though it is not considered important in regulation of that pathway. This enzyme is commonly polymorphic in natural populations of many species and especially so in butterflies of the genus *Colias*. Glycolytic flux, in steady state or in transient state, can limit *Colias*'s capacity for flight activity and thereby affect a number of fitness components (Watt, 1983). Phosphoglucose isomerase variants in *Colias* exhibit a complex array of biochemical differences, primarily in V_{max}/K_m ratios, K_m, and thermal stabilities. K_m is the dissociation constant (a measure of substrate affinity) of an enzyme, and V_{max} is the maximum reaction rate that is obtained when substrate is in excess for the reaction (see

Chapter 5). V_{max}/K_m then is an overall measure of "activity." According to Watt (1983), values of K_m, V_{max}/K_m and thermostability differ among genotypes in a manner that is important in energy metabolism; there is positive selection for those genotypes whose catalytic phenotypes maximize the PGI reaction in glycolytic flux. Individuals with these genotypes can better maximize both the response and the capacity of glycolytic flux, especially under suboptimal or stressful thermal conditions. The interaction between the genotype-specific biochemistry of this polymorphism and the thermal characteristics of the environment leads to predictable patterns of genetic polymorphism in natural populations. For example, differences in flight performance are temperature and genotype dependent. Those PGI genotypes with relatively higher V_{max}/K_m ratios are expected to initiate suboptimal flight under colder morning conditions, fly over a broader span of the day, and maintain higher activity levels under marginal weather conditions. Watt (1983) has shown that genotypes differ in the time of flight onset and breadth of daily flight-time in a manner that is directly proportional to the V_{max}/K_m ratios of the PGI genotypes. In short, biochemical differences among PGI genotypes have measurable effects upon flight physiology of *Colias* butterflies.

The role of glycerol-3-phosphate dehydrogenase (GPDH) in the flight metabolism of *Drosophila melanogaster* is more simple and direct than the PGI isomerase system in *Colias*, since GPDH is one of two enzymes that have an essential role in the α-glycerol phosphate cycle, a pathway that fuels the energy requirements of flight. In *D. melanogaster*, allele frequencies differ with latitude in a similar direction on three continents: the $Gpdh^F$ allele decreases in frequency as latitude increases (Johnson and Schaffer, 1973; Oakeshott et al., 1982; Oakeshott, McKechnie, and Chambers, 1984). This observation suggests that GPDH allozymes may function differentially with temperature and thereby have different effects upon flight performance. [A number of authors have investigated the biochemistry of the GPDH enzyme polymorphism (i.e., Miller, Pearcy, and Berger, 1975; Bewley, 1978; McKechnie, Kohane, and Phillips, 1981; Bewley, Niesel, and Wilkins, 1984), but reported results are often contradictory (reviewed by Zera et al., 1985).]

Barnes and Laurie-Ahlberg (1986) tested the importance of GPDH genotypes to power output during tethered flight. Power output depended on both rearing and flight temperature, but the $GPDH^{SS}$ allotype exhibited a 2 to 4% greater power output than the $GPDH^{FF}$ allotype under a variety of conditions. The temperature effects on power output were consistent with the geographical and seasonal genetic variations observed in nature.

Nearly all marine invertebrates utilize a cytosolic free amino acid pool to maintain cell volume (i.e., remain isosmotic) in response to environmental salinity changes. During acclimation to increased salinity, there is a rapid increase in the concentration of free amino acids; the increase is effected by changes in specific residues, principally alanine and glycine (molluscs) or proline (crustaceans). During acclimation to lower salinity, the free amino acid pool decreases in concentration as a result of the catabolism and excretion of amino acids as ammonia and amines.

Burton and Feldman (1982) have demonstrated the effect of a glutamate pyruvate transaminase (GPT) polymorphism on cell volume regulation in the copepod *Tigriopus californicus*. GPT catalyzes the final step of alanine synthesis, which is the most important amino acid in the early stages of response to hyperosmotic stress in copepods (Bishop, 1976). Two GPT alleles are common in *Tigriopus* populations, Gpt^F and Gpt^S. Gpt^F exhibits significantly higher specific activity (Burton and Feldman, 1983) in both the alanine-synthesizing and the alanine-catabolizing directions of the catalyzed transamination. Under conditions of hyperosmotic stress, individual adult copepods of genotype $Gpt^{F/F}$ and $Gpt^{F/S}$ accumulate alanine (though not glycine or proline) faster than $Gpt^{S/S}$ individuals. The slower rate of response to hyperosmotic stress in $Gpt^{S/S}$ individuals results in a significantly higher rate of mortality of larvae of that genotype when they are exposed to hyperosmotic conditions. In short, the GPT polymorphism results in genetic variation for transamination to alanine, and this catalytic variation has a significant effect, not only upon cell volume regulation, but also upon survival.

Among multidisciplinary studies of a single enzyme polymorphism, that on aminopeptidase-I in the bivalve mollusc *Mytilus edulis* stands as an important example of the biochemical and physiological consequences of genetic variation. It is a good example of how mechanisms of adaptation can be better understood by combining the traditionally different approaches of genetics and physiology.

The aminopeptidase-I polymorphism consists of three common electrophoretic alleles, or electromorphs; these have been designated Lap^{98}, Lap^{96}, and Lap^{94}. The frequency of the Lap^{94} allele is approximately 0.55 in oceanic waters on the East Coast of the United States in the region south of Cape Cod. The frequency of this allele declines abruptly to approximately 0.15 to the north of Cape Cod and at the entrance to all studied estuaries south of Cape Cod (Koehn, Milkman, and Mitton, 1976; Koehn, Hall, Innes, and Zera, 1984). The diminution of Lap^{94} and joint increases in the frequencies of Lap^{98} and Lap^{96} occur over a few kilometers, resulting in steep allele frequency clines in those geographic areas of gene frequency change. These changes are correlated with environmental changes, primarily salinity. This observation, the correlated change in genetic composition with the environment, suggested the existence of strong differentiating forces acting on the *Lap* locus; the enzyme became a prime candidate for detailed investigations of the biochemical and physiological bases of environmentally determined changes in genetic composition.

The aminopeptidase enzyme catalyzes the hydrolysis of neutral and aromatic N-terminal amino acids of di-, tri-, and tetrapeptides (Young, Koehn, and Arnheim, 1979). The production of free amino acids by the action of the enzyme, coupled with the observation (above) that allele frequencies differ with environmental salinity, suggested a role for the enzyme in regulating cellular free amino acid pools. This is indeed the case. Moore, Koehn, and Bayne (1980) demonstrated cytochemically and immunocytochemically the presence of the enzyme in tissues and subcellular organelles associated with protein catabolism, specifically lysosomes. A clear and regular effect of salinity changes upon aminopeptidase activity has been demonstrated (Koehn, 1978; Bayne, Moore, and Koehn, 1981); increases in environmental salinity induce immediate three- to fourfold increases in lysosomal aminopeptidase activity, whereas decreasing environmental salinity results in decreased enzyme activity. Amino acids resulting from salinity-induced changes in the rate of protein turnover in cell-free lysosomes (Bayne et al., 1981) are important cytosolic solutes for cell volume regulation (see above).

With the foregoing knowledge, Koehn and Siebenaller (1981) demonstrated that aminopeptidase-I genotypes with and without the Lap^{94} allele exhibited significantly different enzyme activity per unit enzyme protein; genotypes with the Lap^{94} allele exhibited a 20% greater specific activity than alternate genotypes.

Genotype-specific catalytic efficiencies produce significant differences in the rate of cell volume regulation (Hilbish, Deaton, and Koehn, 1982; Hilbish and Koehn, 1985a). During hyperosmotic stress, individuals carrying the Lap^{94} allele accumulate intracellular free amino acids more rapidly than do other genotypes. This difference in accumulation rate was also demonstrated by the interruption of hyperosmotic adjustment, which resulted in a higher rate of amine and ammonia excretion by Lap^{94} genotypes.

The Lap^{94} allele exhibited the highest relative specific activity, produced the highest rate of hyperosmotic cell volume regulation, and is found in highest relative frequency in those populations experiencing oceanic salinity. The opposite (i.e., lower activity, lower rate of regulation, and lower frequency) is true for non-Lap^{94} genotypes. In natural populations, larvae carrying the Lap^{94} allele are annually transported into low-salinity environments; as a consequence, these individuals suffer higher relative rates of mortality. This mortality results from a relatively higher rate of both use and loss of nitrogen reserves during energetically stressful periods (Hilbish and Koehn, 1985b). Energetic stress occurs in late summer-early fall when warm temperatures cause high metabolic rates, but food availability (i.e., phytoplankton) is low, and animals are thereby required to utilize stored energy reserves. This stress results first in decreased physiological condition (Koehn, Newell, and Immermann, 1980) and ultimately in death. Both stress and death rates are greater for individuals with the Lap^{94} allele.

The foregoing studies of the aminopeptidase-I polymorphism in *Mytilus edulis* illustrate a number of points germane to the topic of this chapter. Genetic variation of enzymes can significantly influence a physiological phenotype, as a consequence of the biochemical diversity represented by genetic variation. From the physiologist's perspective, a significant proportion of the variation among individuals in a physiological parameter (in this case, free amino acid concentrations) can be attributed to variation of a single gene. The link between genetics and physiology in this case is intimate and direct; to understand fully the biological significance of phenotypic variation, both genetic and physiological information is required. Also, the degree to which genetic (and thereby physiological) variation is important in adaptation is critically dependent upon environmental circumstances. In this case, those circumstances that maximize the effect of environmental salinity variation on survivorship are a combination of salinity, temperature, and food availability (a more detailed description of this interaction may be found in Hilbish and Koehn, 1985b).

Dennis Powers and his collaborators have amassed considerable evidence for the importance of lactate dehydrogenase polymorphism in adapting the teleost fish *Fundulus heteroclitus* to differing thermal environments. The two common alleles of the heart type, or lactate dehydrogenase-B, differ in their temperature-dependent catalytic properties, and these are correlated with various aspects of blood chemistry, hatching times and swimming performance. A detailed description of this system may be found in Chapter 5.

Genetic and physiological variations of multiple loci

The preceding section has illustrated some examples of the way in which the combination of genetic and physiological information can be of importance for providing an understanding of biological variations. However, the results of studies of single enzyme polymorphisms, such as aminopeptidase-I in *Mytilus edulis* or lactate dehydrogenase in *Fundulus heteroclitus*, cannot be directly applied to one another or to other genetic systems in other species. However, a number of recent studies have suggested that the proportion of sampled genes in a heterozygous state among multiple enzyme-encoding loci can significantly influence physiological performance, specifically the physiological energetics of growth. This observation has captured the attention of both geneticists and physiologists as it provides an opportunity to understand the precise way in which genetic variations influence the physiological energetics of individual organisms.

Not unlike examples described earlier, the physiological energetic effects of enzyme polymorphism were suggested first by findings in population genetics. A strongly positive correlation has been reported between the degree of multiple-locus heterozygosity of an individual, measured at a series of electrophoretically detectable enzyme loci, and growth rate, measured as size and/or mass at a given age. Although variation is great in the biology of studied organisms, as well as in the experimental designs of each investigation, this relationship has now been demonstrated in a great diversity of organisms, including plants (Schaal and Levin, 1976; Mitton et al., 1981; Ledig, Guries, and Bonefeld, 1983), some mammalian species (Makaveev, Venev, and Baulov, 1977; Baker and Manwell, 1977), and salamanders (Pierce and Mitton, 1982), as well as several marine invertebrates that include oysters of the genus *Crassostrea* (Singh and Zouros, 1978; Zouros, Singh, and Miles, 1980; Fujio, 1982), the mussel *Mytilus edulis* (Koehn and Gaffney, 1984; Diehl, Gaffney, and Koehn, 1986), the gastropod *Thais haemastoma* (Garton, 1984), and the clams *Mulinia lateralis* (Garton, Koehn, and Scott, 1984) and *Macoma balthica* (Green, Singh, Hicks, and McCuaig, 1983). The relationship between individual heterozygosity and growth rate appears to be widespread in nature, possibly a characteristic of all outbred populations, though a number of factors can affect the degree to which the relationship can be experimentally demonstrated (cf. Gaffney and Scott, 1984).

Growth is a conversion of energy to somatic tissue. The rate of growth is governed, in large part, by the biochemistry of the enzymes that catalyze energetic conversions and by how these regulate metabolic pathways of energy production. Variations in growth rate among individuals can be the consequence of variation among individuals in available ration and/or variation in the efficiencies with which food (or stored resources) can be utilized for somatic growth. The balanced energy equation of Winberg (1956) is familiar to all physiologists as a formalized expression of how energy is gained and lost by an organism. Production, P, is equivalent to net assimilated energy and is equal to $C - (F + U + R)$, where C is the amount of food consumed, and F, U, and R are energetic losses through feces, dissolved organics, and respiration, respectively. The relationship is sometimes expressed as "scope-for-growth," where $SG = Ab - (R + U + F)$; SG is a measure of net energy balance, which may be positive or negative, and Ab is net assimilated ration. In the absence of food, as may occur in an experimental context, both C (or Ab) and F are zero. Since U is only a small proportion of excretion costs in energy terms, all other things being equal, a reduction in R will increase the difference between net assimilated ration and respiration, leading thereby to relatively higher growth rates (Bayne, Widdows, and Thompson, 1976). In a similar fashion, P can be increased not only by a reduction in R, but also by a net increase in C (or Ab). When P (or scope-for-growth) is expressed as a mass-specific parameter, comparisons among individuals reflect differences in net energy balance.

Koehn and Shumway (1982) first suggested that more heterozygous individuals in outbreeding populations might achieve increased growth rates by virtue of individual differences in net energy balance, whereby more heterozygous individuals would enjoy greater scope-for-growth. In that study, concerned with the American oyster, *Crassostrea virginica*, an inverse relationship between multiple-locus heterozygosity and mass-specific rates of oxygen consumption was described. Individual variation in heterozygosity, measured at only five loci, explained more than 80% of the variation in standard metabolic rate. The authors hypothesized that the greater metabolic efficiency associated with heterozygosity would be reflected in greater scope-for-growth, thus explaining the higher growth rates of more heterozygous individuals.

Since the initial study by Koehn and Shumway (1982), enzyme heterozygosity has been shown (or strongly suggested) to covary negatively with mass-specific respiration rate in many additional species, including the molluscs *Thais haemastoma* (Garton, 1984), *Mulinia lateralis* (Garton et al., 1984), *Mytilus edulis* (Diehl, Gaffney, McDonald, and Koehn, 1985), fish of the genus *Salmo* (Allendorf, personal communication), and the salamander *Ambystoma tigrinum* (Mitton, Carey, and Kocher, 1986).

Although a precise genetic and metabolic mechanism for the way in which heterozygosity influences metabolic rate is not yet known, the degree of individual heterozygosity at electrophoretically detectable enzyme loci has clear and biologically significant effects upon an individual's energy status. Yet, heterozygosity-correlated growth rates do not appear to derive merely from the channeling into growth of energy saved directly from decreased costs of metabolism. Rather, energy derived from metabolic savings can be used to fuel increased feeding rates, thereby further enhancing net positive energy balance. For example, Garton (1984) demonstrated a significant positive relationship between heterozygosity and feeding rate; both the enhanced feeding rate and the enhanced metabolic efficiency (i.e., decreased respiration rate) contributed to the positive relationship between heterozygosity and scope-for-growth. Feeding rate was much more important than respiration rate in determining the net energy status of an individual. In the clam *Mulinia lateralis*, feeding rate was reported to be positively associated with heterozygosity (though not significantly), but Holley and Foltz (personal communication) have demonstrated a significant effect of heterozygosity upon feeding rate in the bivalve *Rangia cuneata*, under certain environmental conditions.

Individual differences in energy status can be expected to have a pervasive effect upon the biology of a species, though at present, there is little information (relevant to this chapter) beyond that described in the foregoing paragraphs. Individual differences in energy status could well produce differences in fecundity (e.g., Rodhouse, McDonald, Newell, and Koehn, 1986), reproductive effort, time of reproduction, certain behaviors, and many other aspects of an organism's life history. This is a productive area for future investigation.

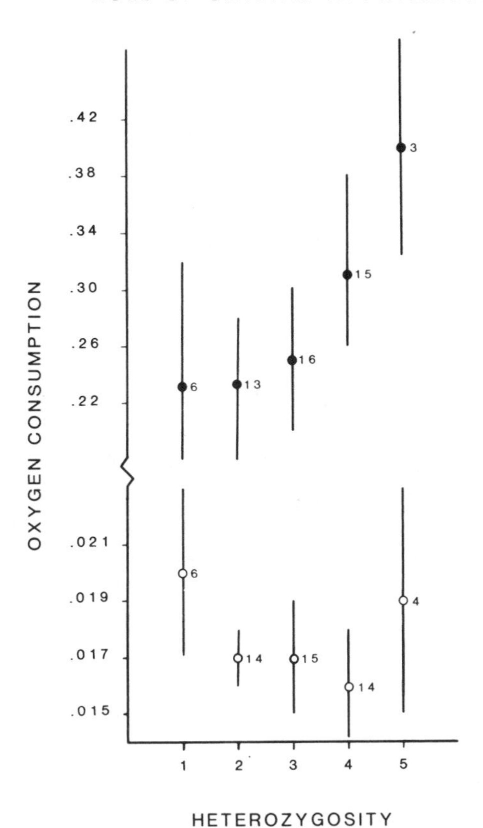

FIGURE 8.3 The relationships between individual heterozygosity (mean ± standard error) and mass-specific standard (open circles) and active (closed circles) rates of oxygen consumption in the tiger salamander *Ambystoma tigrinum*. Sample sizes are indicated adjacent to each group mean. The negative relationship between standard rates and heterozygosity is similar to that described in the text for other species. These relationships suggest that more heterozygous individuals enjoy lower energetic demands of standard metabolism, but greater scope-for-activity during periods of active metabolic demand. (Data are from Mitton et al., 1986.)

Environment effects upon physiological performance are well known. At present, there is little information on the way in which the environment can affect heterozygosity and energy status, though several observations suggest that this can be important. For example, in those experimental circumstances where ration is nonlimiting, heterozygosity has little or even no effect upon growth rate (Diehl et al., 1986). The benefits of reduced metabolic rate appear to be more pronounced during times of low or negative energy balance than under more equitable energy conditions. The lower respiration rates enjoyed by more heterozygous individuals seem to be associated more with resistance to mass loss than with capacity for mass gain (Diehl et al., 1986). In this context, it is interesting to note that Green et al. (1983) showed a positive correlation between growth and heterozygosity in an intertidal population of the clam *Macoma balthica*, whereas an adjacent subtidal population exhibited no such relationship.

The study of the effects of heterozygosity on energy status have focused nearly exclusively upon routine and standard rates of metabolism; under both circumstances, metabolic demands decrease with increasing individual heterozygosity. Mitton et al. (1986) have made this observation for the salamander *Ambystoma tigrinum*, but have further shown that the relationship between heterozygosity and *active* metabolic rate is positive, not negative (Figure 8.3). This result would imply that individual heterozygosity can significantly affect "scope-for-activity"; more heterozygous individuals may enjoy higher energy status under routine conditions and greater metabolic output during periods of active metabolic demand. This finding is potentially important to our general understanding of how heterozygosity influences physiological performance. Additional studies of heterozygosity and active metabolic rate are needed. The specification of precise genetic and metabolic mechanisms whereby enzyme heterozygosity determines energy status is an extraordinarily difficult challenge for future research in this area. Nevertheless, whether such mechanisms are ultimately discovered will not significantly alter the view that the genotype of an individual can have a measurable and biologically significant effect upon physiological performance. The pervasive importance of energy status upon all aspects of physiological performance dictates that the future of physiological ecology must involve attention to the importance of genetic variation.

References

Baker, C. M. A., and Manwell, C. (1977) Heterozygosity of the sheep: polymorphism of 'malic enzyme', isocitrate dehydrogenase ($NADP^+$), catalase and esterase. *Aust. J. Biol. Sci.* 30:127–140.

Barnes, P. T., and Laurie-Ahlberg, C. C. (1986) Genetic variability of flight metabolism in *Drosophila melanogaster*. III. Effects of GPDH allozymes and environmental temperature on power output. *Genetics* 113:267–294.

Bayne, B. L., Widdows, J., and Thompson, R. J. (1976) Physiology: II. In *Marine Mussels: Their Ecology and Physiology*, ed. B. L. Bayne, pp. 207–260. Cambridge University Press.

Bayne, B. L., Moore, M. N., and Koehn, R. K. (1981) Lysosomes and the response by *Mytilus edulis* L. to an increase in salinity. *Mar. Biol. Lett.* 2:193–204.

Bewley, G. C. (1978) Heat stability studies at the α-glycerophosphate dehydrogenase locus in populations of *Drosophila melanogaster. Biochem. Genet.* 16:769–775.

Bewley, G. C., Niesel, D. W., and Wilkins, J. R. (1984) Purification and characterization of the naturally occurring allelic variants of *sn*-glycerol-3-phosphate dehydrogenase in *Drosophila melanogaster. Comp. Biochem. Physiol.* 79B:23–32.

Bishop, S. H. (1976) Nitrogen metabolism and excretion: regulation of intracellular amino acid concentrations. In *Estuarine Processes*, ed. M. Wiley, pp. 414–429. New York: Academic Press.

Burton, R. S., and Feldman, M. W. (1982) Changes in free amino acid concentrations during osmotic response in the intertidal copepod *Tigriopus californicus. Comp. Biochem. Physiol.* 73A:441–445.

Burton, R. S., and Feldman, M. W. (1983) Physiological effects of an allozyme polymorphism: glutamate-pyruvate transaminase and response to hyperosmotic stress in the copepod *Tigriopus californicus. Biochem. Genet.* 21:239–251.

Diehl, W. J., Gaffney, P. M., McDonald, J. H., and Koehn, R. K. (1985) Relationship between weight-standardized oxygen consumption and multiple-locus heterozygosity in the mussel, *Mytilus edulis.* In *Proceedings of the 19th European Marine Biology Symposium*, ed. P. E. Gibbs, pp. 531–536. Cambridge University Press.

Diehl, W. J., Gaffney, P. M., and Koehn, R. K. (1986) Physiological and genetic aspects of growth in the mussel *Mytilus edulis.* I. Oxygen consumption, growth, and weight loss. *Physiol. Zool.* 59:201–211.

Fujio, Y. (1982) A correlation of heterozygosity with growth rate in the Pacific oyster, *Crassostrea gigas. Tohoku J. Agric. Res.* 33:66–75.

Gaffney, P. M., and Scott, T. M. (1984) Genetic heterozygosity and production traits in natural and hatchery populations of bivalves. *Aquaculture* 42:289–302.

Garton, D. W. (1984) Relationship between multiple locus heterozygosity and physiological energetics of growth in the estuarine gastropod *Thais haemastoma. Physiol. Zool.* 57:530–543.

Garton, D. W., Koehn, R. K., and Scott, T. M. (1984) Multiple locus heterozygosity and the physiological energetics of growth in the coot clam, *Mulinia lateralis*, from a natural population. *Genetics* 108:445–455.

Green, R. H., Singh, S. M., Hicks, B., and McCuaig, J. (1983) An Arctic intertidal population of *Macama balthica* (Mollusca, Pelecypoda): genotypic and phenotypic components of population structure. *Can. J. Fish. Aquat. Sci.* 40:1360–1371.

Hilbish, T. J., and Koehn, R. K. (1985a) Dominance in physiological and fitness phenotypes at an enzyme locus. *Science* 229:52–54.

Hilbish, T. J., and Koehn, R. K. (1985b) The physiological basis of natural selection at the *Lap* locus. *Evolution* 39:1302–1317.

Hilbish, T. J., Deaton, L. E., and Koehn, R. K. (1982) Effect of an allozyme polymorphism on regulation of cell volume. *Nature* 298:688–689.

Johnson, F. M., and Schaffer, H. E. (1973) Isozyme variability in species of the genus *Drosophila.* VII. Genotype-environment relationships in populations of *D. melanogaster* from the eastern United States. *Biochem. Genet.* 10:149–163.

Koehn, R. K. (1978) Physiology and biochemistry of enzyme variation: the interface of ecology and population genetics. In *Ecological Genetics: The Interface*, ed. P. Brussard, pp. 51–72. New York: Springer-Verlag.

Koehn, R. K., and Gaffney, P. M. (1984) Genetic heterozygosity and growth rate in *Mytilus edulis*. *Mar. Biol.* 82:1–8.

Koehn, R. K., and Shumway, S. E. (1982) A genetic/physiological explanation for differential growth rate among individuals of the American oyster, *Crassostrea virginica* (Gmelin). *Mar. Biol. Lett.* 3:35–42.

Koehn, R. K., and Siebenaller, J. F. (1981) Biochemical studies of aminopeptidase polymorphism in *Mytilus edulis*. II. Dependence of reaction rate on physical factors and enzyme concentration. *Biochem. Genet.* 19:1143–1162.

Koehn, R. K., Milkman, R. D., and Mitton, J. B. (1976) Population genetics of marine polecypods. IV. Selection, migration and genetic differentiation in the blue mussel *Mytilus edulis*. *Evolution* 30:2–32.

Koehn, R. K., Newell, R. I. E., and Immermann, F. (1980) Maintenance of an aminopeptidase allele frequency cline by natural selection. *Proc. Natl. Acad. Sci. U.S.A.* 77:5385–5389.

Koehn, R. K., Zera, A. J., and Hall, J. G. (1983) Enzyme polymorphism and natural selection. In *Evolution of Genes and Proteins*, ed. M. Nei and R. K. Koehn, pp. 115–136. Sunderland, Mass.: Sinauer.

Koehn, R. K., Hall, J. G., Innes, D. J., and Zera, A. J. (1984) Genetic differentiation of *Mytilus edulis* in eastern North America. *Mar. Biol.* 79:117–126.

Ledig, F. T., Guries, R. P., and Bonefeld, B. A. (1983) The relation of growth to heterozygosity in pitch pine. *Evolution* 37:1227–1238.

Makaveev, T., Venev, I., and Baulov, M. (1977) Investigations on activity level and polymorphism of some blood enzymes in farm animals with different growth energy. II. Correlations between homo- and heterozygosity of some protein and enzyme phenotypes and fattening ability and slaughter indices in various breeds of fattened pigs. *Genet. I Seleck. (Sofia)* 10:229–236.

McKechnie, S. W., Kohane, M., and Phillips, S. C. (1981) A search for interacting polymorphic enzyme loci in *Drosophila melanogaster*. In *Genetic Studies of Drosophila Populations*, ed. J. B. Gibson and J. G. Oakeshott, pp. 121–138. Canberra: Australian National University.

Miller, S., Pearcy, R. W., and Berger, E. (1975) Polymorphism at the α-glycerophosphate dehydrogenase locus in *Drosophila melanogaster*. I. Properties of adult allozymes. *Biochem. Genet.* 13:175–188.

Mitton, J. B., Knowles, T., Sturgeon, K. B., Linhart, Y. B., and Davis, M. (1981) Associations between heterozygosity and growth rate variables in three western forest trees. In *Isozymes of North American Trees and Forest Insects*, ed. M. T. Conkle, pp. 27–34. Berkeley, Cal.: Pacific Southwest Forest and Range Experiment Station.

Mitton, J. B., Carey, C., and Kocher, T. D. (1986) The relation of enyzme heterozygosity to standard and active oxygen consumption and growth rate of tiger salamanders, *Ambystoma tigrinum*. *Physiol. Zool.* 59:574–582.

Moore, M. N., Koehn, R. K., and Bayne, B. L. (1980) Leucine aminopeptidase (aminopeptidase-I), N-acetyl-β-hexosaminidase and lysosomes in the mussel, *Mytilus edulis* L., in response to salinity changes. *J. Exp. Zool.* 214:239–249.

Oakeshott, J. G., Gibson, J. B., Anderson, P. R., Knibb, W. R., Anderson, D. G., and Chambers, G. K. (1982) Alcohol dehydrogenase and glycerol-3-phosphate dehydrogenase clines in *Drosophila melanogaster* on different continents. *Evolution* 36:86–96.

Oakeshott, J. B., McKechnie, S. W., and Chambers, G. K. (1984) Population genetics of the metabolically related *Adh, Gpdh* and *Tpi* polymorphisms in *Drosophila melanogaster*. I. Geographic variation in *Gpdh* and *Tpi* allele frequencies in different continents. *Genetica* 63:21–29.

Pierce, B. A., and Mitton, J. B. (1982) Allozyme heterozygosity and growth in the tiger salamander, *Ambystoma tigrinum. J. Hered.* 73:250–253.

Rodhouse, P. G., McDonald, J. H., Newell, R. I. E., and Koehn, R. K. (1986) Gamete production, somatic growth and multiple-locus enzyme heterozygosity in *Mytilus edulis. Mar. Biol.* 90:209–214.

Schaal, B. A., and Levin, D. A. (1976) The demographic genetics of *Liatris cylindracea* Michx. (Compositae). *Am. Natur.* 110:191–206.

Singh, S. M., and Zouros, E. (1978) Genetic variation associated with growth rate in the American oyster (*Crassostrea virginica*). *Evolution* 32:342–352.

Watt, W. B. (1977) Adaptation at specific loci. I. Natural selection on phosphoglucose isomerase of *Colias* butterflies: biochemical and population aspects. *Genetics* 87:177–194.

Watt, W. B. (1983) Adaptation at specific loci. II. Demographic and biochemical elements in the maintenance of the *Colias* PGI polymorphism. *Genetics* 103:691–724.

Watt, W. B. (1985) Bioenergetics and evolutionary genetics: opportunities for new synthesis. *Amer. Natur.* 125:118–143.

Watt, W. B., Cassin, R. C., and Swan, M. S. (1983) Adaptation at specific loci. III. Field behavior and survivorship differences among *Colias* PGI genotypes are predictable from in vitro biochemistry. *Genetics* 103:725–739.

Watt, W. B., Carter, P. A., and Blower, S. M. (1985) Adaptation at specific loci. IV. Differential mating success among glycolytic allozyme genotypes of *Colias* butterflies. *Genetics* 109:157–175.

Winberg, G. C. (1956) Rate of metabolism and food requirement of fishes. *Translation Series*, no. 194, 253 pp. Fisheries Research Board of Canada.

Young, J. P. W., Koehn, R. K., and Arnheim, N. (1979) Biochemical characterization of "LAP," a polymorphic aminopeptidase from the blue mussel, *Mytilus edulis. Biochem. Genet.* 17:305–323.

Zera, A. J., Koehn, R. K., and Hall, J. G. (1985) Allozymes and biochemical adaptation. In *Comprehensive Insect Physiology, Biochemistry and Pharmacology*, vol. 10, ed. G. A. Kerkut and L. I. Gilbert, pp. 633–674. New York: Pergamon Press.

Zouros, E., Singh, S. M., and Miles, H. E. (1980) Growth rate in oysters: an overdominant phenotype and its possible explanations. *Evolution* 34:856–867.

Discussion

ARNOLD: I wanted to speak to the contrast between quantitative genetics and isozyme function studies. These two approaches should not be viewed as antagonists. They are complementary approaches that address different questions. Using the isozyme function studies, one can take enzymes that play a known role and make some predictions about which components of fitness might be affected by those enzymes. The disadvantage is that a physiological ecologist, interested in a particular attribute, may not be able to get all the way up to that attribute from that particular gene product. In contrast, using quantitative genetics, one can find out something about the inheritance of an attribute without knowing particular gene products. It is the virtue of quantitative genetics that we can use that very crude genetic information to provide the necessary ingredients for evolutionary models, even though we leave unresolved the mechanisms of gene action.

KOEHN: There are some kinds of problems for which only that approach can be taken. For morphology or life history characters, we cannot presently define those specific enzymes that play a functional role. In physiology, however, we can make some fairly intelligent guesses. It would be interesting to combine these two approaches (e.g., single gene and quantitative genetics), particularly in physiology. I think that you will never know a lot about the genetic mechanism if you use a quantitative approach, and that it has some very severe limitations. On the other hand, if you look at studies like Powers's work on LDH or our work on LAP [leucine aminopeptidase], and make the simplistic extrapolation of what we say, it is as if we are dealing with an important physiological character whose variation depends only upon genetic variation of a single gene. That simply is not true. No single approach is perfect, or everyone would be using it.

POWERS: I agree completely. When we found that there was any difference at all at the organismal level among LDH genotypes, we were totally amazed.

BENNETT: I would like to compliment both of you for having associated your genes and your isozymes with physiological properties. There is a temptation to look for correlations of heterozygosity with growth or metabolic rate or whatever, in large numbers of species, and not to make those mechanistic linkages. In that case you may just have correlations with enzymes for which it is difficult to understand how they might be connected with organismal physiological variables.

KOEHN: It is one thing to ask whether or not there is a correlation of average heterozygosity with some trait. It is a very different question to ask how individual genes contribute to that overall correlation. That question needs a different experimental approach incorporating a much larger sample size.

So far as I know, no data set has yet been collected that can adequately address that particular question.

FUTUYMA: In the first half of Koehn's talk, he has tied down a physiological effect and a fitness effect of a particular enzyme. In the second half, he has shown a correlation with a bunch of different enzymes and, if I may say so, no evidence of any functional relationship between those particular gene loci and growth rate.

KOEHN: You may say so.

FEDER: I am intrigued by the different pathways of analysis taken in these cases. Koehn chose a genetic system for analysis based upon the environmental challenges faced by his subject organism, and then analyzed the gene product in terms of an a priori expectation as to its adaptive significance. Powers began with an interesting pattern of genetic variation and then, with no a priori expectation of the gene product's function, deduced what its consequences might be at the organismal level in the field.

UMMINGER: Is there any generalizable property of the degree of variability with the type of enzyme? Is an enzyme that is central to cellular function likely to show less variability than enzymes that are less important to the livelihood of the cell? Or is there any relationship to whether it is a regulatory enzyme?

KOEHN: In 1978, Walt Eanes and I wrote a paper on this which was concerned with molecular structure and variability. Few of the many ideas on this hold up terribly well. Most highly correlated with the degree of polymorphism are aspects of structure rather than function. The one parameter that will explain the most variance among loci is the size of the enzyme subunit. Quaternary structure will also explain some, and function may also. I do not think there are sufficient data even today to partition the variance in polymorphism among these various structural and functional parameters. But the bigger the subunit, the more polymorphic it is going to be.

POWERS: The attention paid to some of the classical regulatory enzymes, like phosphofructokinase or hexokinase, is very small because, when people do not find variation, they normally do not report that fact. So the literature is heavily biased. In the animals in which I personally have looked for variation in phosphofructokinase or hexokinase, the variation is very low or nonexistent. For some systems there may be a restriction on certain regulatory enzymes. In large proteins where you see variable regions, one wonders whether this variability is important or not. Much of the evolutionary literature says that variability is essentially neutral. Yet some of the immunoglobulins, for example, have a constant region that may be important for maintaining the structure and a variable region that may be important for the specificity. Are variability and nonvariability both important?

ARNOLD: Would Dr. Koehn comment on the interpretation of the correlation between heterozygosity and measure of whole-animal performance and fitness components? These could be interpreted as an indication that there is selection for heterozygosity per se, or they could simply reflect inbreeding and other aspects of population structure. What is your feeling on this?

KOEHN: I find problems with the inbreeding explanation. For one thing, the organisms in which this relationship has been most dramatically demonstrated, like mussels, oysters, and various bivalve species, are the kinds of species that you would expect to have very large effective population sizes. They disperse like crazy and are highly fecund. The inbreeding explanation depends on deleterious recessives linked with the genes that we are sampling. There are some serious problems with that. If you sample five genes, to what extend do those five genes represent, in rank order of genic heterozygosity, the rank order of genomic heterozygosity? The answer is virtually zero. This means that the effect has to be due to either the genes that we are actually sampling or genes right in their neighborhood. I think there are a lot of things that one can do to test this experimentally. The answer depends on establishing a functional biochemical genetic mechanism, much as we have done with individual genes (e.g., LDH and LAP).

9 Genetic correlation and the evolution of physiology

STEVAN J. ARNOLD

Correlation ... has no doubt played a most important part [in shaping structures during evolution], and a useful modification of one part will often have entailed on other parts diversified changes of no direct use.

C. Darwin (1859, p. 199)

Statement

Genetic coupling between physiological traits can be studied by analyzing the individual variation that resides within populations. Genetic coupling or correlation can deflect the course of adaptive evolution and cause temporary maladaptation in physiology and other traits. Genetic models indicate that maladaptive phases of physiological evolution could last many thousands of generations.

Introduction

Variation within natural populations has been neglected by physiologists despite its evolutionary importance (Endler, 1986; Chapter 7). Comparative physiology has usually focused on contrasts between populations, species, and higher taxa, and variation within populations has been overlooked. Yet, as Darwin realized, heritable differences within populations are molded by natural selection to produce geographic and specific differences. The study of intrapopulational variation can reveal ongoing processes of selection as well as genetic constraints on the microevolutionary transformation of populations. My aim is to discuss the insights we might get by quantitative genetic analysis of variation in physiology. The discussion is necessarily speculative because several interesting genetic avenues have not been explored. My focus will be on coupling between physiological traits. Genetic coupling, commonly measured as genetic correlation, may constrain the evolution of physiological traits and can lead to counterintuitive evolutionary outcomes.

In the following sections, I begin by describing two different approaches to studying the genetics of physiology. Next I outline the essential concepts of heritability and genetic correlation which underlie one of those approaches, namely quantitative genetics. I briefly discuss the evolutionary importance of those concepts and move on to a discussion of the different manifestations of genetic correlation. This discussion relies on many hypothetical examples because so little quantitative genetic work has focused on physiological traits. Nevertheless, genetic correlations in physiology seem likely in many circumstances. Next I consider some detailed examples (again, mostly hypothetical) of how genetic correlations might have a major impact on physiological evolution. Finally I argue that we need more quantitative genetic studies of physiology and describe some possible directions and practical considerations.

Background

Two complementary genetic approaches

Genetic studies of physiology fall into two general categories that differ in outlook: the gene-to-physiology viewpoint and the physiology-to-genetics perspective. The two approaches are complementary and answer different questions, yet they have hardly ever been used in combination. The gene-to-physiology approach begins with the observation of natural variation of a particular gene locus, commonly assayed by protein electrophoresis, and then shows the effects of that variation on physiology and fitness (Chapters 5 and 8). Watt, Carter, and Donohue (1986), for example, review a particularly elegant example of this approach. They trace the impact of phosphoglucose isomerase variation in butterfly populations on flight capacity, survivorship, and mating success. The gene-to-physiology approach is aesthetically pleasing because it can reveal an entire pathway from gene to fitness. More importantly, the approach yields the information necessary to model change in the frequency of a gene that affects physiology.

In contrast, the physiology-to-genetics approach focuses on physiological attributes that may be affected by numerous genes rather than on the physiological effects of a particular gene. The aim is to understand how the evolution of one physiological trait will affect the evolution of other traits. For example, deliberate selection for large body size in house mice results in larger mice, but these mice also have smaller interscapular pads of brown adipose tissue (Sulzbach and Lynch, 1984). This example illustrates the general principle that evolutionary modification of one element in a physiological system can have reverberating effects on other elements in the system. Such reverberations can be anticipated by studying resemblance among relatives in physiological traits. Thus the physiology-to-genetics approach can succeed at its goal of predicting interactions during evolution without tracing pathways to particular genes. The approach does not tell us which particular genes are associated with variation in physiology or even how many genes affect that variation.

Likewise the gene-to-physiology approach does not tell us how many genes affect a particular physiological trait. Furthermore, because the main goal is to trace a particular pathway from genotype to phenotype, the gene-to-physiology approach does not tell us whether physiological traits are likely to interact and affect one another's evolution.

Quantitative genetics and the importance of genetic correlation

Quantitative genetics is the genetic discipline that uses what I have called the physiology-to-genetics approach. The discipline focuses on the inheritance of traits whose variation is affected by many genes and by the environment. Many physiological attributes are probably polygenic, so quantitative genetics is a natural analytical tool.

The basic idea in quantitative genetics is to predict the attributes of offspring from the characteristics of their parents. A hypothetical example of resemblance in body size between offspring and parents in a population of mice is shown in Figure 9.1. Each point represents the average body size of the offspring plotted against the average body size of their parents, all sizes measured in adulthood. In this example, offspring strongly resemble their parents in body size. If we randomize environments in both generations, so as to eliminate environmental causes of parent-offspring resemblance, we can show with a little algebra that the observed resemblance is due to the additive properties of the many genes that might affect body size (Fisher, 1918). The slope of the best-fit line that predicts offspring values from those of their parents is known as *heritability*. The slope turns out to be mathematically equivalent to a measure of variation in the genetic values of the parents, referred to as *genetic variance* for body size.

The plot of offspring against their parents also enables us to visualize the consequences of selection. Suppose the plot represents all the potential parents and their hypothetical offspring but only a fraction of the potential parents actually survive to breed and produce offspring. In particular, imagine that only the pairs of parents in the top 20% of average body masses actually breed. We can readily deduce the consequences of this selection. In Figure 9.2, we have the same plot of offspring-parental resemblance as in Figure 9.1. The vertical dotted line marks the average body size of the actual, selected parents. The distance between these lines measures the strength of selection on body size, shown with the heavy arrow. The dotted horizontal line shows the body size of offspring expected from the average of all potential parents. This is the body size that would prevail in the next generation, and in all succeeding generations, if no selection were imposed. The solid horizontal line shows the size of offspring expected from the selected parents. These offspring are larger than the offspring of all potential parents, as shown by

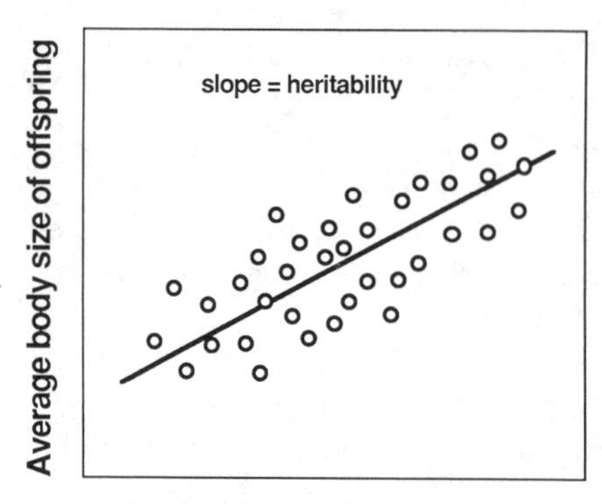

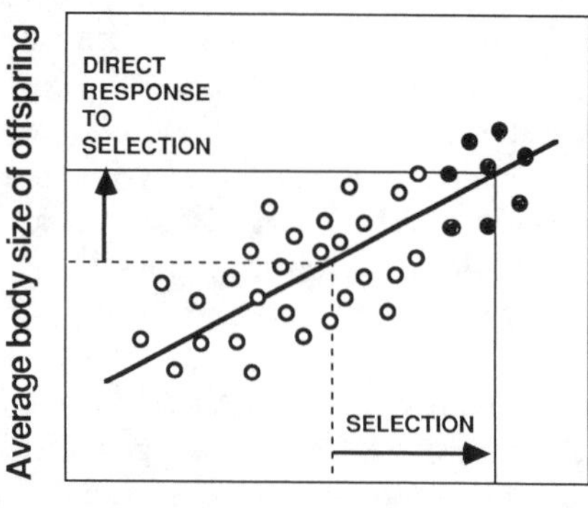

FIGURE 9.1 (Left) The average body size of offspring plotted against the average body size of their parents. Each point in this hypothetical plot represents the average value for offspring and the average for their two parents. The slope of the least squares regression of offspring on parents estimates the heritability of the trait.

FIGURE 9.2 (Right) Direct response to selection on body size. A subset of the hypothetical parents actually survive and produce offspring. These surviving parents are indicated with solid dots. The strength of selection on body size (horizontal arrow) can be visualized as the distance between the average body size of all potential parents (vertical dotted line) and the average body size of the actual parents (vertical solid line). The direct response to selection is a function of the strength of selection and the heritability of the trait.

the vertical arrow. This change in the body size of the next generation is known as the "direct response to selection" on body size. If there were no heritability of body size, the slope of the best-fit line would be zero, and there could be no response to selection.

Plots assaying offspring-parental resemblance can also be used to visualize genetic coupling between characters. Figure 9.3 shows the body size of progeny plotted against a second hypothetical trait, size of brown adipose pads, of parents. Trends in such plots reflect genetic coupling if common environmental effects are absent or have been eliminated by design. In Figure 9.3, positive genetic effects on body size tend to be coupled with negative genetic effects on size of adipose pad, resulting in an overall negative trend in the data (more precise genetic meanings of coupling will be discussed later). The slope of the best-fit line through the data reflects covariance between genetic values for the two traits or "genetic covariance." In Figure 9.3, the genetic values for adipose pad size are expressed by parents whereas genetic values for body size are expressed in offspring. The "genetic correlation" between

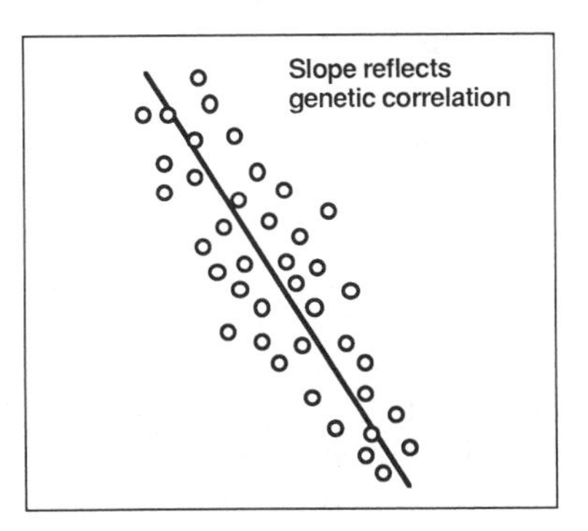

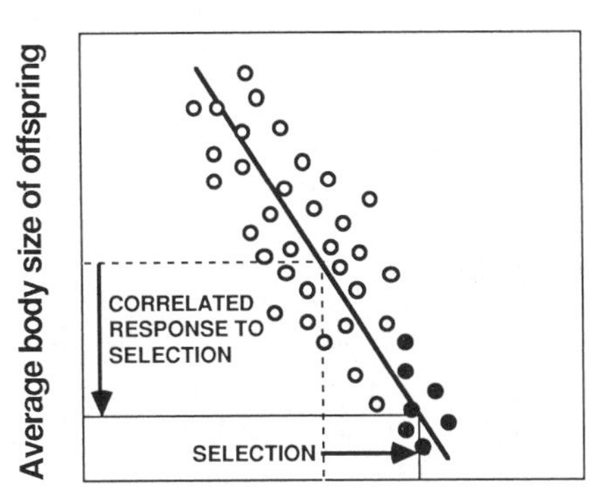

FIGURE 9.3 (Left) Average body size of offspring plotted against a second hypothetical character of their parents, size of brown adipose pads. The trend in the data reflects genetic coupling or genetic correlation between the two traits. In this case the genetic correlation is negative.

FIGURE 9.4 (Right) Correlated response of body size to selection on adipose pads. Actual parents of the next generation are indicated with solid dots. Selection favoring large adipose pads (horizontal arrow) causes a decrease in the body size of offspring because of the negative genetic correlation between the two traits.

two traits is simply a standardized genetic covariance that varies within the limits −1 to +1. In Figure 9.3, the genetic covariance is negative.

As a consequence of genetic covariance, selection on one trait will cause evolution in other traits. In Figure 9.4 we have the same offspring-parent data as in Figure 9.3. Imagine that selection is imposed on adipose pads, so that only the parents in the upper 20% of pad sizes survive and produce offspring. Because of the negative genetic correlation, these selected parents will yield progeny with smaller body sizes (solid horizontal line) than the average of potential parents (their expected progeny is shown with a dotted horizontal line). The change in body size in the next generation due to selection on adipose pads is referred to as a "correlated response to selection" on adipose pads.

Quantitative genetics has been an important conceptual framework for plant and animal breeders. Quantitative genetics has been useful because it enables the breeder to summarize the inheritance of phenotypic traits and to predict how those traits will respond to selection. For these same reasons, quantitative genetics can be useful to the physiologist who is interested in the evolution of physiology. The phenotypic resemblance between offspring and parents, as described in Figures 9.1 and 9.3, is only a rough summary of the underlying genetic system, but it captures the essential information needed to predict the consequences of selection. In recent years, quantitative genetics has been increasingly used by evolutionary theorists. The main thrust of this work has been to see how multiple phenotypic traits will respond to long-term patterns of selection (Lande, 1976, 1979, 1980a).

The importance of genetic correlations for evolution has been revealed by applied and theoretical lines of inquiry. Breeders and experimentalists have had to take genetic correlations into account time and time again in trying to interpret the results of deliberate selection (Dickerson, 1955; Robertson, 1980; Falconer, 1981). In addition, recent theoretical models have shown that genetic correlations can have major, sometimes nonintuitive effects on the course of evolution (e.g., Lande, 1979, 1980a). The essential point, revealed by both approaches, is that genetically coupled traits will not evolve independently. Thus to understand the evolution of physiological systems, we must consider the possibility of genetic coupling between elements in the system.

Sources of genetic correlation

The genetic coupling between traits, described by a genetic correlation, can arise from two sources: pleiotropy and linkage disequilibrium (Falconer, 1981). A single gene may affect many traits in an organism, a phenomenon called "pleiotropy." Two traits may be genetically correlated because of the summed action of many genes with pleiotropic effects. The contributions of alleles at one locus to a genetic correlation are shown in Figure 9.5. The ubiquity of multiple effects by single mutations suggests (Figure 9.5) that pleiotropy is the rule rather than an exceptional mode of gene action (Wright, 1968). "Linkage disequilibrium" refers to correlation between the allelic effects of different loci (Crow and Kimura, 1970). So, for example, a negative genetic correlation between body size and fat pad size might arise because alleles at one locus increase body size and those alleles occur most frequently in conjunction with alleles at a second locus that tend to decrease size of fat pads (Figure 9.6). The contributions of pleiotropy and linkage disequilibrium to a genetic correlation can be separated with an elaborate series of crosses. Both such crosses and computer simulations indicate that pleiotropy, rather than linkage disequilibrium, is usually the major determinant of genetic correlation (Bulmer, 1974). In routine work in quantitative genetics, where the main issue is response to selection, the two sources of correlation in genetic effects are simply lumped together and summarized as a genetic correlation.

Phenotypic correlations versus genetic correlations

If we measure two traits, say body size and size of brown adipose pads, in each individual in a population, we could readily compute the correlation between the traits. Such similar correlations, based on individual values

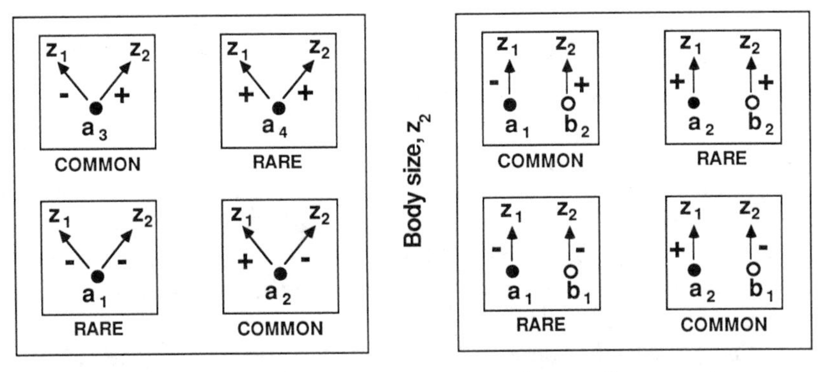

FIGURE 9.5 (Left) Genetic correlation due to pleiotropy. Four alleles at a hypothetical locus exist in a population and are denoted a_1, a_2, a_3, and a_4. Each affects both size of adipose pads, denoted z_1, and body size, denoted z_2. Two of the alleles, a_2 and a_3, are common in the population. Because these alleles have effects of different sign on the two traits, they produce a negative genetic correlation between size of adipose pads and body size.

FIGURE 9.6 (Right) Genetic correlation due to linkage disequilibrium. Two alleles at a particular hypothetical locus have different effects on body size. The a_1 allele tends to decrease size of adipose pads while the a_2 allele tends to increase size of adipose pads. Two alleles at a second hypothetical locus have contrasting effects on body size. If genotypes with contrasting signs (a_1b_2 and a_2b_1) are common in the population while genotypes with complementary signs (a_1b_1 and a_2b_2) are rare, a negative genetic correlation will be produced.

within a generation, are known as "phenotypic correlations." Why couldn't we use phenotypic correlations to assess the degree of genetic coupling between traits? The answer is that phenotypic correlations reflect both genetic and environmental effects on the traits. In extreme cases the genetic and environmental effects may be opposite in sign (Falconer, 1981). For example, mice with more access to food may be both larger in size and have larger adipose pads for nutritional reasons. If such nutritional effects are large, the phenotypic correlation between the traits may be positive even though the genetic correlation is negative.

Detecting and measuring genetic correlations

Genetic correlation can be assessed by assays of phenotypic resemblance among known relatives or by selection experiments. For example, we could measure one trait in parents, measure the other trait in offspring, and assess their correlation (Figure 9.3). This procedure will work so long as parents and offspring do not share environmental features that lead to nongenetic correlation of the characters. Alternatively, offspring can be reared apart from parents, so that environmental causes of offspring-parent resemblance are disrupted. Many other combinations of relatives can give comparable genetic information [e.g., sets of progeny, each set with the same mother and father (full sibs) or with the same father but different mothers (parental half-sibs)]. A large part of the literature in quantitative genetics is devoted to the problem of devising experimental breeding plans that give efficient and reliable estimates of genetic variances and covariances (Cockerham, 1963; Falconer, 1981; Becker, 1984).

Deliberate selection on experimental populations can also reveal genetic correlation and assess its magnitude (Robertson, 1980; Falconer, 1981). The basic procedure is to select on one trait, usually for a few to many generations, and see if other traits respond to that selection. Such experiments can produce correlated responses to selection and give dramatic evidence for genetic coupling. For example, the practical question of whether selection for rapid growth in domestic animals is likely to have deleterious effects on reproduction has been investigated with selection experiments (e.g., Bradford, Barkley, and Spearow, 1980). Often selection experiments include unselected control lines, so that random or systematic effects of the environment can be disentangled from response to selection. A selected line gives only a single estimate of genetic correlation so replicate selection lines are needed in order to estimate standard errors (Hill, 1980).

Varieties of genetic correlation

The aim of the following survey is to suggest some physiological situations in which genetic correlation will have interesting evolutionary consequences, rather than to attempt an inventory of quantitative genetic studies of physiology. Thus a common feature in the examples that follow is the possibility of antagonistic responses to selection, which arise when selection acts on genetically coupled traits and produces conflicting results. In most of the examples, I will consider homologous traits that are likely to show positive genetic correlation due to pleiotropy. For example, body size at one age may be genetically the same character as body size at a later age, and consequently the two traits may show a large positive genetic correlation. Antagonism in adaptation can arise when selection acts in different directions on the two traits, for example, when large size is favored at one age and small size at another age. Likewise, antagonism can occur when selection acts in the same direction on two traits that have a negative genetic correlation (Figures 9.2 and 9.4).

Correlation between characters expressed at the same age

An important generalization, emerging from studies of correlation in morphology, is that homologous traits, traits in close spatial proximity and traits that are functionally linked, often show positive phenotypic correlations and sometimes they show positive genetic correlations as well (Olson and Miller, 1958; Lande, 1980b; Cheverud, 1982). Because so little quantitative genetic work has been done with physiology, we do not have comparable generalizations for physiological traits.

Sometimes unexpected genetic correlations occur, as in the following behavioral example. Chemoreception is a major sensory modality used by garter snakes to recognize prey species. Even naive, newborn garter snakes will give an active tongue-flicking response to cotton-tipped swabs that are laden with prey odors. Sometimes such swabs elicit prey attack. Arnold (1981a, 1981b) tested the tongue-flicking response of newborn snakes to the odors of several natural prey species (slugs, leeches, salamanders, anurans, and fish). Litters from the same population differ significantly in their chemoreceptive responses to virtually all of these prey. This variation is probably a consequence of heritable differences because Burghardt (1971) showed that the tongue-flicking responses of newborn snakes are not influenced by the mother's diet during gestation. Arnold (1981b) estimated genetic correlations between responses to different prey by analyzing covariation in the average scores of litters. Not surprisingly, responses to different species of anurans showed positive genetic correlations. Litters that showed strong responses to treefrog odor also showed strong responses to tadpole odor. Some genetic correlations, however, were unexpected. In particular, litters that reacted strongly to slug odor also reacted strongly to leech odor. The genetic correlation between responses to these unrelated prey was very high (nearly .9) and may reflect the sharing of a critical chemical cue. Thus suppose that a chemical substance on the surface of slugs triggers tongue-flicking and prey attack. If there is heritable variation in responsiveness to that substance and if the same substance is found on leeches, we would expect a genetic correlation between chemoreceptive responses to these two prey.

Correlation between characters expressed at different ages

Many studies have estimated genetic correlations between the same morphometric character measured at different ages. Generally, genetic correlations are higher when the ages are close together (e.g., Leamy and Cheverud, 1984). For example, in a laboratory rat population, the genetic correlation between mass at two and four weeks of age is nearly 1, but between the ages of two and forty-seven weeks it has fallen to .4 (Cheverud, Rutledge, and Atchley, 1983).

Another prospect is that one type of character expressed at one age may be correlated with another type of character expressed at a different age. For example, in holometabolous insects, a larval trait might be genetically correlated with a phenotypically dissimilar trait in adults (e.g., if the traits share developmental or metabolic pathways so that pleiotropy gives rise to genetic correlation). For example, Palmer and Dingle (1986) detected a positive genetic correlation between adult wing length and developmental time by selecting on wing length in an hemipteran. By implication, then, evolution of increased flight performance through selection for longer wings might frustrate the evolution of a short larval period in this species.

Correlation between characters expressed in males and females

Homologous traits in males and females often show high genetic correlation. For example, the correlation between body size in males and females may be .9 or higher in birds, mammals, and *Drosophila* (Lande, 1980a). (All the studies to date have been done on species that show only modest sexual dimorphism in size.) The result is not surprising when we realize that new mutations will tend to give comparable effects in males and females. Consequently, when selection acts differently on the sexes, the possibility for antagonistic responses to selection is very great.

Sexual dimorphism is an indication that selection acts differently in males and females (Darwin, 1874). For example, in most anurans, males actively call to attract mates, while females are silent and rarely vocalize. Recent studies have shown that male vocalization is energetically expensive, resulting in as much as a twenty-five-fold elevation of metabolic rate (Bucher, Ryan, and Bartholomew, 1982; Taigen and Wells, 1985). In a hylid frog, this sexual difference in energy expenditure is paralleled by a striking sexual dimorphism in size of trunk musculature and in activity levels of oxidative enzymes (Taigen, Wells, and Marsh, 1985). Citrate synthase activity in male trunk musculature is seventeen times greater than in females. Recent theoretical models suggest that positive genetic correlation between the sexes in trunk musculature and in citrate synthase activity could have delayed the evolution of sexual dimorphism in these traits. Those models provide a genetical perspective on the evolution of sexual differences in physiology and are discussed in the next section.

Correlation between characters expressed in different environments

Some of the genetic difficulties that thwart adaptation to alternative environments can be analyzed by measuring genetic correlation between traits expressed in different environments (Falconer, 1952, 1981; Robertson, 1959; Via and Lande, 1985). The simple step of defining a trait, say milk yield, in two environments as separate characters has helped to solve practical problems in animal breeding, and it opens new ecological vistas as well. Suppose we select for increased milk yield in dairy cows maintained on good pastures and succeed in improving the breed in this environment. Will the breed show improved milk yield when also maintained on a poor pasture or will their milk yield be worse than that of the original stock? This question can be approached by asking whether there is a positive or a negative genetic correlation between milk yield in the two environments. A high positive correlation indicates that the same genes promote milk yield in both nutritional environments, while a negative correlation indicates that selective improvement in one environment may cause a genetic loss in milk yield in the other environment.

Genetic correlation between traits in alternate environments is a current issue in studies of how arthropods are able to adapt to each of several host plant species (Via, 1984, 1986; Futuyma and Peterson, 1985; Lofdahl, 1985). Genetic correlation between arthropod performance on different host plants has been investigated by selection experiments and by analysis of covariation within populations. Gould (1979) raised herbivorous mites for many generations on two separate host plant species: one that yielded good mite growth and one that was mite-resistant. The line on the mite-resistant host improved in performance (developmental rate and survivorship) over a twenty-one-month period. However, by transferring mites back to the permissive host, Gould found that its performance there had slightly deteriorated. The result suggests a negative genetic correlation between performance on the two hosts. In contrast, Rausher (1984) did not find a negative genetic correlation between host plant performances within tortoise beetle populations. A large number of such case studies will be needed before we will know whether genetic trade-offs are important in the evolution of host plant specialization in arthropods.

Correlation between direct genetic and maternal genetic effects

In many organisms with maternal care, the behavior and physiology of the mother can profoundly affect the offspring's phenotype. Milk supply or feeding at the nest affects offspring growth, for example. Because growth rate is of great economic importance in domestic mammals, much experimental and theoretical work has focused on the genetic analysis of maternal effects (Dickerson, 1947; Willham, 1963, 1972; Falconer, 1965; Cheverud, 1984a; Riska, Rutledge, and Atchley, 1985). The resulting concepts and findings are directly applicable to the problem of analyzing the evolution of a physiological trait that experiences or produces a maternal effect.

The critical first step is to realize that the maternal effect itself may have a genetic basis. Thus milk supply may be heritable to some degree. Consequently, when we consider offspring growth rate, we can imagine two classes of genetic effects: direct effects due to genes that act in the zygote and affect its growth, and maternal effects that act in the mother and affect the offspring's growth by changing the offspring's maternal environment. We also have the possibility that the two kinds of genetic effects, direct and maternal, might be correlated. Experimental work with mammals, for example, has shown that such correlations exist. Furthermore, when the correlation between direct and maternal genetic effects is positive, selection on maternal performance can amplify the effects of selection on offspring growth. However, when the correlation is negative, the evolution of maternal performance may be antagonistic to the evolution of offspring growth, and vice versa.

Negative correlation between direct genetic and maternal genetic effects on body mass and growth have been found in both mice and cattle (Riska et al., 1985). In some cases the negative correlation may result from a trade-off between offspring size and offspring numbers. Thus, larger female mice tend to have larger litters but smaller offspring (Falconer, 1965). Because of the trade-off between offspring size and numbers, a genetic tendency for larger size will be associated with a negative pleiotropic effect on offspring size (mediated through the maternal effect of body size on litter size and hence on offspring size). Antagonism between the two kinds of genetic effects cannot be mediated through litter size in all cases, however, because the negative genetic correlation has been found in some cases even when litter sizes were held constant (Riska et al., 1985).

Correlation between direct genetic and maternal genetic effects is conceivable in many circumstances and could have an important impact on the evolution of maternal (or paternal) care. For example, in viviparous natricine snakes, body temperature during pregnancy affects the scalation and number of vertebrae of developing offspring (Fox, 1948; Fox, Gordon, and Fox, 1961; Osgood, 1978). A quantitative genetic perspective on maternal effects raises the issues of (1) whether gravid females show phenotypic and genetic variation in thermoregulation, (2) whether natural variation in maternal thermoregulation (contrasted with the dramatic temperature differences imposed in perturbation experiments) affects offspring thermoregulation and vertebral numbers, and (3) whether there is a genetic coupling between maternal thermoregulation and number of vertebrae. A correlation between direct genetic effects on thermoregulation and maternal genetic effects on number of vertebrae would cause selection on vertebral numbers to affect the evolution of thermoregulation. Conversely, selection on thermoregulation would affect the evolution of number of vertebrae.

Carey (1980) has stressed the need for studies of how wide-ranging bird species meet the challenge of greater water loss through the eggshell at higher elevations. Within species, water loss appears to be buffered against the effects of altitude. Carey proposed four kinds of maternal effects as candidates for adaptations to elevation: (1) a change in porosity of the shell, which reduces water vapor conductance through the eggshell; (2) an increase in initial water content; (3) an increase in shell thickness; and (4) parental behavior that changes the egg microenvironment and thereby reduces water loss. Phenotypic effects of the zygotic genome might also affect water loss.

For example, properties of the chorioallantoic membranes, which are constructed by the embryo, might affect water loss. Returning to our theme of genetic correlation, we can ask whether direct genetic effects on embryonic membranes are correlated with maternal genetic effects on the eggshell and its environment. With positive correlation, maternal adaptation might reinforce zygotic adjustments to high elevation, but with negative correlations, maternal and zygotic adjustments might work against each another and delay adaptation.

Maternal effects complicate the detection and measurement of genetic variances and covariances. The complication arises because we are interested in genetic effects that act at two stages of the life cycle (in offspring and in mothers). Consequently, to analyze the inheritance of maternal effects one needs a breeding design that produces mothers of known relationship as well as offspring of known relationship (Eisen, 1967; Rutledge, Robinson, Eisen, and Legates, 1972).

Evolutionary consequences of genetic correlation

The major detrimental effect of genetic correlation is to delay adaptation. Departures from evolutionary optima caused by genetic correlation may be temporary, but they may last for hundreds of thousands of generations. To understand these effects, we need a more precise concept of adaptive antagonism.

Antagonism between direct and correlated responses to selection

When two traits are genetically coupled, selection on one trait may interfere with the adaptive evolution of the other trait. We can readily see how such antagonism can arise by referring to the standard quantitative genetic equations for one generation of response to selection. By convention, z refers to the value of a trait (e.g., thermal tolerance, wing length, citrate synthase activity) in a population; $\bar{z}$ is the average value of the trait in the population, and $\Delta\bar{z}$ is the change in the average value of the trait from one generation to the next, due to natural selection. If selection acts only on a single trait then

$$\Delta\bar{z} = G\beta \tag{9.1}$$

where G is the genetic variance of the trait and β represents the magnitude of natural selection (Falconer, 1981). In other words, the amount of change in a trait due to selection is proportional to the product of *both* the strength of selection *and* the genetic variance. Equation 9.1 is but a mathematical representation of directional natural selection (e.g., Figure 9.2), with which ecological physiologists are familiar. Of course, traits may not behave independently. Thus, if we consider the values of any two correlated traits, z_1 and

z_2, with their respective genetic variances (G_{11} and G_{22}) and selection coefficients (β_1 and β_2):

$$\Delta\bar{z}_1 = G_{11}\beta_1 + G_{12}\beta_2 \quad (9.2)$$

$$\Delta\bar{z}_2 = G_{22}\beta_2 + G_{12}\beta_1 \quad (9.3)$$

where G_{12} is the "genetic covariance" between traits 1 and 2 (Lande, 1979). The general implication of these equations is that the evolutionary change in a trait is the sum of two components. The first is the "direct response to selection" (e.g., $G_{11}\beta_1$ for trait 1), which gives the shift in the mean of z_1 due to selection acting directly on that trait. The second component, termed the "correlated response to selection" (e.g., $G_{12}\beta_2$ for trait 1), is the shift in the mean of z_1 due to selection acting on the correlated character, z_2. Likewise, the response to selection by z_2 is composed of a direct response ($G_{22}\beta_2$) and a correlated response ($G_{12}\beta_1$).

Importantly, depending on the relative magnitude and sign of the direct response and the correlated response, the correlated response may reduce the direct response to natural selection or override it entirely. In either of these situations, the population is in a kind of genetic treadmill, with part or all of the genetic advance in each character being canceled in each generation by selection acting through the correlated character. For example, the direct response to selection on body size (Figure 9.2) is overridden by the correlated response of body size to selection on adipose pads (Figure 9.4). Consequently, a smaller body size evolves even though selection favors large body size.

We have few examples of antagonism between physiological features of animals, for such phenomena have rarely been studied. However, by turning to breeders' attempts to select artificially for multiple traits in agricultural species, we can readily appreciate how important antagonism might be in affecting the outcome of natural physiological adaptation. In some cases, repeated artificial selection fails to alter a trait after an initial large change, not because the genetic variance is small, but because of negative genetic correlations between the traits. In chickens, for example, a negative genetic correlation between egg production and body size frustrated attempts to produce chickens that were both large and laid many eggs (Gyles, Dickerson, Kinder, and Kempster, 1955). Such genetic trade-offs are common in crop species, and Dickerson (1955) has referred to the resulting cancellation of direct and correlated responses as "genetic slippage." Dickerson anticipated current interest in genetic trade-offs by students of life-history evolution by pointing out that deliberate selection for multiple objectives bears many similarities to selection on major fitness components.

Genetic slippage can also arise from antagonistic selection. In the garter snake example described earlier, newborn snakes from two populations were tested for chemoreceptive responses to slugs and leeches (Arnold, 1981a, 1981b). In a population allopatric with slugs, naive newborn snakes hardly react to slug or leech odors, even though leeches are a minor constituent of the natural diet (the principal prey are fish and anurans). In a population sympatric with slugs, naive newborn snakes give strong feeding reactions not only to slugs, the principal natural prey, but also to leeches, even though leeches are never encountered in nature. The enigmatic reaction of these slug-sympatric snakes to leeches might represent a correlated response to selection for slug recognition. I tested this supposition using the data reported in Arnold (1981a, 1981b) and equations given in Lande (1979). The results indicate that the reactions of slug-sympatric snakes to leeches, while strong, are not strong enough to be simply a correlated response. To account for the population difference in reaction to the two prey, and taking into account the strong positive genetic correlation in both populations, I needed to invoke both selection for slug reaction and selection against leech reaction. The results indicated that antagonistic selection on positively coupled traits might be responsible for the geographic differences.

Counterintuitive evolutionary trajectories and temporary maladaptation

In the preceding sections, we considered the possibilities that direct and correlated responses to selection might cancel so that no evolution occurs even though selection favors improvement. We also considered a case, in garter snakes, in which the population seemed to have evolved in the opposite direction to selection on one of two traits. More extreme possibilities exist. It is conceivable that negative genetic correlations could cause evolution in the opposite direction to the force of selection on both traits. Cheverud (1984a) shows that this nonintuitive result is possible if there is a strong negative genetic correlation between direct genetic and maternal genetic effects. For instance, let us return to the earlier example of thermoregulating female snakes that affect the number of vertebrae in their embryos. Even if selection favored a higher body temperature for thermoregulating mothers as well as more vertebrae in their embryos, a sufficiently large negative genetic correlation could cause the evolution of lower maternal temperatures and fewer vertebrae. Such bizarre evolution is possible in the short run, but we need a model of long-term response to selection, as well as single generation equations, to predict ultimate outcomes.

Lande's (1979, 1980a) models for the evolution of genetically coupled traits make some predictions about the time-course and direction of long-term evolution. The general result from Lande's models is that genetic correlation tends to delay adaptation (adaptation being conceived as proximity to the hilltop on the adaptive landscape). Those models examine the effects of genetic correlation on the evolution of trait averages by holding the genetic variations and covariance(s) constant. This assumption of constancy is not outrageous because both theoretical and empirical studies indicate that the

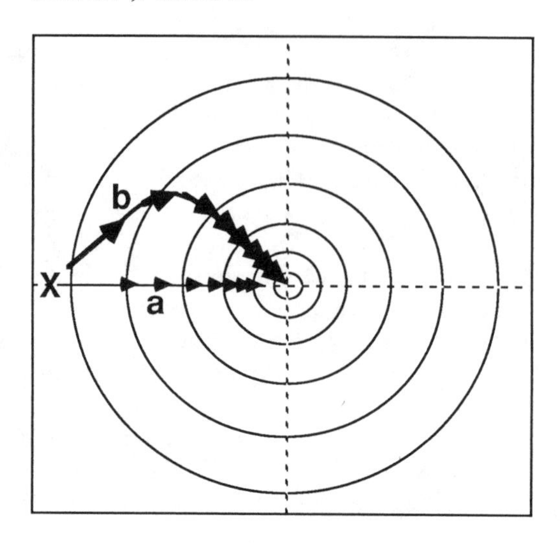

FIGURE 9.7 The evolution of two genetically coupled traits, body size in females and males. A population with a bivariate mean at the position shown by the X (middle left) experiences a changed environment that favors the same female body size but a larger male body size. The new adaptive landscape is represented by the concentric circles which are contours of average fitness in the population. The fitness maximum is located at the center of the concentric circles, and the dotted lines show optimal female and male body sizes. If there is no genetic correlation, the population will evolve along the straight path (a). A strong, positive genetic correlation will cause the population to evolve along the curved trajectory. Arrowheads are drawn at intervals of a few hundred generations. (After Lande 1980a.)

trait means evolve on a much faster time scale than genetic variances and covariances (Lande, 1976). The models treat the common situation in which selection favors some intermediate optimum for each of the two characters. Thus the population is visualized as evolving on a hill-shaped surface in a space whose two horizontal axes are the population averages for the two evolving traits, with average population fitness as the vertical dimension. Mean fitness in the population is at a maximum at the hilltop, and, it can be shown, the population tends to evolve toward that hilltop (Figure 9.7).

Delay in adaptation due to genetic correlation can be appreciated by considering the evolution of sexual dimorphism. Consider the case of body sizes of the two sexes, homologous traits that show high genetic correlation between males and females. Suppose the environment changes so that now selection favors a larger body size in males but the original female body size is still at its optimum (Figure 9.7). In the absence of genetic correlation (and with comparable genetic variances in the two sexes), average body size in the sexes would evolve directly toward the new adaptive peak. That is, average female size would stay constant and average male size would increase. Genetic correlation causes a curved evolutionary trajectory. Initially the population evolves rapidly in the direction of increased male and female body size, for in that direction the population possesses the most genetic variance. During this rapid phase, the females may evolve above optimal body size as the males continually evolve larger, always toward their optimum. The maladaptive evolution of the females toward a larger than optimal size is due to a correlated response. The correlated response to selection for larger males temporarily overwhelms the direct response to selection for females, which would otherwise tend to maintain females at their original body size. Finally, a slow process of sexual differentiation ensues during which males gradually increase in size and females slowly decrease in average size. This differentiation is slow since there is little genetic variability in the direction of sexual dimorphism because of the strong genetic coupling between the sexes (Lande, 1980a). The second phase of sexual differentiation could last tens or even hundreds of thousands of generations (Lande, 1980a). Throughout that phase the average female is larger than the optimum. Thus the effect of genetic correlation is to cause the evolutionary trajectory to curve and to delay adaptation. Dissimilar genetic variances in the two traits can have comparable effects (Leutenegger and Cheverud, 1985).

The important message is that the delay in adaptation may be exceedingly long. Consequently, when considering physiological traits that are likely to be genetically coupled, we must seriously entertain the possibility that our study population has excessive or deficient average trait values. Genetic correlation can cause long-lasting disharmony between the population and its environment.

Populations adapting to two different environments may also experience curved evolutionary trajectories and temporary maladaptation due to genetic correlation. Via and Lande (1985) model the situation in which individuals reside in one of two environments throughout their lives and experience corresponding differences in selective pressures. Adaptation can be much delayed if traits that confer success in the two environments are genetically coupled. In particular, if one of the environments is common (e.g., one host plant of a sedentary, herbivorous insect), adaptation to that environment may dominate the first phase of evolution. During that phase, the population may evolve traits deleterious to success in the second environment (e.g., a second host plant species), because correlated responses to selection in the common environment swamp direct responses to selection in the rare environment. Thereafter, a slow process of adaptation to the rare environment may ensue.

Even when the two environments are equally common, genetic correlation can deflect the course of evolution and promote temporary maladaptation.

Long-term evolutionary solutions

Genetic correlation may cause temporary maladaptation, but other traits may ameliorate or even remove the constraints imposed by genetic coupling. Some long-term solutions to the examples of genetic constraint that we previously discussed include (1) the evolution of age-specific or sex-limited expression of traits, and (2) the evolution of the ability to discriminate among hosts or prey that differ in their fitness consequences for the predator. Such mechanisms may eventually lessen the impact of genetic correlation or even eliminate its effects. Nevertheless, specificity in expression and modification of behavior and physiology do not evolve overnight. As new ensembles of traits evolve to meet new environmental challenges, the population will repeatedly confront the constraints and consequences of genetic coupling. Modifications that help retrieve the population from the slow phase of adaptation imposed by genetic coupling may themselves consist of genetically correlated traits and so will experience their own slow evolutionary phase.

The evolution of genetic correlation

In the preceding discussions we took genetic correlation as a given and examined its evolutionary consequences. Yet, genetic correlations themselves evolve, so the constraints they impose are unlikely to last forever, a point sometimes overlooked in arguments for the ever-lasting power of developmental constraints. Thus Charlesworth, Lande, and Slatkin (1982) refute S. Gould's (1980) and Alberch's (1980) arguments that developmental constraints can impose long-term evolutionary stasis by pointing to the success of selection experiments in moving population averages of numerous traits far beyond the limits of variation in the initial population. In addition, the long-term effect of selection may be to change the pattern of genetic and developmental constraints (Schmalhausen, 1949; Waddington, 1957; Cheverud, 1984b).

Genetic models indicate that selection can shape genetic correlations. The pattern of genetic correlation (sign and magnitude) represents a balance between selection, mutation, migration, and other factors (Lande, 1976, 1980b, 1984; Turelli, 1985). Because so many factors influence the evolution of genetic correlations, generalizations about when genetic correlations are likely to be weak or strong, positive or negative, will probably emerge from empirical rather than from theoretical work.

Unexplored genetic issues in ecophysiology

There is a great need for quantitative studies of physiological traits. Knowledge of the heritabilities and genetic correlations of physiological traits is critical to our understanding of the evolution of physiology, yet we have extremely few estimates for natural populations. Most of the relevant past work has been focused on applied problems or basic genetic issues other than the evolution of physiology in nature. Thus when quantitative work has included physiological traits, it usually has had the goal of improving commercial livestock breeds rather than elucidating evolutionary processes. Much of the most sophisticated empirical work in quantitative genetics has been designed to explore basic genetic issues (e.g., how does the heritability of a trait change as a function of the age when the trait is measured?). To accomplish such genetic goals, arbitrary traits are often used (e.g., bristle numbers and wing variation in *Drosophila*). As a consequence, the results seem far from concerns of the ecophysiologist. The perceived distance is an illusion. Much can be learned from the genetic literature about conceptualizing and executing genetic studies of physiology. At the same time, studies of physiology can bring a much needed ecological and evolutionary focus to quantitative genetics.

Physiology is an attractive field for quantitative genetic work for a number of reasons. First, the functional significance of many traits is well understood. The traits are not mere markers of an underlying genetic system, but play specific, known roles in adaptive processes. Second, comparative work is often available and can reveal which traits are evolutionarily liable. Third, many physiologists work with animals sampled directly from nature so that specific natural reference populations can be established for genetic work. In combination, these attributes mean that by studying the inheritance of evolutionarily liable physiological traits we can assess the joint roles of selection and inheritance in pursuing evolutionary problems. In other words, the genetic results can be placed in a specific ecological and evolutionary framework.

Heritability is an important issue confronting the physiologist. Because so little work has been done, heritability remains an active, unresolved issue. Most geneticists would expect physiological traits to show heritable variation in a natural population, but this expectation is merely an educated guess derived from more extensive work on morphological traits. For physiological traits there is virtually no data base, so we do not know whether genetic material for natural selection to act on is abundant or sparse. The extensive experience of animal breeders shows that heritabilities of linear dimensions (conformation) are often in the high range (70% or higher). Such traits respond rapidly to selection and are presumably capable of rapid evolution in nature. We have no comparable generalizations for physiology.

Genetic correlation is a second unresolved issue. We can expect many physiological traits to be genetically coupled because pleiotropic gene action is so common (Wright, 1968). Despite the fact that genetic correlation could play a major role in the evolution of physiological systems, the relevant correlations are a virtually unexplored field of study. For example, a trade-off between burst speed and capacity to sustain running performance might be a reasonable expectation because different, perhaps conflicting physiological systems might support these two kinds of performance. The genetic manifestation of the trade-off is a negative genetic correlation and thus is the parameter that will tell us whether there will be an evolutionary trade-off. The genetic basis of such trade-offs, revealed in genetic correlation, is unexplored.

Genetic correlations with potentially major evolutionary effects are particularly promising candidates for study. Thus high genetic correlations cause correspondingly large deflections of evolutionary trajectories (Figure 9.7). We have already reviewed several instances in which strong genetic coupling is likely: between traits at successive ages; between sexually homologous traits; between the same trait expressed in different environments. A further criterion for genetic correlations having major evolutionary effects is that of antagonism between direct and correlated responses to selection. Thus the genetic correlations just listed are likely to be both strong and positive. If selection favors, say, large trait values at one age but small values at the next oldest age, then trait differentiation will be much slowed because of antagonism in adaptation. In general, the genetic system will have a major impact on adaptation when genetic correlations and selective pressures differ in sign.

The minimal requirements for doing quantitative genetic work are replicated sets of known relatives from a specific population. The requirement of a specific, natural reference population can be easy to satisfy and is critical to interpreting the results. If animals are pooled from a variety of localities, the genetic parameters that are estimated may reflect a totally artificial population entity unlike any particular, local population in nature. The sampled animals should be randomly sampled from a local, interbreeding population to maximize the value of the results. Ability to breed the animals through a succession of generations is desirable but not necessary. For many difficult-to-breed animals it is nevertheless practical to collect series of gravid females and arrange for egg-laying or birth of broods under uniform laboratory conditions. Heritabilities and genetic correlations can be estimated for such animals by analyzing variation within and among broods or by regression of progeny averages on mothers' trait values (Falconer, 1981). Roughly speaking, dozens of families, each composed of several individuals, are needed if heritabilities and genetic correlations lie in the low range. The capacity to breed animals of known parentage permits more powerful estimation as well as selection experiments.

In contemplating a first project in quantitative genetics, it may be reassuring to consult with a colleague who is already familiar with the formidable jargon and technical literature. The complexities of the field are more apparent than real. Likewise, consultation with a statistician may be useful in planning to estimate heritabilities and genetic correlations. How many sets of mothers and offspring will be needed? Falconer (1981) and Becker (1984) are extremely useful introductory texts.

Acknowledgments

I am grateful to A. F. Bennett, M. E. Feder, L. D. Houck, H. B. Shaffer, and D. Townsend for critical comments that much improved the manuscript. I thank J. M. Cheverud, R. Lande, and participants in the symposium for stimulating discussions. The preparation of this paper was supported by the National Science Foundation.

References

Alberch, P. (1980) Ontogenesis and morphological diversification. *Am. Zool.* 20:653–667.

Arnold, S. J. (1981a) Behavioral variation in natural populations. I. Phenotypic, genetic and environmental correlations between chemoreceptive responses to prey in the garter snake, *Thamnophis elegans. Evolution* 35:489–509.

Arnold, S. J. (1981b) The microevolution of feeding behavior. In *Foraging Behavior: Ecological, Ethological and Psychological Perspectives*, ed. A. Kamil and T. Sargent, pp. 409–453. New York: Garland Press.

Becker, W. A. (1984) *Manual of Quantitative Genetics*, 4th ed. Pullman, Wash.: Academic Enterprises.

Bradford, G. E., Barkley, M. S., and Spearow, J. L. (1980) Physiological effects of selection for aspects of efficiency of reproduction. In *Selection Experiments in Laboratory and Domestic Animals*, ed. A. Robertson, pp. 161–175. Slough, U.K.: Commonwealth Agricultural Bureaux.

Bucher, T. L., Ryan, M. J., and Bartholomew, G. A. (1982) Oxygen consumption during resting, calling and nest building in the frog *Physalaemus pustulosus. Physiol. Zool.* 55:10–22.

Bulmer, M. G. (1974) Linkage disequilibrium and genetic variability. *Genet. Res. Camb.* 23:281–289.

Burghardt, G. M. (1971) Chemical-cue preferences of newborn snakes: influence of prenatal maternal experience. *Science* 171:921–923.

Carey, C. (1980) Adaptation of the avian egg to high altitude. *Am. Zool.* 20:449–459.

Charlesworth, B., Lande, R., and Slatkin, M. (1982) A neo-Darwinian commentary on macroevolution. *Evolution* 36:474–498.

Cheverud, J. M. (1982) Phenotypic, genetic and environmental morphological integration in the cranium. *Evolution* 36:499–516.

Cheverud, J. M. (1984a) Evolution by kin selection: a quantitative genetic model illustrated by maternal performance in mice. *Evolution* 38:766–777.

Cheverud, J. M. (1984b) Quantitative genetics and developmental constraints on evolution by selection. *J. Theor. Biol.* 110:155–171.

Cheverud, J. M., Rutledge, J. J., and Atchley, W. R. (1983) Quantitative genetics of development: genetic correlations among age-specific trait values and the evolution of ontogeny. *Evolution* 37:895–905.

Cockerham, C. C. (1963) Estimation of genetic variances. In *Statistical Genetics and Plant Breeding*, pp. 53–94. NAS-NRC Publication 982. Washington, D.C.: National Academy of Sciences – National Research Council.

Crow, J. F., and Kimura, M. (1970) *An Introduction to Population Genetics Theory.* New York: Harper and Row.

Darwin, C. (1859) *The Origin of Species by Means of Natural Selection or the Preservation of Favored Races in the Struggle for Life.* London: Murray.

Darwin, C. (1874) *The Descent of Man and Selection in Relation to Sex*, 2nd ed. London: Murray.

Dickerson, G. E. (1947) Composition of hog carcasses as influenced by heritable differences in rate and economy of gain. *Iowa Agric. Exp. Stat. Res. Bull.* 354:489–524.

Dickerson, G. E. (1955) Genetic slippage in response to selection for multiple objectives. *Cold Spring Harbor Symp. Quant. Biol.* 20:213–224.

Eisen, E. J. (1967) Mating designs for estimating direct and maternal genetic variances and direct-maternal genetic covariances. *Can. J. Genet. Cytol.* 9:13–22.

Endler, J. (1986) *Natural Selection in The Wild.* Princeton, N.J.: Princeton University Press.

Falconer, D. S. (1952) The problem of environment and selection. *Am. Natur.* 86:293–298.

Falconer, D. S. (1965) Maternal effects and selection response. In *Genetics Today*, vol. 3, *Proc. 11th Int. Cong. Genet.*, ed. S. J. Geerts, pp. 763–774. Oxford: Pergamon Press.

Falconer, D. S. (1981) *Introduction to Quantitative Genetics*, 2nd ed., New York: Longman.

Fisher, R. A. (1918) The correlation between relatives on the supposition of Mendelian inheritance. *Trans. R. Soc. Edinburgh* 52:399–433.

Fox, W. (1948) Effect of temperature on development of scutellation in the garter snake, *Thamnophis elegans atratus. Copeia* 1948:252–262.

Fox, W., Gordon, C., and Fox, M. A. (1961) Morphological effects of low temperature during the embryonic development of the garter snake, *Thamnophis elegans. Zoologica* 46:57–71.

Futuyma, D. J., and Peterson, S. (1985) Genetic variation in the use of resources by insects. *Annu. Rev. Entomol.* 30:217–238.

Gould, F. (1979) Rapid host range evolution in a population of the phytophagous mite *Tetrahychus urticae* Koch. *Evolution* 33:791–802.

Gould, S. J. (1980) Is a new and general theory of evolution emerging? *Paleobiology* 6:119–130.

Gyles, N. R., Dickerson, G. E., Kinder, G. B., and Kempster, H. L. (1955) Initial and actual selection in poultry. *Poultry Sci.* 34:530–539.

Hill, W. G. (1980) Design of quantitative genetic selection experiments. In *Selection Experiments in Laboratory and Domestic Animals*, ed. A. Robertson, pp. 1–13. Slough, U.K.: Commonwealth Agricultural Bureaux.

Lande, R. (1976) The maintenance of genetic variation by mutation in a polygenic character with linked loci. *Genet. Res. Camb.* 26:221–235.

Lande, R. (1979) Quantitative genetic analysis of multivariate evolution, applied to brain: body size allometry. *Evolution* 34:402–416.

Lande, R. (1980a) Sexual dimorphism, sexual selection and adaptation in polygenic characters. *Evolution* 34:292–305.

Lande, R. (1980b) The genetic covariance between characters maintained by pleiotropic mutations. *Genetics* 94:203–215.

Lande, R. (1984) The genetic correlation between characters maintained by selection, linkage and inbreeding. *Genet. Res. Camb.* 44:309–320.

Leamy, L., and Cheverud, J. M. (1984) Quantitative genetics and the evolution of ontogeny. II. Genetic and environmental correlations among age-specific characters in random bred house mice. *Growth* 48:339– 353.

Leutenegger, W., and Cheverud, J. M. (1985) Sexual dimorphism in primates: the effects of size. In *Size and Scaling in Primate Biology*, ed. W. L. Jungers, pp. 33–50. New York: Plenum Press.

Lofdahl, K. L. (1985) A quantitative genetic analysis of habitat selection behavior in the cactus-breeding species *Drosophila mojavensis.* Ph.D. thesis, University of Chicago.

Olson, E., and Miller, R. (1958) *Morphological Integration.* Chicago: University of Chicago Press.

Osgood, D. W. (1978) Effects of temperature on the development of meristic characters in *Natrix fasciata. Copeia* 1978:33–37.

Palmer, J. O., and Dingle, H. (1986) Direct and correlated responses to selection among life-history traits in milkweed bugs (*Oncopeltus fasciatus*). *Evolution* 40:767–777.

Rausher, M. D. (1984) Tradeoffs in performance on different hosts: evidence from within- and between-site variation in the beetle *Peloyala guttata. Evolution* 40:767–777.

Riska, B., Rutledge, J. J., and Atchley, W. R. (1985) Covariance between direct and maternal genetic effects in mice, with a model of persistent environmental influences. *Genet. Res. Camb.* 45:287–297.

Robertson, A. (1959) The sampling variance of the genetic correlation coefficient. *Biometrics* 15:469– 485.

Robertson, A. (1980) *Selection Experiments in Laboratory and Domestic Animals.* Slough, U.K.: Commonwealth Agricultural Bureaux.

Rutledge, J. J., Robinson, O. W., Eisen, E. J., and Legates, J. E. (1972) Dynamics of genetic and maternal effects in mice. *J. Anim. Sci.* 35:911–918.

Schmalhausen, I. F. (1949) *Factors of Evolution, the Theory of Stabilizing Selection.* Philadelphia: Blakiston.

Sulzbach, D. S., and Lynch, C. B. (1984) Quantitative genetic analysis of temperature regulation in *Mus musculus.* III. Diallel analysis of correlation between traits. *Evolution* 38:541–552.

Taigen, T. L., and Wells, K. D. (1985) Energetics of vocalization by an anuran amphibian (*Hyla versicolor*). *J. Comp. Physiol.* 155:163–170.

Taigen, T. L., Wells, K. D., and Marsh, R. L. (1985) The enzymatic basis of high metabolic rates in calling frogs. *Physiol. Zool.* 58:719–726.

Turelli, M. (1985) Effects of pleiotropy on predictions concerning mutation-selection balance for polygenic traits. *Genetics* 111:165–195.
Via, S. (1984) The quantitative genetics of polyphagy in an insect herbivore. II. Genetic correlation in larval performance within and across host plants. *Evolution* 38:896–905.
Via, S. (1986) Genetic covariance between oviposition preference and larval performance in an insect herbivore. *Evolution* 40:778–785.
Via, S., and Lande, R. (1985) Genotype-environment interaction and the evolution of phenotypic plasticity. *Evolution* 39:505–522.
Waddington, C. H. (1957) *The Strategy of the Genes.* London: Allen and Unwin.
Watt, W. B., Carter, P. A., and Donohue, K. (1986) Females' choice of 'good genotypes' as mates is promoted by an insect mating system. *Science* 233:1187–1190.
Willham, R. L. (1963) The covariance between relatives for characters composed of components contributed by related individuals. *Biometrics* 19:18–27.
Willham, R. L. (1972) The role of maternal effects in animal breeding. III. Biometrical aspects of maternal effects in animals. *J. Anim. Sci.* 35:1288–1293.
Wright, S. (1968) *Genetic and Biometric Foundations,* vol. 1: *Evolution and the Genetics of Populations.* Chicago: University of Chicago Press.

Discussion

POWERS: You said that the effect of temperature on vertebral number is an environmental factor and yet you are looking at its genetic component. Could you elaborate a little on why you are addressing it and how you are addressing it relative to environmental versus genetic factors?

ARNOLD: What I was showing is the reasonable prospect, on the basis of studies on other vertebrates, that there is a maternal effect of temperature on vertebral numbers in garter snakes. There are two extreme possibilities. One is that this maternal effect is completely nongenetic. Another is that there is some heritable component to the maternal effect, that is to say, that there is some heritable variation in temperatures that females are selecting during pregnancy. If that is the case, then vertebral number could have two heritable components: a component due to genes that affect vertebral numbers directly, and a second category of genes that affect maternal thermoregulation and influence indirectly the trait in question. So we have the possibility of a complicated genetic basis for vertebral numbers, and also the possibility of a genetic correlation between vertebral numbers and maternal performance. Then the two traits could evolve as a consequence of variation in genes affecting vertebral number directly, genes affecting maternal thermoregulation directly, and genes that have pleiotropic effects on both traits. When we consider the large number of traits that are influenced by maternal performance or paternal performance, this kind of situation might be quite general, particularly in viviparous vertebrates.

FLORANT: If Siamese cats are kept at cold temperatures, the paws and the ears of the kittens are much darker than in controls. This seems parallel to the garter snakes. I can see why Siamese cats might want darker paws if their mother was in a cold environment, but I do not understand why a garter snake would want more vertebrae in a warmer environment.

ARNOLD: An issue that I did not even try to address was the evolution of the norm of reaction and why we might have a "U"-shaped norm of reaction in so many species of teleost fish, for example. There is a zone of stability: at the bottom of the "U," these populations can experience quite an excursion of temperatures and not see an impact on vertebral numbers. In garter snake populations, we know there is stabilizing selection on vertebral numbers: that individuals that deviate in either direction from the population mean suffer a decrement in fitness. My hypothesis would be that if there is a "U"-shaped curve in garter snakes, that zone of stability might have evolved to coincide with the stabilized mean of the vertebral number distribution. That is a hypothesis I am using to motivate the field and laboratory work.

FUTUYMA: This is reminiscent of the theory of canalization, the notion that there has been selection to mold a developmental system that will produce the "right" phenotype in the usual range of environmental conditions. The plateau at the bottom of this "U" could be the target that development is shooting toward, but the high vertebral number at an extreme temperature might be just a nonadaptive developmental anomaly.

ARNOLD: Yes, that is possible.

FUTUYMA: It is a case where you would like to know whether the entire norm of reaction is an adaptation, or if the ends of that norm of reaction are a pathology of some kind.

ARNOLD: I am not dealing with the difficult problem of how the norm of reaction evolves. If we take that as a given for the moment, we can imagine that thermoregulation in pregnant females might evolve in relation to that norm of reaction. Given that there is stabilizing selection on the vertebral numbers, there might be selection on maternal thermoregulation to move in on that zone of stability.

BENNETT: I would like to point out that the functional consequences of these morphological features can easily be tested. In the first pass, if one thinks of locomotion as being a function of vertebral number, it is easy to ask whether animals with more vertebrae locomote better in cold environments than in warm ones.

BARTHOLOMEW: Is what you are studying different from examining the nature of the fit between the maternal behavior of a mammal and the physiological state of the infant at birth?

ARNOLD: No, I think that you could use the same framework for viewing the evolution of maternal performance of any kind and target offspring phenotypes. In this case, this approach makes us look at thermoregulation in a perspective that I had not entertained before. I think the framework of quantitative genetics gives us an agenda of issues that will help us focus attention on different aspects of the problem.

FUTUYMA: A reaction norm is a geneticist's jargon for the form of the relationship between some feature of the organism and the environmental conditions in which the organism has been reared or kept for some time. A quantitative geneticist may approach this by taking, for example, performances at various temperatures as being in themselves characters, and ask how, from individual to individual, these characters are correlated with one another. If you plot different genotypes, would you find that they show different reaction norms? This could help you determine whether or not the entire reaction norm constitutes an adaptation. Do you interpret every part of the curve as being adaptive? Or do you assume that perhaps some of this curve is simply a correlated nonadaptive consequence of other parts of the curve?

BENNETT: Certainly the extremes of the curve are nonadaptive: if you freeze it or boil it, the system is going to come apart and those are not adaptations but structural consequences. When people do these studies, they try to phrase them in terms of natural environmental exposure and what the animal really experiences.

HUEY: Your question borders on an issue that we have not addressed, which is, if you get selection on one part of the performance curve, how does that affect other parts of the curve? If you get selection for higher tolerance, does that drag along the lower parts of the curve; or, if you get selection on the variance, how does that affect traits?

JACKSON: I think that it is going to depend on the type of function you are looking at. Some functions are simple reactions to change in temperature, such as heart rate. Upon those may be built the effect of temperature on other physiological control mechanisms that may affect heart rate, such as the autonomic nervous system. I think the issues of the direct effect of temperature, regulatory mechanisms existing within the animal, adaptations, and longer-term changes are all going to play a part.

DAWSON: There is in some cases a correlation between, for example, thermal preferenda and lethal temperature in lizards. Paul Licht's work showed that one set of enzymes, the myofibrillar ATPases, exhibit thermal differentiation, but alkaline phosphatase had such a high thermal denaturation point that it did not show any diversification. It relates to specific characteristics and how centrally they are involved, being either permissive or nonpermissive.

BISHOP: We were working on a bivalve and we calculated that there is about ten percent protein turnover. If the organism trapped all of the free amino acids and accumulated them, it could survive, and you need really no increase in the LAP [leucine aminopeptidase] activity. Then Koehn found that the animals with genetically higher LAP activities would adapt more rapidly in the face of osmotic stress. However, because they had higher LAP activities, they had a higher protein turnover activity when they were starved, which actually selected against the animal. So when you talk about an adaptive character, it depends upon all of the stresses that the animals face.

10 The misuse of ratios to scale physiological data that vary allometrically with body size

GARY C. PACKARD AND
THOMAS J. BOARDMAN

Introduction

Allometry is the zoological discipline that is concerned with the manifold consequences of variation in body size (Gould, 1966). Important variation in body size may result from growth and other changes occurring during the ontogeny of a single individual, or it may result from static differences among adult animals in a single population or among adult animals belonging to different species (Gould, 1966; Fleagle, 1985). The consequences of this variation in size are manifested in almost every facet of the morphology, physiology, and ecology of the animal(s) in question (Peters, 1983; Calder, 1984; Schmidt-Nielsen, 1984).

We were invited by the editors of this volume to undertake a review of the field of allometry and to tender our recommendations for future research on this topic. However, several excellent introductions to allometry were published quite recently (Peters, 1983; Calder, 1984; Schmidt-Nielsen, 1984), and the ink is barely dry on several thoughtful critiques of the concepts and methods of allometry (e.g., Smith, 1984a, 1984b; Fleagle, 1985; Donhoffer, 1986). Rather than simply recapitulate the main points of these contributions, we shall focus instead on a pervasive problem attending allometric variation in physiological data, namely the problem of how to scale experimental data correctly for variation in body size. We should mention, however, that the problem we perceive is not limited to physiological studies, but extends to biochemical, morphological, and ecological investigations as well.

Most physiological functions vary in level or intensity with body size of the animal(s) being studied. When a plot of some variable of interest (e.g., oxygen consumption, or extractable body fat, or the combined mass of the gonads) against some measure of body size (e.g., body mass, or snout-to-vent length) yields a straight line passing through the origin of a graph with linear coordinates, the variable varies *isometrically* with body size (Figure 10.1). Thus, a doubling of body size results in a doubling also in the level or intensity of the variable of interest. When the line is curvilinear or when it does not pass through the origin, however, the variable varies *allometrically* with body size (Figure 10.1). In such instances, a doubling in body size does not

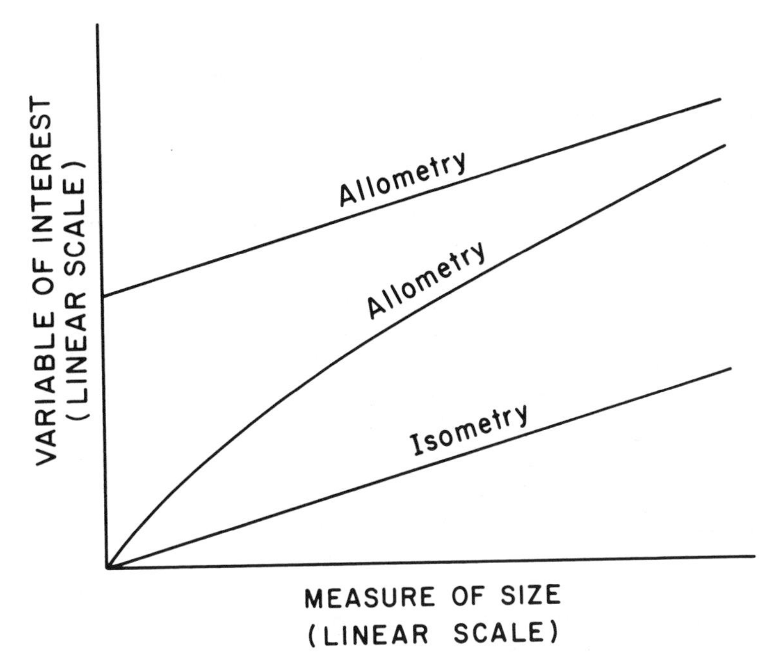

FIGURE 10.1. Three examples illustrate the distinction we make between allometric variation and isometric variation. Allometric variation occurs when a plot of the variable of interest (Y) against a measure of body size (X) yields a straight line intersecting the ordinate at some point other than zero, or when the plot forms a smooth curve (irrespective of the intercept). In contrast, isometric variation occurs only when the variable of interest is related to the measure of body size by a straight line passing through the origin, that is, when the line conforms with the equation $Y = bX$. Thus, the ratio Y/X does not vary with body size in cases of isometry, but the ratio does vary with body size in cases of allometry.

result in exactly a doubling in level or intensity of the variable of interest, and the variable therefore changes in its proportion to body size. We could have distinguished isometric variation from allometric variation using plots of data transformed into their common logarithms (e.g., Huxley, 1932), but logarithmic transformations frequently are unnecessary (Thompson, 1942) and they may even be misleading (Smith, 1984a). Given the restrictive definition for isometry, it is not surprising that the size-related variation in most physiological variables is allometric (Smith, 1984a).

Zoologists generally recognize that variation in body size of organisms is a potential source of confounding variation in the data they gather. Accordingly, many workers executing designed experiments assign animals to treatments so as to assure that the groups do not vary appreciably with respect to either the mean or the distribution of body size (i.e., the experiments are "balanced"). An implicit assumption in such investigations is that allometric (or even isometric) variation in the variable of interest is of little concern, because animals in the several treatment groups are so carefully matched with respect to size (Finney, 1957). As we shall see, however, part of the variation in the physiological variable of interest usually stems from variation in size of the animals in the different samples, even when the experimental groups are carefully matched with respect to size. Physiological measurements gathered in such experiments frequently lack precision, and real treatment effects often go undetected (Cochran, 1957; Finney, 1957; Cox, 1958).

Other workers (especially those performing descriptive studies on organisms differing in body size) try to compensate for effects of size by dividing the variable of interest by the measure of body size, thereby forming a proportion, percentage, or size-specific index. Such ratios are commonplace in the modern literature, and frequently represent the primary form for data being presented (e.g., mass-specific metabolism, percent body fat, gonosomatic index). Although ratios afford a satisfactory adjustment when the variable of interest varies isometrically with body size, such an adjustment often is inadequate when allometric variation prevails (Cochran, 1957; Gould, 1966). Indeed, several workers have shown that the use of ratios to scale data can lead investigators to draw improper conclusions from their experiments (Tanner, 1949; Dinkel, Wilson, Tuma, and Minyard, 1965; Gonor, 1972; Atchley, 1978; de Vlaming, Grossman, and Chapman, 1982; Blem, 1984; Reist, 1985; Strauss, 1985).

A superior method for increasing precision in planned experiments and for minimizing (or eliminating) confounding effects of varying body size in observational studies is the analysis of covariance (Cochran, 1957; Finney, 1957; Cox, 1958). Although many zoologists seem to be intimidated by the formidable name given to this analytical tool, they probably are familiar with the two procedures on which the analysis of covariance is based: regression and analysis of variance (see Fisher, 1932, p. 259). We shall illustrate the use of analysis of covariance in several simple yet realistic examples, and shall compare the outcome with results of statistical tests on the original data and on ratios. Our intent is to emphasize practical considerations, and to point out how inappropriate methods can lead to improper conclusions. Readers wishing to avail themselves of statistical theory or details of computation should consult standard accounts (e.g., Snedecor and Cochran, 1980; Steel and Torrie, 1980; Sokal and Rohlf, 1981).

Example 1: Increasing precision in planned experiments

Figure 10.2a is a bivariate plot showing the distribution of hypothetical data gathered in a planned experiment comparing two treatments (i.e., open cir-

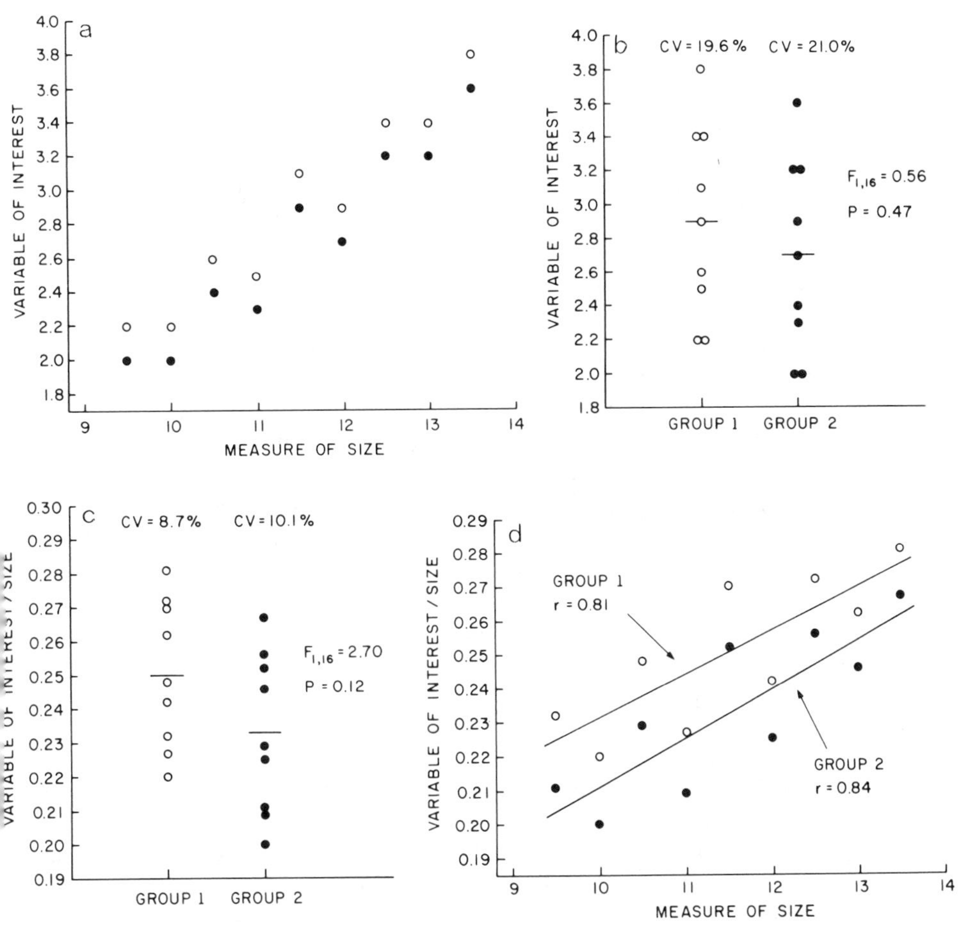

FIGURE 10.2 (a) Bivariate plot of hypothetical data illustrating the importance of proper scaling in planned experiments that are balanced with respect to body size of animals in the various treatment groups. The variable of interest is any physiological process that varies allometrically with body size. Sample size is nine for both group 1 (open circles) and group 2 (closed circles). (b) Values for the variable of interest are displayed in a simple dot-plot. Horizontal lines represent means of 2.9 and 2.7 for groups 1 and 2, respectively. CV = coefficient of variation. The *F*-ratio is from an analysis of variance comparing means. (c) Ratios were formed by dividing the variable of interest by the measure of body size. Horizontal lines represent means of 0.250 and 0.233 for groups 1 and 2, respectively. The *F*-ratio is from an analysis of variance comparing means. (d) Bivariate plots of the ratios against the measure of body size. Product-moment correlation coefficients (*r*) are significantly different from zero and indicate that both sets of ratios vary with body size.

cles vs. closed circles). The variable of interest here is any physiological character that varies allometrically in magnitude or intensity with body size of the organism being studied. For example, the physiological response could be the rate of uptake of water by fully hydrated frogs (assessed as changes in their mass), and the treatments could involve injecting animals with antidiuretic hormone (ADH) and with saline. The measure of size is any accepted measure of body size, such as standard mass for the frogs of interest here. Measurements of size generally are taken before the animals are assigned to treatment groups, but could be taken later if the measurements are unaffected by the experimental manipulations (Cochran, 1957; Cox, 1958). The distribution of body sizes is the same for animals in both of the groups illustrated here, so the experiment is balanced. The values presented in this bivariate plot admittedly were chosen for ease in computation and to illustrate a point, but the patterns of response shown here have numerous parallels in real data sets published in the literature.

Examination of the graph reveals that responses of individuals in each of the experimental groups varied in essentially a linear manner with size (Figure 10.2a). A linear relation facilitates computations that we shall perform later, but is not an essential condition for our general argument or for this example (Cochran, 1957; Cox, 1958). The responses of animals comprising group 1 were slightly higher, on average, than those of individuals comprising group 2, but the overlap of values is so great as to render conclusions equivocal. Further examination of the data is required.

When values for the variable of interest are presented in a dot-plot (Figure 10.2b), we see that the mean for this variable is 2.9 for animals in group 1 and 2.7 for those in group 2. Variation about these means is substantial, and coefficients of variation are about 20%. When the data are examined by analysis of variance, the resultant *F*-ratio is small enough that the null hypothesis cannot be rejected. Indeed, we probably would conclude instead that the means do not differ at all. Some workers would argue, of course, that the samples were too small to demonstrate the existence of significant treatment effects. However, assuming momentarily that the difference between means of 0.2 represents a real treatment effect, samples of 130 would be needed to be 80% certain of demonstrating the difference at a probability level of .05 (Sokal and Rohlf, 1981). Such samples are prohibitively large for most zoologists, so increasing the size of samples is not a realistic solution to the problem.

We next compute ratios by dividing each value for the variable of interest by the corresponding value for body size, in an effort to compensate for variation in body size within each of the samples. When we present the values in dot-plots (Figure 10.2c), we discover that the apparent precision of our measurements has been increased, because the coefficients of variation now are approximately 10%. Despite the increase in precision, however, the *F*-

ratio from an analysis of variance still is too small to provide compelling support for existence of a treatment effect (Figure 10.2c). Additionally, effects of variation in body size have not been removed completely from the data, because the ratios still are highly correlated with the measure of body size (Figure 10.2d). Correlations between such ratios and the measure of size are the rule rather than the exception (Tanner, 1949; Atchley, Gaskins, and Anderson, 1976; Atchley, 1978; Atchley and Anderson, 1978; Strauss, 1985), so the correlations we observe here are not peculiar to our hypothetical data.

As an alternative to examining ratios, we now perform an analysis of covariance on the original data. At the first step in this analysis, straight lines are fit to data in each of the experimental groups (Figure 10.3a). If the slopes of these lines do not differ significantly, an "average" slope (weighted for possible differences in sample size) is computed. This average slope then is used to adjust values for the variable of interest to some common value for body size (which usually is the grand mean for body size of the animals used in the study). Values for small animals are therefore adjusted (or scaled) upward, and those for large animals are adjusted downward (Figure 10.3b). Variation in the data stemming from variation in body size is thereby removed, but the means are not affected.

The adjusted values for the variable of interest then are displayed in a dot-plot and compared by an analysis of variance (Figure 10.3c). Variation in the data has been reduced substantially by the adjustment procedure (coefficients of variation are less than 6%), and the power of the statistical analysis has been increased accordingly. Indeed, the difference between means of 0.2 now is significant at a probability level of .02. Thus, we would conclude from this analysis that the physiological responses of the animals differed significantly between the two experimental groups.

In summary, use of ratios to scale data in example 1 for variation in body size increased the apparent precision of measurements on the physiological variable, but the statistical analysis of the ratios still lacked sufficient power to discriminate between responses of animals in the two treatment groups. In contrast, the analysis of covariance led to an even greater increase in precision of the measurements, and covariance had the added advantage of removing completely the influence of body size on the variable of interest. The analysis of covariance consequently had sufficient power that real effects of the treatments were detected statistically.

Example 2: Assessing treatment effects in planned experiments

A second application of the analysis of covariance is closely related to the first example, and explores possible treatment effects in planned experiments that are balanced with respect to the size of animals in different treatment

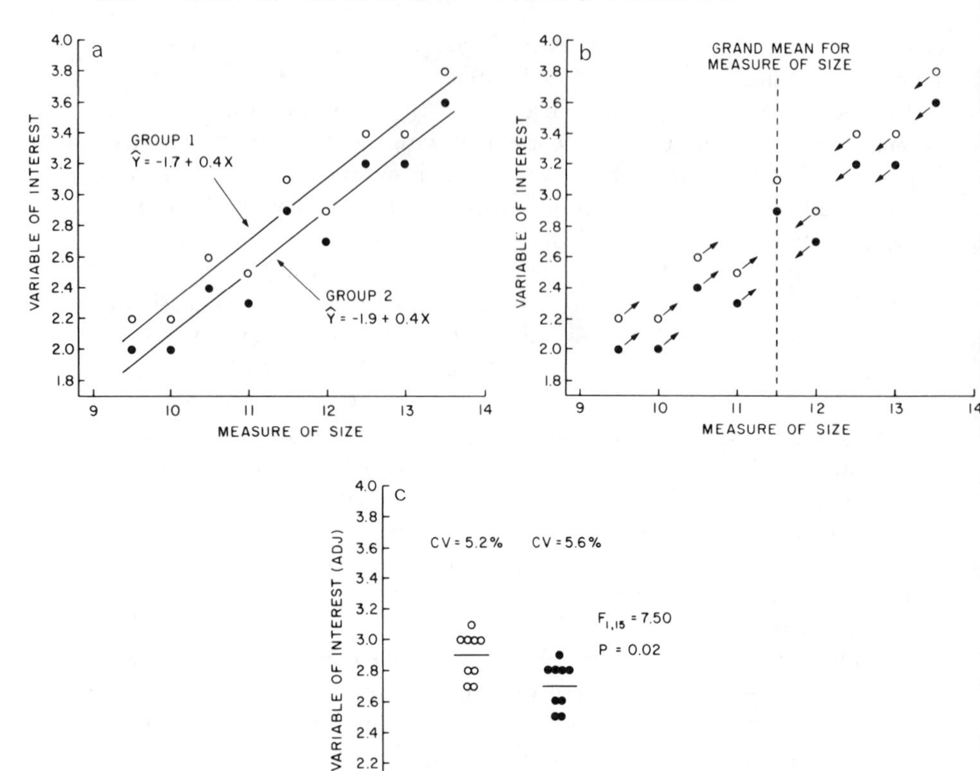

FIGURE 10.3 (a) Straight lines were fit to data in Figure 10.2a by the method of least squares. Such curve-fitting is a preliminary step in the analysis of covariance. (b) The slopes of the lines in Figure 10.3a do not differ significantly, so an "average" slope is calculated. This average slope is used to adjust values for the variable of interest to the grand mean for body size, that is, to the vertical line in the graph. Thus, values are scaled upward for small animals and downward for large animals. (c) Individual values for the variable of interest were adjusted by the regression procedure (= covariance procedure). Horizontal lines represent means of 2.9 and 2.7 for groups 1 and 2, respectively. CV = coefficient of variation. The *F*-ratio is from an analysis of variance comparing means. One degree of freedom was lost from the residuals in this analysis as a result of estimating the slope of the line used to adjust data to the grand mean for body size.

groups. Hypothetical data displayed in a bivariate plot allow us to explore this problem further (Figure 10.4a).

As before, each of two groups of nine animals is subjected to some experimental manipulation, and a response of interest is recorded. The manipulation could entail exposing two groups of some passerine bird to different photoperiods and then measuring their consumption of food. The individual responses (i.e., food consumed) are plotted against a measure of body size (probably body mass) taken at the outset of study (Figure 10.4a). The relation between the variable of interest and body size is apparent, but it again is not clear whether the treatments have affected the response of interest.

When data for the physiological variable are summarized in a dot-plot (Figure 10.4b), we can see that the means for both groups are 2.9. An analysis of variance on these data yields an *F*-ratio of zero, which is a strong indication that the means do not differ significantly. However, no consideration has been given as yet to possible confounding effects introduced by variation in size of animals within each of the samples.

In an effort to remove effects of variation in body size, we form ratios by dividing each value for the physiological variable by the corresponding measure of body size. A dot-plot of the ratios reveals that means for the two groups are very similar (Figure 10.4c). Although the variance for group 1 is greater than that for group 2 ($F_{(8,8)} = 6.95$, $p = .007$), this difference probably would go undetected in most studies because few investigators plot their data in the way that they have been plotted here. Analysis of variance (which works well even in instances like this where variances differ slightly) produces an *F*-ratio that is so small that the null hypothesis cannot be rejected, and we probably would draw the same conclusion here that we drew from the analysis of the original data, namely that the physiological response did not differ between the two treatment groups.

If we examine the data by analysis of covariance, however, we discover that the slope of the line describing data from group 1 is significantly different from (steeper than) the slope of the line characterizing data from group 2 (Figure 10.4d). Variation in body size had a more pronounced effect on expression of the variable of interest in group 1 than in group 2. Thus, the experimental manipulations caused animals in group 1 to manifest a higher dependence between the physiological variable and the measure of body size than was true for animals in group 2. This is a clear indication of a significant treatment effect in the experiment (Cochran, 1957).

Adjusted values could be computed for animals in the two groups (by using the different slopes for the two lines as the basis for separate adjustments), but the means would be identical in this case and therefore would not be especially useful. There are other instances, however, when we might want to compare adjusted means to determine whether one treatment elicited a general increase or decrease in the level of response relative to the other treat-

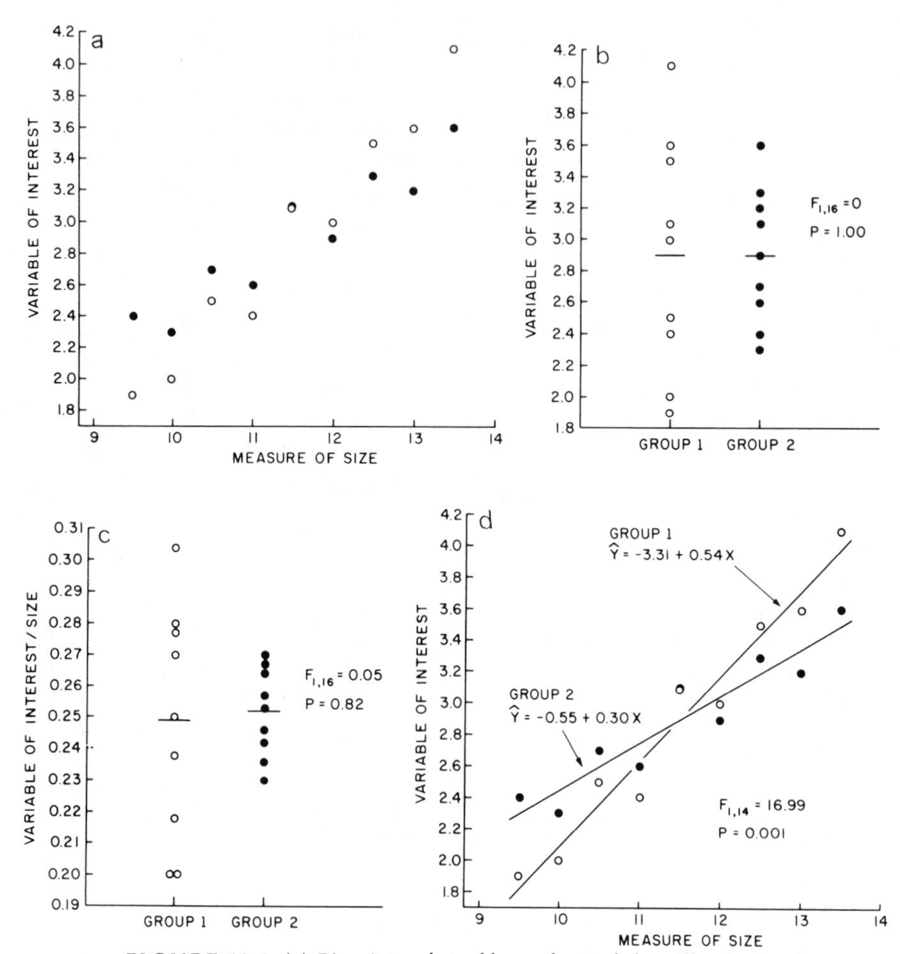

FIGURE 10.4 (a) Bivariate plot of hypothetical data illustrating the importance of proper analysis in planned experiments that are balanced with respect to body size of animals in the various treatment groups. The variable of interest is any physiological process that varies allometrically with body size. Sample size is nine for both group 1 (open circles) and group 2 (closed circles). (b) Values for the variable of interest are displayed in simple dot-plots. Horizontal lines represent means of 2.9 for individuals in both of the experimental groups. The *F*-ratio is from an analysis of variance comparing means. (c) Ratios were formed by dividing the variable of interest by the measure of body size. Horizontal lines represent means of 0.249 and 0.252 for groups 1 and 2, respectively. The *F*-ratio is from an analysis of variance comparing means. (d) Straight lines were fit to data in Figure 10.4a by the method of least squares. The *F*-ratio is from an analysis of covariance comparing the slopes of the lines.

ment (or control). Thus, adjustment by covariance may be appropriate even when lines with different slopes are used to scale data in different samples.

In summary, uncritical use of ratios to scale data for the physiological variable in example 2 led us to conclude erroneously that the experimental manipulations had no important effect on the variable of interest. Use of the analysis of covariance, however, led us to conclude correctly that the pattern of response was altered significantly by the experimental treatments.

Example 3: Assessing confounding effects in descriptive studies

A third application of the analysis of covariance is in descriptive studies in which size of animals differs among treatment groups. Such a circumstance commonly arises in studies comparing animals at different seasons, or animals from different genetic strains, different populations, or different species. For example, an investigator may be interested in comparing the blood oxygen capacity of frogs from high and low altitude populations that differ in average body size in an effort to elucidate adaptations of the oxygen transport system to high altitude.

A plot of hypothetical data suggests that the variable of interest is not affected appreciably by body size, although size of animals differs appreciably between the two groups (Figure 10.5a). If the former impression is correct, values for the variable of interest do not need to be scaled for variation in body size. Many zoologists omit the critical step of plotting and examining their data, however, and automatically divide the variable of interest by body size in an effort to compensate for the allometric (or isometric) variation that is assumed to exist. When such ratios are examined in the present case, we find that the mean for group 1 is significantly larger than that for group 2 (Figure 10.5b). We might conclude from this analysis that the groups differ with respect to the variable of interest once the confounding effects of body size have been removed.

This conclusion conflicts with impressions gained earlier while examining the bivariate plot of the raw data (Figure 10.5a). If we perform an analysis of covariance on the data instead of examining ratios, we discover at the first step that lines fit to each of the data sets have slopes that do not differ significantly from zero. This finding confirms the correctness of our initial impressions, and indicates that the final analysis of variance is appropriately performed on the unadjusted values. The F-ratio in this analysis is zero, which strongly indicates that the data were drawn from the same statistical distribution (Figure 10.5c).

Thus, ill-considered use of ratios in example 3 led us to conclude incorrectly that scaled responses of animals in group 1 were higher than those of animals in group 2. The difference arising when data were expressed as ratios resulted entirely from the difference in body size between groups, and not

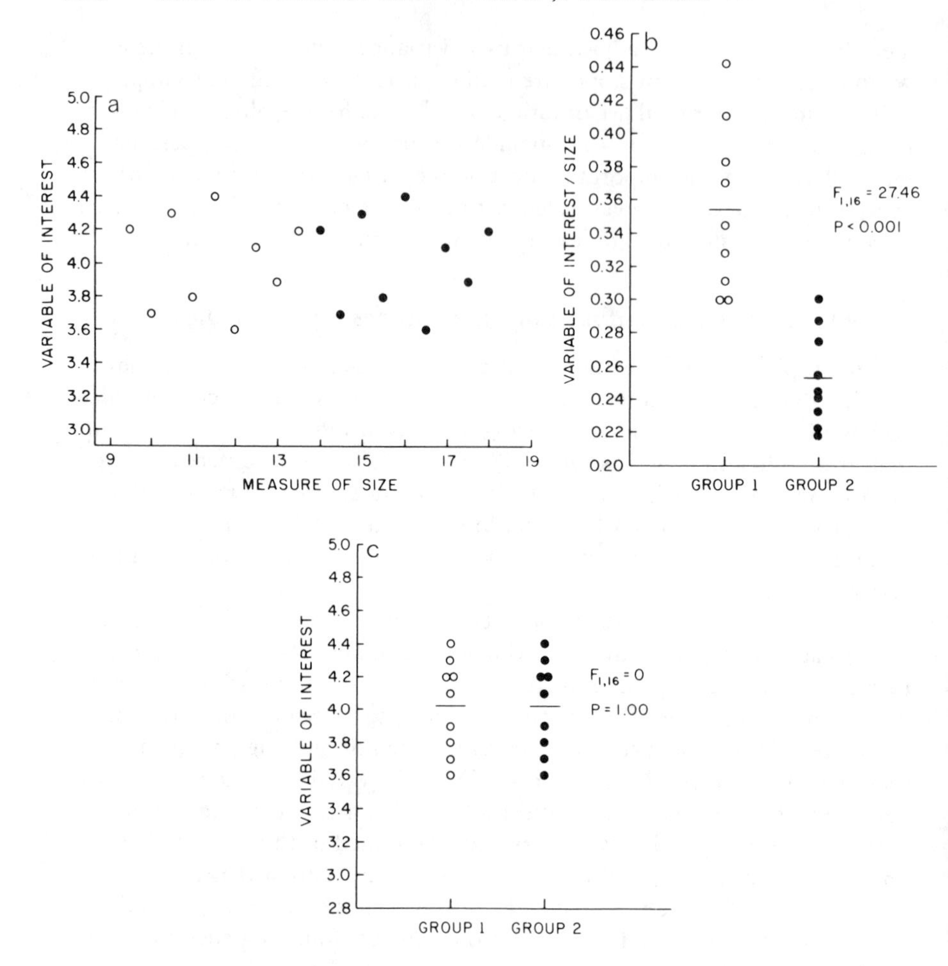

FIGURE 10.5 (a) Bivariate plot of hypothetical data illustrating the importance of proper analysis in descriptive studies comparing groups of animals differing in size. The variable of interest is any physiological process that does not vary in level or intensity with body size. Sample size is nine for both group 1 (open circles) and group 2 (closed circles). (b) Ratios were formed by dividing the variable of interest by the measure of body size. Horizontal lines represent means of 0.354 and 0.253 for groups 1 and 2, respectively. The *F*-ratio is from an analysis of variance comparing means. (c) Regression analysis indicated that values for the variable of interest do not need to be adjusted for variation in body size, so data are displayed here in simple dot-plots. Horizontal lines represent means of 4.02 for individuals in both groups. The *F*-ratio is from an analysis of variance comparing means.

from any difference in the physiological variable. In contrast, use of the analysis of covariance led us to conclude properly that responses of animals in the two groups were indistinguishable.

Example 4: Removing confounding effects of body size

A fourth application of the analysis of covariance concerns removing effects related to differences in size of animals comprising the study groups. This again is a situation that might arise in studies comparing animals at different times of year, comparing individuals from different genetic strains or populations of a single species, or comparing individuals from different species. For example, measurements might be taken of urine production under standard conditions by representatives of two species of fish differing in average body size.

A bivariate plot of hypothetical data reveals that the variable of interest increases in level or intensity with body size of the animals in question, and that individuals in group 1 are smaller than those in group 2 (Figure 10.6a). The response of interest seems to be higher, on the average, in animals in group 2 than in those in group 1, and this impression is confirmed by an analysis of variance on unadjusted values (Figure 10.6b).

This difference between groups in the variable of interest could be a simple manifestation of differences in size rather than a consequence of some fundamental difference in the physiological variable. In an effort to address this question, we divide individual values for the physiological variable by corresponding measures for body size, and then display the resulting ratios in a dot-plot (Figure 10.6c). An analysis of variance on the ratios indicates that the means are quite similar, and suggests that physiological responses of animals in the two groups do not differ once the confounding effects of body size have been removed.

As an alternative to using ratios, however, we submit the original data to an analysis of covariance. At the first step in this analysis, we fit straight lines to each of the sets of data and determine that the slopes of the two lines are similar (indeed, they are identical in the present example; Figure 10.6d). The "average" slope of 0.4 is used to adjust values for individuals in group 1 upward to the grand mean for body size (i.e., to 13.75 on the abscissa), while values for animals in group 2 are scaled downward. The adjusted values then are displayed in a dot-plot, and the data are submitted to an analysis of variance. Adjusted values in group 1 clearly are higher than adjusted values in group 2 (Figure 10.6e). Thus, we would conclude from this analysis that the groups differ in the variable of interest when confounding effects of body size have been factored out of the study. This conclusion again contrasts with that based on an examination of ratios.

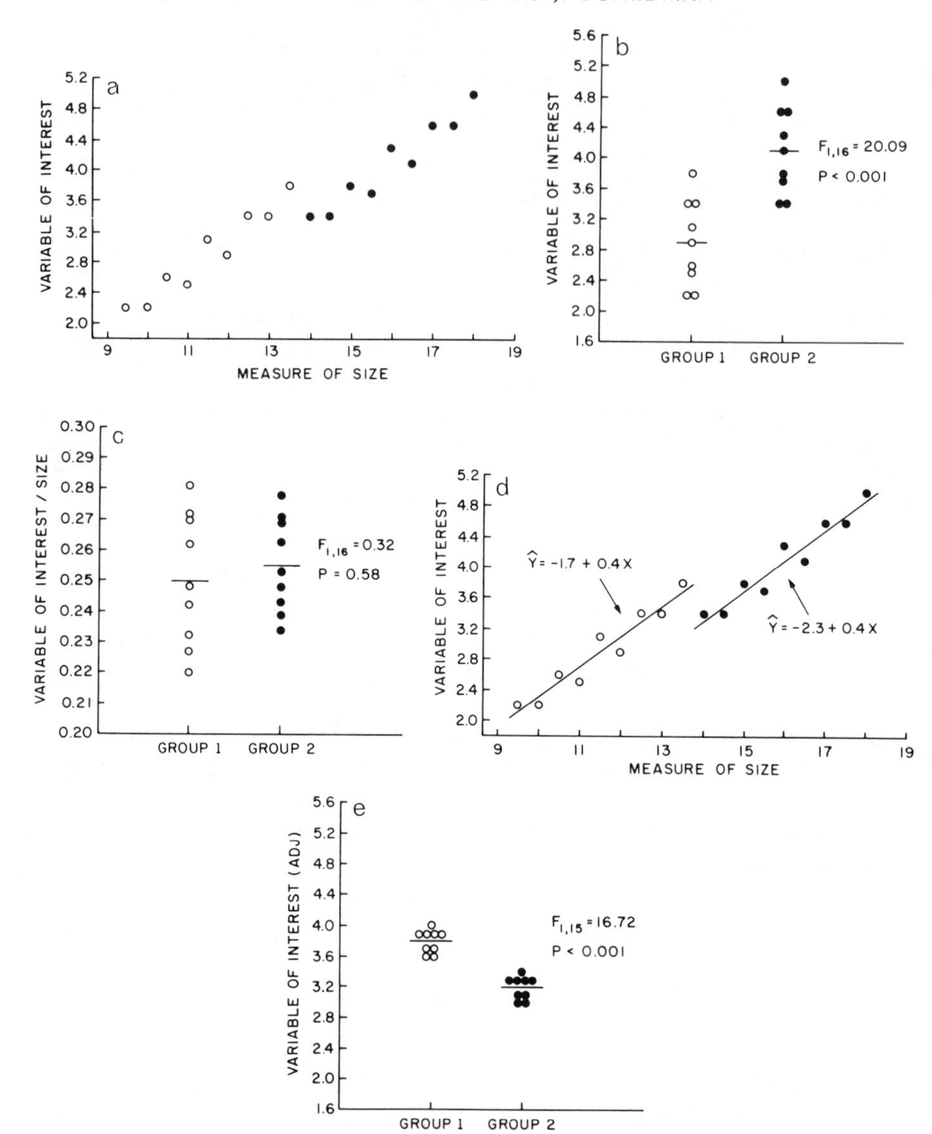

FIGURE 10.6 (a) Bivariate plot of hypothetical data illustrating the importance of proper scaling in removing confounding effects of allometry from descriptive studies comparing groups of animals differing in size. The variable of interest is any physiological process that varies allometrically with body size. Sample size is nine for both group 1 (open circles) and group 2 (closed circles). (b) Values for the variable of interest are displayed here in simple dot-plots. Horizontal lines represent means of 2.9 and 4.1 for groups 1 and 2, respectively. The *F*-ratio is from an analysis of variance comparing means. (c) Ratios were

A potential problem with this application of the analysis of covariance concerns adjusting the variable of interest to a value for body size that falls beyond the limits of size for animals in both groups (Cochran, 1957). The implicit assumption in such adjustments is that the same linear relation between the variable of interest and the measure of body size prevails over the range of sizes between samples as prevails over the ranges within samples. Such extrapolations are not likely to introduce problems when the grand mean for body size is near to the boundaries for the upper or lower limits of size for animals in different groups, as in the present example, but may introduce major problems when extrapolations are to a grand mean that is well beyond the limits of size for all of the groups under study (e.g., in a comparison of oxygen consumption by mice and elephants). In cases such as the latter, the animals are so different that meaningful statistical comparisons of the variable of interest probably are obviated anyway.

Thus, use of ratios to scale data in example 4 led us to conclude incorrectly that the physiological responses of animals in group 1 were similar to those of animals in group 2. In contrast, analysis of covariance led to the proper conclusion that responses of animals in group 1 actually were higher than those of animals in group 2 once the confounding effects of body size had been removed.

Example 5: Removing confounding effects of body size

Our last example is similar to the preceding example in that it also deals with confounding effects stemming from differences in body size among groups being studied. Such a case could arise in a study of seasonal variation in oxygen consumption by some species of small mammal exhibiting an annual cycle in deposition of fat and, therefore, in body mass.

A bivariate plot of hypothetical data again suggests that the physiological variable of interest increases in level or intensity with body size of the ani-

CAPTION TO FIGURE 10.6 *(cont.)* formed by dividing the variable of interest by the measure of body size. Horizontal lines represent means of 0.250 and 0.255 for groups 1 and 2, respectively. The *F*-ratio is from an analysis of variance comparing means. (d) Straight lines were fit to data in (a) by the method of least squares. Note that the lines have identical slopes, but different intercepts. (e) Individual values for the variable of interest were adjusted by the covariance procedure. Horizontal lines represent means of 3.8 and 3.2 for groups 1 and 2, respectively. The *F*-ratio is from an analysis of variance comparing means. One degree of freedom was lost from the residuals in this analysis as a result of estimating the slope of the line used to adjust data to the grand mean for body size.

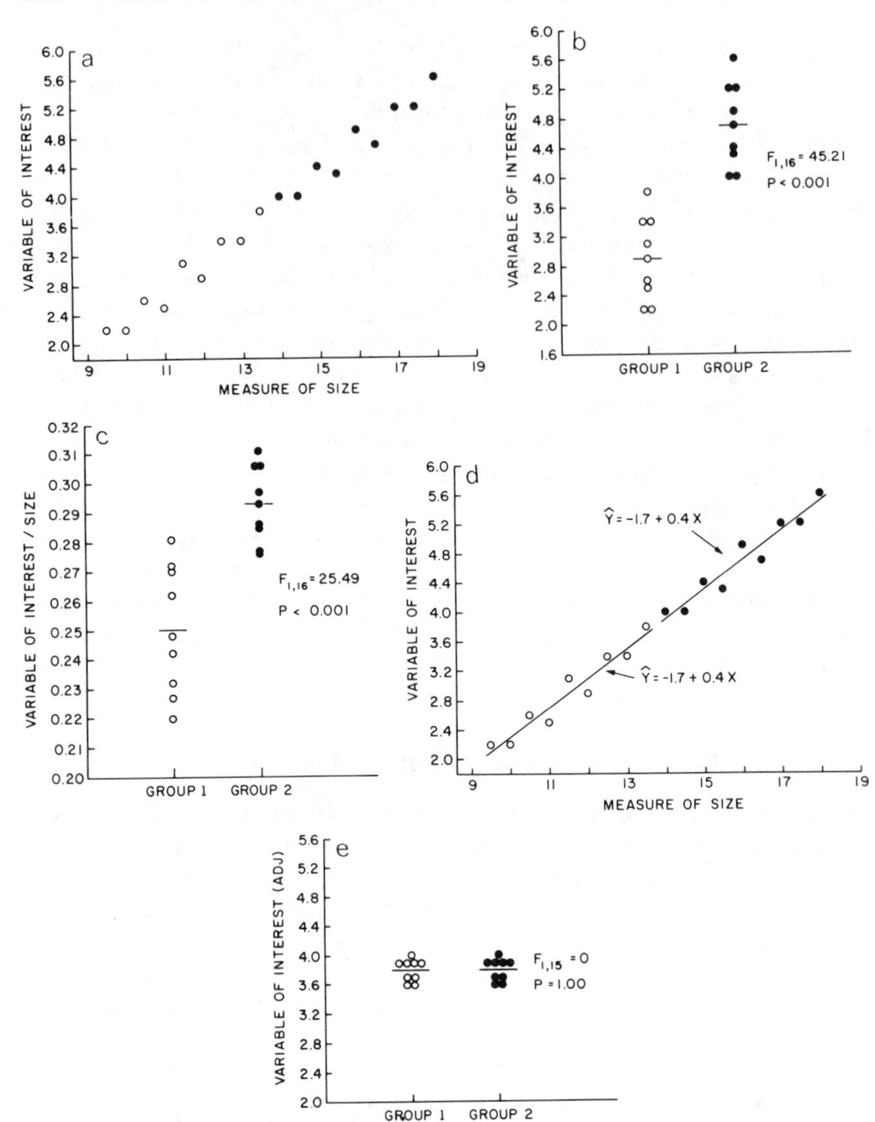

FIGURE 10.7 (a) Bivariate plot for a second set of hypothetical data illustrating the importance of proper scaling in removing confounding effects of allometry from descriptive studies comparing groups of animals differing in size. The variable of interest is any physiological process that varies allometrically with body size. Sample size is nine for both group 1 (open circles) and group 2 (closed circles). (b) Values for the variable of interest are displayed here in simple dot-plots. Horizontal lines represent means of 2.9 and 4.7 for groups 1 and 2, respectively. The *F*-ratio is from an analysis of variance comparing means. (c)

mals under investigation (Figure 10.7a). When the data are displayed in a dot-plot and examined by an analysis of variance, we find that values for animals in group 2 clearly are significantly higher, on the average, than values for animals in group 1 (Figure 10.7b).

What happens, however, when effects of body size are removed? In an attempt to answer this question, we divide each value for the physiological variable by the corresponding value for size, thereby forming ratios, percentages, or size-specific measures of the physiological response. These ratios are then presented in dot-plots and examined statistically (Figure 10.7c). An analysis of variance indicates that values are greater for animals in group 2 than for those in group 1 (Figure 10.7c). We might conclude from this analysis that the groups differ in important ways in the variable of interest, even after effects of body size have been removed.

As an alternative to examining ratios, we again perform an analysis of covariance on the original data. Straight lines are fit to the two sets of data at the first step in this analysis (Figure 10.7d). When we find that slopes of these lines do not differ significantly (they are, in fact, identical in the present example), we use an "average" slope to adjust values for animals in group 1 upward to the grand mean for body size and to adjust data for group 2 downward. The adjusted values are displayed in a dot-plot and are subsequently examined by an analysis of variance (Figure 10.7e). Means for the two groups clearly do not differ once confounding effects of body size have been removed. Furthermore, differences between groups in unadjusted values for the physiological variable stem entirely from the differences between groups in body size. These conclusions again contrast with those based on a consideration of ratios.

Thus, use of ratios to scale physiological data in example 5 led to the erroneous conclusion that responses of animals in group 1 were lower than those of animals in group 2. Use of analysis of covariance, however, led to the proper conclusion that responses of animals in group 1 were identical to those of animals in group 2 once the effects of size had been removed.

CAPTION TO FIGURE 10.7 *(cont.)* Ratios were formed by dividing the variable of interest by the measure of body size. Horizontal lines represent means of 0.250 and 0.293 for groups 1 and 2, respectively. The *F*-ratio is from an analysis of variance comparing means. (d) Straight lines were fit to data in (a) by the method of least squares. Note that the lines have identical slopes and intercepts. (e) Individual values for the variable of interest were adjusted by the covariance procedure. Horizontal lines represent the same mean of 3.8 for individuals in both groups. The *F*-ratio is from an analysis of variance comparing means. One degree of freedom was lost from the residuals in this analysis as a result of estimating the slope of the line used to adjust data to the grand mean for body size.

Conclusions

One criticism that doubtless will be directed at the preceding examples is that the data are not "real" and that the ensuing analyses therefore are flawed. However, we deliberately resorted to artificial data sets to keep the analyses simple and easy to follow. The shortcomings of ratios that we have illustrated in this essay have been shown repeatedly to plague real data sets reported in the literature (e.g., Tanner, 1949; Dinkel et al., 1965; Gonor, 1972; Atchley, 1978; de Vlaming et al., 1982; Blem, 1984; Reist, 1985; Strauss, 1985).

A second criticism is that well-planned experiments do not require "fancy" statistics to reveal patterns and trends, and that we are using a big hammer to crack very small nuts. Implicit in this criticism is the belief that poorly designed experiments are the only ones that require more than a modicum of statistical analysis. We answer this criticism by pointing to example 1, which illustrates an experiment having a good design. Despite the soundness of the design for the experiment, neither an analysis of unadjusted values (Figure 10.2b) nor an examination of ratios (Figure 10.2c) had sufficient power to detect real treatment effects. "Noise" introduced into the data by variation in body size was so great as to overwhelm the "signals" elicited by the experimental manipulations. Analysis of covariance reduced the "noise" and made it possible for us to discriminate between the "signals."

A third possible criticism is that we have overstated the magnitude of the problem surrounding uncritical use of ratios to scale data for variation in body size. However, we examined tables and graphs appearing in *Physiological Zoology* in 1985 (volume 58) and discovered that 42 of 76 papers published that year used some kind of ratio to scale data for variation in size or surface area. Similarly, 26 of 54 physiological papers published in *Journal of Experimental Zoology* (volumes 233–236) used ratios in this way, and 45 of 96 papers appearing in *Journal of Comparative Physiology* [*B*] (volume 155) expressed some data in ratio form. The most common use for ratios was to express oxygen consumption in mass-specific units, but other common uses included adjusting data on ion transport for differences in surface area, expressing ventilation rates in mass-specific units, and "normalizing" different measures for cellular activity to a unit mass of protein in the samples. Certainly not all of these papers misused ratios, but we have no way of determining which ones did and which ones did not. Nevertheless, when approximately 50% of papers appearing in the current literature make use of ratios to scale data for variation in size, it is hard for critics to argue convincingly that we have overstated the magnitude of the problem.

Finally, a fourth possible criticism is that we are advocating use of a statistical manipulation that is not accessible to most investigators. This criticism is simply unfounded. Analysis of covariance is widely available in program packages such as BMDP, SAS, and SPSS, all of which can be readily accessed on several of the more popular microcomputers as well as on mainframe systems. However, the quality of the "canned" routines varies, so users must exercise caution. For example, the analysis of covariance performed by program BMDP1V includes a comparison of regression slopes for the several treatment groups, but this critical information is not provided in the analysis of covariance performed by program BMDP2V from the same series.

It is easy to understand why ecological physiologists use ratios to adjust data that vary allometrically with body size. First, some very competent biometricians encouraged several generations of biologists to use ratios in this way (see Simpson, Roe, and Lewontin, 1960). Second, ratios are easy to compute by hand or with the aid of a calculator, and therefore were of considerable appeal to biologists laboring in the period before the advent of computers. And finally, it simply is hard to imagine that ratios can be misleading, because they are so simple to form and (seemingly) easy to comprehend.

There nonetheless is reason to doubt the utility of ratios. In each of the preceding examples, we first plotted values for the variable of interest against corresponding values for the measure of body size. We then examined the plot visually, and formed preliminary impressions concerning both the scaling of the variable against body size and possible differences between experimental groups. Finally, we analyzed the data statistically, but in three different ways. First, the unadjusted values for the variable of interest were examined by analysis of variance. Next, ratios were formed by dividing the variable of interest by the measure of body size, and the ratios were examined by analysis of variance. Finally, values for the variable of interest were examined by analysis of covariance, where body size was the potential covariate. In every instance, conclusions based on the analysis of covariance (1) were consistent with initial impressions based on visual examination of the bivariate plots and (2) differed from conclusions based on analyses of ratios.

The use of ratios will not lead investigators invariably to draw improper conclusions. Forming ratios is adequate for scaling data in those few instances where the variable of interest varies isometrically with body size (Cochran, 1957; Thorpe, 1976), and where the coefficient of variation for data in the numerator is substantially greater than that for data in the denominator (Anderson and Lydic, 1977). Also, differences between experimental groups in the variable of interest sometimes are so large that detrimental effects of ratios are overwhelmed, and analysis of ratios leads to the same conclusions as does analysis of covariance (Blem, 1984). The critical problem is in deciding when the use of ratios is justified and when it is not. In our experience, most of the steps that must be followed in assessing the propriety of using ratios must be followed also in performing an analysis of covariance. Having gone through these steps, we see little advantage in doing an analysis on ratios – even when the use of ratios is justified – and therefore would continue with the analysis of covariance (see also Albrecht, 1978; Strauss, 1985).

Most of the published studies with which we are familiar do not present plots of unadjusted data against body size, and authors seldom indicate in the text that bivariate plots were examined as a first step in the analysis of their data. Accordingly, readers cannot determine readily whether use of ratios was defensible, or whether use of ratios caused the authors to draw inappropriate conclusions. Unfortunately, ratios are misused so often that readers may be forced to disregard these studies (and their conclusions) altogether. Disregarding published studies that rely on ratios certainly is serious, because considerable time, effort, and expense may have gone into gathering the data. Additionally, objectives of the research may have been important, design of the investigations may have been sound, and the data themselves may have been flawless. What can be done in the future to prevent the waste that results from the uncritical use of ratios? We offer the following recommendations. First, we urge investigators to plot their raw data and to examine them visually, because this process is likely to provide the very best protection against drawing inappropriate conclusions from any data set. Next, we recommend that use of ratios be discouraged in favor of more reliable methods of analysis. We clearly prefer the analysis of covariance in this regard (see also Cox, 1958; Snedecor and Cochran, 1980; Steel and Torrie, 1980), but a few workers recommend analysis by partial correlation because of unresolved problems attending the fitting of regressions in covariance analysis (Sokal and Rohlf, 1981). Finally, we encourage full publication of data (not merely summaries), perhaps as appendices in microprint, so that future generations of biologists can fully evaluate both the data and the conclusions emerging from earlier work.

Acknowledgments

We thank A. F. Bennett, D. Bowden, M. E. Feder, M. LaBarbera, K. Miller, and M. J. Packard for reviewing drafts of this paper, and D. Carlson for preparing the line drawings. Preparation of the manuscript was supported in part by NSF Grant DCB 83–08555.

References

Albrecht, G. H. (1978) Some comments on the use of ratios. *Syst. Zool.* 27:67–71.

Anderson, D. E., and Lydic, R. (1977) On the effect of using ratios in the analysis of variance. *Biobehav. Rev.* 1:225–229.

Atchley, W. R. (1978) Ratios, regression intercepts, and the scaling of data. *Syst. Zool.* 27:78–83.

Atchley, W. R., and Anderson, D. (1978) Ratios and the statistical analysis of biological data. *Syst. Zool.* 27:71–78.
Atchley, W. R., Gaskins, C. T., and Anderson, D. (1976) Statistical properties of ratios. I. Empirical results. *Syst. Zool.* 25:137–148.
Blem, C. R. (1984) Ratios in avian physiology. *Auk* 101:153–155.
Calder, W. A. III. (1984) *Size, Function, and Life History.* Cambridge, Mass.: Harvard University Press.
Cochran, W. G. (1957) Analysis of covariance: its nature and uses. *Biometrics* 13:261–281.
Cox, D. R. (1958) *Planning of Experiments.* New York: Wiley.
de Vlaming, V., Grossman, G., and Chapman, F. (1982) On the use of the gonosomatic index. *Comp. Biochem. Physiol.* 73A:31–39.
Dinkel, C. A., Wilson, L. L., Tuma, H. J., and Minyard, J. A. (1965) Ratios and percents as measures of carcass traits. *J. Anim. Sci.* 24:425–429.
Donhoffer, S. (1986) Body size and metabolic rate: exponent and coefficient of the allometric equation. The role of units. *J. Theor. Biol.* 119:125–137.
Finney, D. J. (1957) Stratification, balance, and covariance. *Biometrics* 13:373–386.
Fisher, R. A. (1932) *Statistical Methods for Research Workers*, 4th ed. Edinburgh: Oliver and Boyd.
Fleagle, J. G. (1985) Size and adaptation in primates. In *Size and Scaling in Primate Biology*, ed. W. L. Jungers, pp. 1–19. New York: Plenum Press.
Gonor, J. J. (1972) Gonad growth in the sea urchin, *Strongylocentrotus purpuratus* (Stimpson) (Echinodermata: Echinoidea) and the assumptions of gonad index methods. *J. Exp. Mar. Biol. Ecol.* 10:89–103.
Gould, S. J. (1966) Allometry and size in ontogeny and phylogeny. *Biol. Rev.* 41:587–640.
Huxley, J. S. (1932) *Problems of Relative Growth.* London: Methuen.
Peters, R. H. (1983) *The Ecological Implications of Body Size.* Cambridge University Press.
Reist, J. D. (1985) An empirical evaluation of several univariate methods that adjust for size variation in morphometric data. *Can. J. Zool.* 63:1429–1439.
Schmidt-Nielsen, K. (1984) *Scaling: Why Is Animal Size So Important?* Cambridge University Press.
Simpson, G. G., Roe, A., and Lewontin, R. C. (1960) *Quantitative Zoology*, rev. ed. New York: Harcourt, Brace.
Smith, R. J. (1984a) Allometric scaling in comparative biology: problems of concept and method. *Am. J. Physiol.* 246:R152–R160.
Smith, R. J. (1984b) Determination of relative size: the "criterion of subtraction" problem in allometry. *J. Theor. Biol.* 108:131–142.
Snedecor, G. W., and Cochran, W. G. (1980) *Statistical Methods*, 7th ed. Ames, Iowa: Iowa State University Press.
Sokal, R. R., and Rohlf, F. J. (1981) *Biometry*, 2nd ed. San Francisco: W. H. Freeman.
Steel, R. G. D., and Torrie, J. H. (1980) *Principles and Procedures of Statistics*, 2nd ed. New York: McGraw-Hill.

Strauss, R. E. (1985) Evolutionary allometry and variation in body form in the South American catfish genus *Corydoras* (Callichthyidae). *Syst. Zool.* 34:381–396.
Tanner, J. M. (1949) Fallacy of per-weight and per-surface area standards, and their relation to spurious correlation. *J. Appl. Physiol.* 2:1–15.
Thompson, D. W. (1942) *On Growth and Form*, new ed. Cambridge University Press.
Thorpe, R. S. (1976) Biometric analysis of geographic variation and racial affinities. *Biol. Rev.* 51:407–452.

Discussion

FEDER: I would like to make two points about the specific regression model you used. Typically, when we take two variables, "*X*," say body mass, and "*Y*," say ventilation rate, we do a regression that implies that we know which variable is the independent one and which is the dependent one. But we may not be certain which is independent and which is dependent. If you use "*X*" as the dependent variable and "*Y*" as the independent variable, you get another relationship. There are statistical techniques, major axis regression and reduced major axis regression, that give you a slope without making any assumptions as to which variable is dependent and which is independent. For reduced major axes, this slope is simply the slope of the "*Y*" upon "*X*" regression, multiplied by the correlation coefficient. Unless you can specify without any doubt which is the independent and which is the dependent variable, the proper model may not be common least squares linear regression but reduced major axis, which can give you a very different result if the correlation coefficient is not high. My second point is that if, as we typically do, we regress log "*Y*" on log "*X*," the regression is then effectively fitted to the median of "*X*" and "*Y*," not the mean. If you want to get back to the mean of "*X*" and "*Y*," you must make specific corrections, which are hardly ever made in the animal ecological physiology literature but in general should be. The "take-home message" is that much of the use of regression in the analysis of ecophysiological data may be questionable.

PACKARD: The problem is identifying the variable of interest, the one that you suspect to be influenced by some other variable. If your interest is in "*Y*," then you regress "*Y*" against "*X*" to correct "*Y*" for variation in "*X*." If your interest is in "*X*," then you plot it against "*Y*" to adjust "*X*" for variation in "*Y*."

HUEY: This issue has been debated a lot in the last few years, particularly in the behavioral literature. However, Joe Felsenstein has pointed out that if you are trying to remove the effects of body size, the appropriate model is linear regression, because it minimizes the variance about the regression line on the vertical axis. Reduced major axis will in fact confound your measurements.

FEDER: Michael LaBarbera [University of Chicago] would assert that the reduced major axis model makes fewer assumptions about the structure of the data. The typical regression model assumes that "X" is fixed and measured without error, which is not always the case. Second, assigning independence and dependence to particular variables assumes that you know which is the cause and which the effect. Third, I think we have to distinguish between the proper removal of a confounding variable, and the analysis of data where we don't really know which is cause and which is effect. In the latter case, reduced major axis would probably be the route of choice.

DAWSON: Do you mean median or geometric mean?

FEDER: I believe that it is the median.

PACKARD: I think the most important thing in working with data is to plot the raw, unadjusted values, and examine them. If then you proceed with an analysis that leads you to conclusions to which you have not been led by your examination of the data, there is something wrong with it. If you do something with reduced major axes and that seems to be a more appropriate conclusion than the one based on the analysis of covariance procedure, then I would recommend that you accept that conclusion.

BENNETT: I would like to move on to the ratio problem. A difficulty that permeates the literature is that of mass-specific values, where metabolic rate is expressed per gram, or whatever. This is a ratio analysis that you consider to be probably inadequate. How do you feel, however, about some newer approaches that involve regression of the variable on body mass, determining the allometric power exponent of the variable, and then dividing the value by the mass raised to that allometric exponent, effectively eliminating body mass – for example, expressing oxygen consumption in milliliters of oxygen consumption per gram to the power 0.75? Do you feel this is a more acceptable technique for dealing with mass dependence?

PACKARD: In general, no, because most people are using an exponent, here 0.75, derived from earlier studies, and that is not valid.

BENNETT: No, no. I refer to using an exponent that is empirically determined from your own data on mass.

PACKARD: If they are doing that, then that would be perfectly reasonable. I do not think most people are using an exponent suggested by their own data. In fact, that is one of the powers of the analysis of covariance, that the adjustments are made on a regression based on those data, not on some prior experiments.

SCHEID: Often the exponent that is derived interspecifically is used to make an intraspecific correction, for example, in the plot of log standard metabolic rate against log body mass in the mammals. That may be completely wrong, because intraspecific mass dependence is normally completely different. You said that if you use your own intraspecific data to generate the regression, you can use this correction. But is that true? Just translate the interspecific vs. intraspecific difficulty into an interindividual vs. intraindividual problem. In other words, you used different individuals of different body size to get your regression and exponent, but is it really correct that any individual animal would behave with that exponent if it grew or gained mass or lost mass? It could be that any individual would have a completely different mass dependence than the group.

PACKARD: I think you have indicated one of the many problems with the mouse-to-elephant curves that have been constructed. In many allometric studies, investigators have automatically performed the logarithmic transformation without first examining the unadjusted values and untransformed data, but the use of logs without prior examination of unaltered data can be very misleading.

FEDER: Regression coefficients on mass are often different, depending on the level of analysis. There are many instances in which, for oxygen consumption and body size, the regression coefficient computed by using the mean for a single adult for each species is greater than that obtained by using individuals of different ages and different sizes within a species. One simply cannot assume that interspecific allometry will have the same pattern as intraspecific allometry.

LINDSTEDT: I will make two comments. The first is that in striving for objectivity, let us not throw out the use of biological intuition. We are still biologists, and have a feeling for how the system works. Second, I think we are asking much too much of these regression equations if we think that we can use them to make predictions of very small differences between organisms. They explain a broader pattern of design among diverse groups of organisms, but once that pattern is described, I think we need to look at the variance with a different tool. I do not think we can use regression equations to make the predictions between animals that are close on a line. There is too much variance.

DAWSON: Going back to Scheid's point, in studying seasonal effects and changes in body composition, I found that there are animals with different masses but similar metabolic rates, and you would actually effect a distortion if you normalized these data with a ratio. You can be so concerned with normalization that sometimes you lose a little biology. Also, if you are really going to get so precise, it is incumbent to know about the body composition.

If you are adding or taking away fat, metabolically that may not have much meaning. Protein-specific rates would be appropriate for some purposes. Determination of nuclear DNA might be used to distinguish between hypertrophy and hyperplasia.

BENNETT: I disagree about the lack of utility of taking mass effects into account within populations or within groups of animals. You have a very insidious problem of false correlations between variables which then appear to have some kind of physiological connection, when each of them is simply correlated with body mass. Consequently it is important to factor out mass. I find allometry a useful analytical tool to look at broad ranges, but I think it is just as useful for getting rid of false positive correlations within a species.

11 Interindividual comparisons: a discussion

DOUGLAS J. FUTUYMA

FUTUYMA: Since I am not a physiological ecologist, let me react as an evolutionary biologist to the presentations in this section. I see physiological ecology as one of several fields, including functional morphology and sociobiology, that constitute the study of adaptation. In these fields, which ask how adaptations have come about, what the nature of adaptation is, and what the ecological and evolutionary implications of adaptations are, two major themes have arisen repeatedly and have arisen also in our discussion here. One is the very great importance of explicit phylogenetic analysis whenever one approaches a comparative study of species. Huey's comments in Chapter 4 summarize this important theme very well. The other familiar theme is the importance and utility of studying variation within species. These two themes represent what seem to me the two major divisions of evolutionary biology as a whole: the study of the actual history of evolution, in which phylogenetic analysis plays an important role, and the study of mechanisms of evolution, in which analysis of variation within populations is central. In both areas, there exist unsolved problems that permeate evolutionary biology, which find a particular manifestation within physiological ecology but can be studied equally well from the morphological, behavioral, or biochemical point of view.

One is the study of constraints on adaptation: it has become realized that studying constraints on adaptation, factors that may block particular paths of evolution from being taken, may be at least as interesting as discovering the existence of adaptation. For example, we all know of the cases of rapid adaptation of a number of species of plants to toxic soils impregnated by heavy metals, both in this country and in England. But only a few species, out of many hundreds in these areas, achieved this adaptation. It may be more difficult, but it is just as interesting to ask why these other species of plants have not undergone similar genetic change. Similarly, a long-standing problem is posed by geographical distribution: why can't a species adapt enough to spread a little further over an ecological or geographical gradient? Here again, what are the constraints on adapting further? In studying constraints, one wants to study the magnitude and nature of genetic variation in the relevant traits, as well as genetic correlations, including allometric correlations.

There must be many physiological issues in which correlations of this kind are relevant. For example, is basal metabolic rate correlated, genetically or in some very tight functional way, with the capacity for activity metabolism, and could activity explain variation among taxa in basal metabolic rate?

A second major question is how phenotypic differences map onto differences in fitness. Do slight phenotypic differences have measurable benefits and costs in fitness? Because population sizes of many organisms are small, natural selection has to be fairly intense to overcome genetic drift. This again is relevant to issues such as the evolution of ecological specialization and ecological breadth of distribution. For example, a common explanation of why an insect may feed on only a few species of plants is that a genotype that specializes on just one class of toxic secondary compounds will be more efficient. In our own work on genetic variation in a species of moth, we have not been able to find evidence of a genetic trade-off in the ability to deal with chemically different kinds of plants. Ron Pulliam recently showed that species of finches with large bills are indeed more efficient than small-billed finches at husking large seeds, but they are no less efficient than small-billed birds in husking small seeds. In others words, there isn't necessarily the kind of trade-off that we assume when accounting for the genesis of morphological diversity. Looking within species for benefits and costs of slight differences in phenotype addresses some fundamental issues in evolution and ecology.

The third major issue is the mechanistic basis of adaptation. For example, to what extent does it suffice to study physiological adaptations at very low levels of biological organization, such as the primary structure of enzymes, and to what extent must we look at more organismal-level features such as changes in morphological structure?

Those are my general reactions, but I would like to raise a few points that seem to be lacking from the discussion so far.

One is the apparent absence within physiological ecology of a priori theory against which to test data. Contrast this, for example, with the neutral theory of gene frequencies in population genetics, or the application of game theory to social behavior. My impression is that in physiological ecology, expectations tend to be based on empirical generalizations from past studies rather than a priori theories of adaptation or engineering design. But I may be wrong.

Second, physiological ecology must have implications outside evolutionary biology per se, especially for ecology, but we have hardly discussed them. There must be imaginative new ways of joining physiological ecology to classic problems such as the distribution and abundance of species, or to the study of interactions among species.

Third, factors that contribute to individual variation other than genetic variations should have interesting implications. It seems to me these are worthy of study themselves. In insects, for example, both maternal and paternal nongenetic effects contribute importantly to variation. And there are classic problems in the study of physiological acclimation, the study of individual plasticity, that are important and have evolutionary and ecological implications that have not been fully explored.

COLLINS: I will start with the idea of adaptation also. In his 1982 book, *The Growth of Biological Thought* [see Chapter 7 for reference], Ernst Mayr distinguished an essentialist way of thinking about a problem from the population level of thinking. The essentialist way tends to minimize variation. Beginning with Darwin, biologists began to appreciate that variation in a character is interesting in itself. Mayr also distinguished the analysis of proximate factors, the study of mechanisms, from the study of ultimate factors, that is, evolutionary issues. Bennett observed that comparative physiologists are principally interested in mechanisms. I suggest also that they usually assume adaptation, perhaps even perfection. I think this reflects a kind of essentialism. On the other hand, if you assume that an organism may not really be perfectly adapted, and then ask how does adaptation come about, you begin to engage in population thinking. Outliers or extreme values may be important. Some traits may simply be dragged along in phylogeny, and animals may not be perfectly adapted.

Population-level thinking requires contact with new areas: phylogenetics, statistics, genetics. Koehn quoted Prosser, who said in 1955 [see Chapter 3 for reference] that physiological ecologists and physiologists ought to make contact with these different areas. Perhaps today there is a new unwillingness to live with some assumptions, especially about adaptation, and there is a renewed desire to understand ultimate factors.

Bennett discussed repeatability in the analysis of individual variability and approaches such as stepwise multiple regression. A host of statistical questions are involved here, a point discussed in detail by Packard. For example, in determining the heritability of organismal or physiological performance, one must worry about statistical design and sample size. But why worry about heritability? Measuring heritability for the sake of just measuring heritability is not a justification. Still, why, in the thirty years since Prosser's remark, has there been so little effort to incorporate genetics into physiological ecology?

Koehn mentioned that physiological ecologists now want to incorporate comparative biochemistry and systematics. Again, we can ask, why? I think ecological physiologists now want to understand the dynamics of adaptation, which is much more complex than we thought in the flush of confidence of the post-Evolutionary Synthesis period of the 1950's and early 1960's. Understanding it in a deep way will require more of a synthetic approach among many disciplines.

Arnold made the point that measuring and analyzing variation will be a difficult task. Again, why use a quantitative genetic approach and not a molecular approach? We should ask, What are the strengths of each, and what will each bring to any particular analysis? I also wonder how good the quantitative genetic theory is. What about the effect of ontogeny, for example, on the stability of the variance-covariance matrix? Can we have a lot of confidence in the predictions?

Packard emphasized proper statistical design and analysis of statistical data. Dawson was concerned about the assumptions of the statistical models used in any particular analysis of variation. Koehn noted that some of the important and interesting physiological differences may be small. But Packard said that unless physiological differences between organisms are large, then not using the proper statistical analysis can lead you to completely wrong results. The message seems to be that if you are going to analyze variability per se, it must be done in an extremely rigorous way from the point of view of statistics.

I will finish by saying that studying variation departs from the essentialist tradition in comparative physiology. Comparative physiologists can be very suspicious of variation, often attributing it to faulty techniques. On the other hand, physiological ecologists are interacting with population biologists who are accustomed to population-level thinking and are often sophisticated in statistical techniques. Physiological ecologists who want to cope with variation may find themselves serving two masters with different expectations: comparative physiologists and population biologists.

BENNETT: I would like to respond to Collins's query about the failure of the community to respond to Prosser's call for interactions with morphology, systematics, etc. We have to remember that in the mid-1950's, the people who were instrumental in founding this field – George Bartholomew, Bill Dawson, Knut Schmidt-Nielsen – were looking at the physiology of camels and kangaroo rats and marine iguanas for the very first time. They were only then describing basic adaptations to the environment. This was the cradle time of our field – not a time to begin building bridges, for we had no place to stand to build from. Now we have filled a lot of those different areas. Now it is time to look outward, largely because we have been so successful in attaining the major goals of our own field. Second, like progress in many other areas, we were limited by the techniques available. Just as beautiful color in painting in the latter part of the nineteenth century had to follow the synthesis of aniline dyes, so our field is dependent upon technical progress. We have talked about quantitative genetics a lot here. Much of the theory was developed in the very late 1970's and 1980's. We have picked up on this pretty quickly. The movement of statistical techniques and molecular techniques into more sophisticated areas is just now beginning to influence us.

BARTHOLOMEW: I wanted to comment on the same thing. In 1955, when Prosser said that, everybody knew it, but there was just nothing that anyone could do about it. Just as here and now, I know that what I need is robust statistics for handling variations, but they are really not available, not in terms that are convenient for the kind of data that I have access to.

KOEHN: I share the two opinions, but I do think there are other factors. One is a genuine hesitancy to step outside the context of familiar problems and technologies, and to try to assimilate new approaches. The reason why Powers and I are here is that we came to a problem in genetics that could be solved only by approaches that incorporated the techniques and orientation of physiological ecology. This was blatantly obvious to most people in the center of the neutralist vs. selectionist storm in the mid-1970's, but most people turned to other problems in evolutionary biology within the context of their field rather than moving into biochemistry and/or physiology. There should be ways to make such transitions as easy as possible.

ARNOLD: Another impediment to synthesis between fields is the risk that you are not going to find anything. Some functional morphologists are very skeptical that these minor variations within populations have any evolutionary significance. In fact, they are skeptical that the differences between species have any significance. The risk in doing this work is that it is not obvious that you are going to find any effects on fitness unless you have huge samples. Collins raised the issue of whether finding heritability in physiological traits is a sufficient justification for research. At this stage it is. Quantitative genetics was not addressed by and was not assimilated into the modern synthesis. You won't find the word "heritability" discussed by Mayr, Simpson, Dobzhansky, Schmalhausen, Rensch, or Huxley. The concepts have only now received renewed attention. It is a risky endeavor. We do not know whether the minor variations are going to be demonstrably heritable in natural populations. For reptiles, I know of *one* published study that shows an approximation to heritability in a natural population. We all believe that the traits that we work on are heritable, but we have a data base that is almost completely zero. I think it would not be good policy to demand elaborate theoretical justifications for measuring genetic variation in natural populations when we do not know anything. We are not talking about an encyclopedia here; we are talking about a tabula rasa.

DAWSON: A criticism of theoretical approaches and predictions is circularity, because when you are talking about the impact of adaptation on fitness, you have to define fitness operationally. You have that problem in many cases, both physiological and otherwise. Given the almost infinite matrix you can work yourself into with a given physiological character and a set of environmental circumstances, operationally how are you going to deal with this? Is there a circularity here or not?

FUTUYMA: This notion that fitness and natural selection are mere tautology has been voiced again and again, but in principle the issue has been dealt with. Elliot Sober, in *The Nature of Selection* [1985; Cambridge, Mass.: MIT Press], deals with this very well. I think it's just a red herring. Now, how you actually demonstrate that a feature is an adaptation, or that variant forms of a species differ in fitness, can be a difficult question in practice. It may involve looking at survivorship and reproductive output of different phenotypes or genotypes, as Koehn does.

COLLINS: I agree with Futuyma. The important point is how you do the analysis. If the organism or part of it is really not perfectly adapted, how would you, after all, decide that it is not? In the final analysis, what leads you to be confident that a trait is an adaptation? Physiological ecologists are especially suited for analyzing adaptation because they understand physiological processes and their relationship to abiotic factors. Maybe they can make strong predictions a priori and then see if the trait really is functioning in that particular way.

FEDER: I want to skip to the absence of null models in physiological ecology. Especially when we are dealing with adaptations at the species and population level, the general analytical paradigm is to assume that a particular feature is an adaptation, and to ascertain in which of several alternative ways it actually works. The more basic question is, Is it adaptive in the first place? When you analyze data at the level of individuals within populations, you can cope with this question, whereas when you restrict your analysis to population, species, or higher levels, you cannot. So I think the problem of tautology is a real one, but in order to address it, one has to move to another level, to another kind of analysis.

ARNOLD: I think that by analyzing minor variations in physiology within populations, and seeking correlations with fitness, we get a particularly direct approach to the issue of adaptation. But I do not think that when we are dealing with the means of species and higher taxa, the analysis of adaptation is closed to us. John Endler, in his new book *Natural Selection in the Wild* [reference in Chapter 9], discusses several approaches. One is looking for correspondence between, say, physiological traits and environmental variables. To the extent that we see such correlations, there is a strong presumption of adaptation. Another approach is to look for parallel clines in species that are straddling the same environmental gradients. Again, one comes away with a strong presumption of adaptation, if not a direct demonstration. So there are tools.

POWERS: I am not so sure that there isn't a null model. In the controversy between the neutralists versus the selectionists, the neutralists' assumption, that all of this genetic variation is of little if any significance in terms of natural selection, is a null model. When you see variations in a trait, you can at least reject the null that there is no physical or biochemical difference associated with the trait. Whether or not it has significance in terms of fitness is another series of null models.

FUTUYMA: I agree in part with Feder's comments, and in part not. I agree that you should not a priori assume that every feature of an organism is adaptive. On the other hand, as Maynard Smith has pointed out, there are some features for which, even if you have no idea what the adaptive significance is, you just know that there has got to be one – in particular, for very complex organs. He points to the ampullae of Lorenzini of elasmobranchs as something whose function was mysterious for decades. It was obvious that they had a function. I think that there is such a thing as biological judgment or intuition. The other point is that, while study of variation within species can be a particularly powerful method of showing that some phenotypic character does have adaptive significance, first of all, that is not the only way to do it, as Arnold said; and second, I am not sure that it will always give you the right answer. This relates to the problems of scale, of the magnitude of the phenotypic difference. To me it is evident that beak differences between birds as different as a swift and a hummingbird have adaptive significance. On the other hand, it is much less obvious that variations in the beak length of a ruby-throated hummingbird are going to show adaptive differences. They may or they may not. There may be some threshold of difference beyond which you find adaptive significance, but below which you do not. I think that there remains a very important role for interspecific comparisons, tempered, however, by the kinds of considerations that Huey raised.

POWERS: I would like to go even further. Many people, myself included, say that these physiological, genetic, and biochemical differences are great, and that maybe they are involved in selection and so forth. But one could say that this is simply fine tuning, and in the grand scheme, the evolution of species and taxa, it has little if any significance.

COLLINS: That is a good point. An important issue in evolutionary theory is the evolution of very different forms through nonmicroevolutionary processes. If the field narrows in on a finer and finer understanding of microevolution and microadaptation, it may be getting very good at that, but the major questions in evolutionary theory may concern other phenomena.

BENNETT: At the risk of stating the obvious, one of the important things that has come from animal physiological ecology is the importance and the centrality of the individual organism and the organismal level of function. As we move into evolutionary and genetic studies, and into more biochemical and mechanistic studies, it is very important that we not lose sight of organismal function and that these studies be done with reference to organismal function. We should not get lost in the enzymes or in the genes, and assume that the cascade with which we are working will completely operate, without verifying it at the whole-animal level. By way of a cautionary tale, several of us looked at two species of lizards in South Africa. They varied considerably in their foraging strategies and in their capacities for activity, for sprint speeds, and for endurance, but there was no hint of those differences in the muscle tissue of the animals in any of the factors that we looked at. You could have taken muscles from one species and stuck them on another. If somebody had simply assumed that locomotion is muscle dependent and had worked on the genetics or the molecular biology of muscle tissues, they would have come to totally wrong conclusions about organismal function. I think that it is very important not to lose sight of where we came from and always do things in reference to organisms.

FUTUYMA: Let me conclude. My own reaction to much of this discussion is that most of it has dealt with conceptual questions, methodological difficulties, and analytical methods that are similar in kind to those dealt with by population biologists. These are the same kinds of issues discussed by people who study life histories, behavior, or morphological variation; what we have experienced today is an extension of these issues to physiological traits. This suggests that interactions between physiological ecologists and population biologists could be profitable. Population biologists deal with many of these issues on an everyday basis. Most of them, however, are not well versed in physiology, and certainly are not accomplished in physiological methodology. Yet the characters they study often have important physiological underpinnings – growth rates, for example – or have physiological functions (e.g., thermoregulatory effects of pigmentation). So population biologists often stop short of physiological analysis when it would be useful, and on the other hand have conceptual and analytical tools that physiological ecologists should find useful as they extend their work in new directions.

PART THREE

Interacting physiological systems

Part Three considers physiology *at the system level within the individual.* Nevertheless, the book has not abandoned a "holistic," strongly ecological perspective. Indeed, as will become evident, the interactions of physiological systems within individuals often can be understood only when the constraints and demands of the individual's environment are also taken into account.

Chapter 12 evaluates the traditional dichotomy between "noninvasive" and "invasive" physiological methodologies in physiological ecology. While noting that each approach has advantages and disadvantages, Warren W. Burggren argues that an integration of current approaches, aided by collaboration and by new technological advances, will facilitate future advances.

In Chapter 13, Peter Scheid advocates the use of physiological modeling in ecological physiology. He urges that models be carefully constructed, so that they are sufficiently simple to be usable yet do not compromise general validity by sacrificing complexity. Chapter 14 explores the concept of symmorphosis, or optimal design, as it applies to the physiology of organ systems. Using examples primarily from gas exchange, Stan L. Lindstedt and James H. Jones discuss whether organ systems adaptively vary so as to maintain effective matching of structure and function.

In Chapter 15, Donald C. Jackson examines a common situation in physiological ecology where the "ideal" physiological responses to complex environmental conditions may require conflicting adjustments in organ systems. How functional priorities are established to resolve these physiological conflicts is discussed in the context of control systems and organismal physiology.

The importance of ontogeny in physiological ecology is the subject of James Metcalfe and Michael K. Stock in Chapter 16. Critical ontogenetic changes in physiological performance are discussed using both mammalian and avian examples. Ontogenetic processes can have a large impact on studies in physiological ecology.

Part Three is concluded by a general discussion led by David J. Randall and Steven H. Bishop, and edited here by Randall. A central theme emerging from this discussion is of the increasing importance that physiological experimentation will have in the previously more "ecologically oriented" areas of evolution and population biology.

12 Invasive and noninvasive methodologies in ecological physiology: a plea for integration

WARREN W. BURGGREN

The problem

A recurrent theme throughout this volume has been "integration," whether in the context of experimental approaches or the data and concepts these approaches provide. This chapter discusses the merits of integrating the currently divergent methodologies within the field of physiological ecology. Many investigators in this field have long recognized a need for such integration, and indeed have achieved it in their own experimental approaches. As a group, however, physiological ecologists often follow established methodological pathways that support, rather than challenge, prevailing hypotheses.

I wish to pursue three specific objectives. First, I will illuminate and categorize a widely recognized (but vaguely defined) dichotomy in the experimental approaches and methodologies of physiological ecology. Specifically, I refer to the "invasive" laboratory-oriented approach, dependent upon experimental manipulation of both internal and external environments of the animal, and the "noninvasive," field-oriented approach, which emphasizes observations of minimally disturbed animals in natural or simulated natural surroundings. Perhaps less easy to achieve will be the second goal, which is to persuade those readers not already in agreement that this methodological dichotomy is, in fact, largely historical, and lacks foundation and justification in theory. The third goal is to suggest ways in which integration of these divergent methodologies might be most rapidly and effectively achieved, for the benefit of the discipline of physiological ecology.

Essays for diagnosis or prognosis of a particular biological field often emerge as a largely subjective account reflecting one's own perceptions, biases, and experiences (cf. Prosser, 1975; Waterman, 1975; Ross, 1981; Greenberg, 1985; Randall, in press), and this particular essay is no exception. My training in system-level organ physiology may allow me to comment on that particular field, but I certainly claim no particular insights into more ecological aspects of physiological ecology. In attempting to describe the essence of methodologies with which I am not especially familiar, I hope that at least my misperceptions might in themselves be informative when regarded as symptomatic of the unfortunate dichotomy in this field.

The two traditional methodologies: definitions

Traditionally, comparative physiological investigations have tended to use one of two major approaches or methodologies. Since each methodology to some extent has its own "rules," experimental designs, jargon, and even preferred statistical analyses, it is hardly surprising (but still unfortunate) that the two methodologies have been wielded by distinct groups of investigators who have failed, by and large, to enter into productive collaborative studies.

First, let me clearly, albeit subjectively, define these approaches (Table 12.1).

Noninvasive approaches

The noninvasive approach is probably the most familiar to more ecologically oriented physiologists. The goal of advocates of this approach is ultimately to understand how the overall evolution of a species is influenced by its ability (or inability) to adapt physiologically to its environment (Table 12.1).

The noninvasive approach typically involves the investigation of animals behaving relatively normally in their natural environment. Less frequently, but with increasing occurrence, simulated environments of varying degrees of realism are used in laboratory settings (Figure 12.1). By observing intact, essentially undisturbed animals in as natural an environment as possible, it is assumed that the resulting data will suffer minimally from experimental perturbation [Figure 12.2 (Top)]. Of course, as discussed later, such an environmentally rigorous experimental approach also places significant restrictions on the extent as well as the type of information that can be gathered.

Typically in noninvasive studies, a variety of important and diverse environmental variables (e.g., temperature, photoperiod, population density, food availability) are carefully monitored in detail on site. Other factors, such as an animal's past thermal history and phenotype, are carefully controlled, if not actually serving as experimental variables.

Noninvasive physiological ecology historically has included the study of (1) *exchange* of respiratory gases, water, ions, and heat between the animal and its surroundings; and (2) *energetics* of predation, foraging, assimilation, reproductive effort, and locomotion. Such studies, the methodology and conceptual approach of which have already been discussed in some detail in previous chapters, have been and will continue to be a cornerstone, if not the foundation, of physiological ecology.

In theory, the study of exchange and energetics is rarely regarded as of intrinsically greater merit than the study of other aspects of ecological physiology. In practice, however, it is a focal point for pragmatic reasons, since

TABLE 12.1 Description of the traditional methodologies in physiological ecology

	Noninvasive	Invasive
Proponents	Ecologically oriented biologists with strong background in behavior, systematics, and evolutionary biology	Physiologically oriented biologists with strong background in biochemistry, physics, and engineering
Goals	Understanding evolutionary processes and events related to environmental adaptation	Understanding evolutionary processes and events related to environmental adaptation
Scope of study	Intraspecific as well as interspecific differences	Primarily interspecific differences
Setting	Field; occasional use of simulated natural environment in laboratory	Laboratory; occasional use of field sites or simulated natural environment in laboratory
Methods	Observations of intact, undisturbed, unrestrained animals exhibiting normal behavior; environmental variables monitored but not necessarily regulated	Observations from animals instrumented and often confined; "recovery" usually allowed after surgery in chronic experiments; experimental environment rigorously controlled
Typical topics	Exchange between animal and its environment, and associated energetics	Exchange and transport within the animal; homeostatic mechanisms; organ systems engineering

Note: See text for detailed description.

many aspects of exchange and energetics can be investigated in considerable depth without necessitating surgical intervention or other major disturbance of the animal's body or normal behavior.

Finally, although understanding evolutionary pathways and events is the goal of both noninvasive and invasive physiological ecology (see below and Table 12.1), the investigators wielding noninvasive approaches appear by and large to have considerably more extensive training in evolutionary biology. Consequently, noninvasive studies much more frequently include consideration of phenotypic/genotypic differences within populations, as well as interpopulation differences within species (see Chapters 4, 5, and 7).

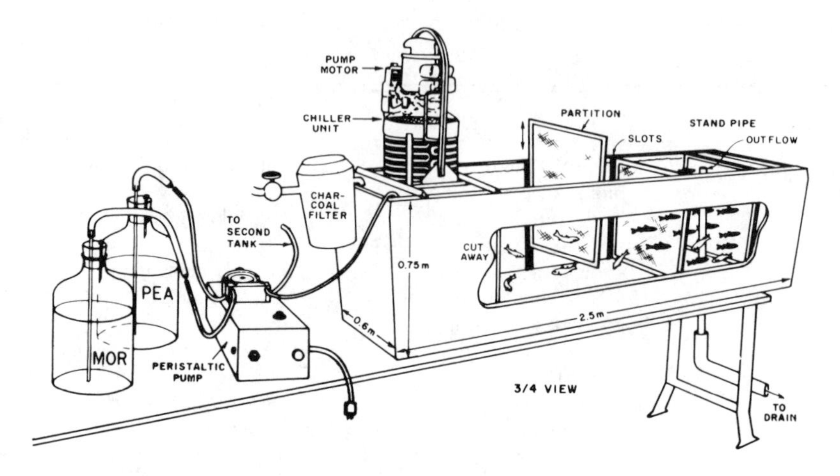

FIGURE 12.1 A laboratory stream habitat constructed for investigating the role of olfactory cues in homing salmon. Chemical agents suspected to have a role in imprinting can be released into the simulated stream environment and the resulting changes in behavior of undisturbed fish recorded. (From Hasler and Scholz 1983.)

Invasive approaches

The invasive experimental approach, which some of us who use it have affectionately (if somewhat self-critically) termed "slash-and-burn physiology" (Juan Markin, unpublished), has also enjoyed a long and successful history. Indeed, the thriving disciplines of comparative physiology and comparative biochemistry were founded upon invasive methodologies (Ross, 1981; Greenberg, 1985).

Like those who use noninvasive approaches, advocates of the invasive methodology are striving to understand physiological adaptation to the environment (Table 12.1). However, users of invasive approaches to problems in physiological ecology bring quite different skills, experimental designs, and analytical approaches to the field.

By its very nature, invasive physiology imposes experimental constraints and demands upon the animals being investigated [Figure 12.2 (Bottom)]. The "invasion" is not only of the animal's internal environment, but also of the behaviors normally at an animal's disposal. Some degree of physical confinement or restraint, if not an actual surgical procedure for sampling or continuous measurement, commonly may be employed at some point in the experiment. Experiments may be "acute" (i.e., short-term experiments performed on anesthetized animals), or they may be "chronic" (i.e., involving longer-term monitoring following recovery from surgical or other experimental manipulation). Typically, physiological data may be accumulated via implanted instrumentation or devices (e.g., muscle potential electrodes for electromyography, bladder intubation for urine collection, transducers for blood flow measurement, etc.). In what might be considered the ultimate in removal from natural environments, in vitro experiments might be performed on excised organs or tissues. Clearly, there is a continuum from invasive physiology to comparative biochemistry.

Because of the heavy dependence upon complex instrumentation, invasive experiments have almost always been restricted to a laboratory rather than a field setting. One consequence of this is that, unlike the natural experimental setting common for non-invasive studies, the laboratory "environment" the animal experiences can be rigorously controlled. This control has become the hallmark of invasive studies (Table 12.1). Typically, all but one (or at most a few) environmental variables are tightly regulated, while the variables of interest are carefully varied in some sort of step- or ramp-function fashion. (In spite of a purported ability to manipulate many aspects of the laboratory environment, multivariant experiments are not commonly used, in spite of the extensive information they potentially can yield.)

Invasive methodology is by its very nature designed to yield information on specific mechanisms that ultimately underlie whole-animal responses to short- or long-term environmental challenges. Transport within the animal, as a natural inward extension of exchange between animal and environment, has been a major subject for investigation. Also, most of what we know of homeostasis and its constituent processes and elements (reflexes, receptors, effectors, feedback, etc.) has been gleaned from invasive rather than noninvasive physiological studies (see Chapters 13 through 16).

As indicated previously, although understanding evolution of physiological processes and adaptation is the ultimate goal of advocates of invasive physiological approaches, only rarely does consideration of differences in phenotype, genotype, or whole populations receive consideration in experimental design (Table 12.1). Indeed, more than ten years ago, Prosser (1975) identified the physiological bases for speciation as a neglected area in comparative physiology that was ripe for investigation.

The two traditional methodologies: advantages and limitations

Invasive and noninvasive methodologies each have many compelling features. In addition, they have their limitations, not to mention outright pitfalls, as will now be explored. Some of these aspects are summarized and compared in Table 12.2 (see p. 258).

Noninvasive approaches

One of the major strengths of the noninvasive approach lies in the fact that data are acquired from intact, essentially undisturbed animals that have not

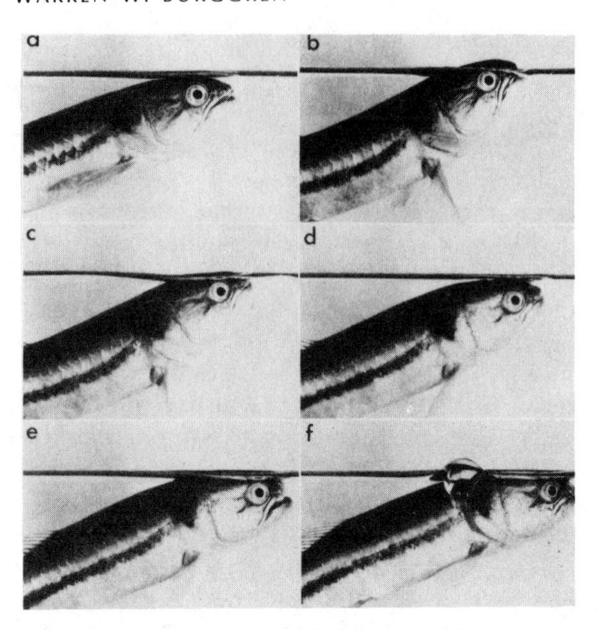
a
b
c
d
e
f

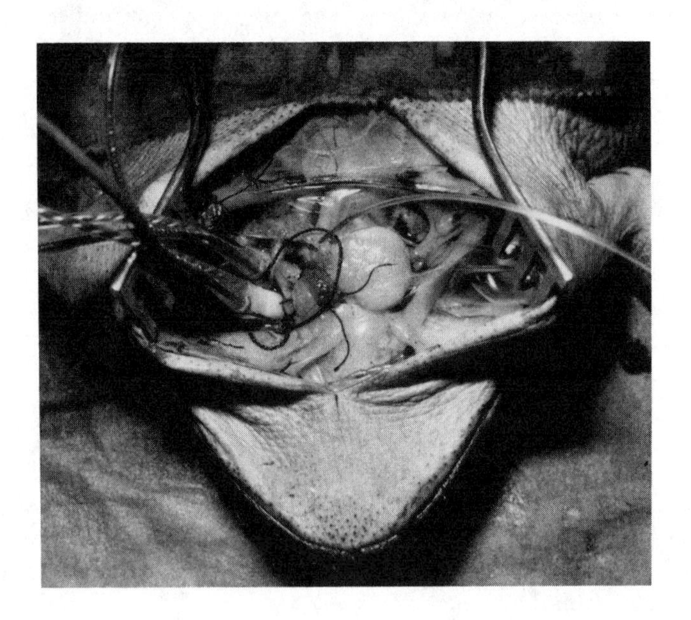

been surgically or otherwise compromised [Figure 12.2 (Top)]. Very importantly, in response to environmental stressors animals in most instances are free to utilize complex behaviors (which in many situations are the first and sufficient responses) as well as physiological responses. Consequently, differences between individuals within an experimental population are less likely to result from confounding effects of restraint, surgical intervention, or disruption of normal behavior. Differences between undisturbed animals can be more convincingly examined in the context of phenotypic variability, the considerable significance of which has already been explored in Chapters 5 and 7.

Another appealing aspect of noninvasive physiology is the general tendency for crucial ecologically related factors (photoperiod, season, thermal history, genotype, reproductive state, to name but a few) to serve as the actual experimental variables. There is no feature or quality of these particular variables that should necessarily lend itself to the noninvasive experimental approach. (My own opinion, clearly evident in Table 12.1, is that noninvasive methodology is most often utilized by ecologically oriented physiologists, whose training may lead them to be more aware of such environmental factors.) Although the variables mentioned above are not necessarily amenable to experimental manipulation, often they can at least be carefully measured in the experimental setting. Consequently, many noninvasive experiments have very successfully exploited powerful multivariate design and statistical analyses depending on correlative observations.

Yet another strength of noninvasive approaches is the "apparent" ease of the actual data collection *relative* to more invasive experiments. My intention is not to draw the ire of physiological ecologists who have, for example, spent seemingly endless weeks in desert blinds cataloging lizard basking behavior, or suffered frostbite while measuring environmental variables appropriate to winter hibernation in ground squirrels. To clarify, it seems that many noninvasive studies frequently are extremely complex in design but technically "simple" (though potentially very time-consuming) in execution when compared with invasive studies. Contrast, for example, invasive studies involving some types of in situ neurophysiological recording, where an experimental success rate of 1 in 2, 5, or even 10 preparations may be

FIGURE 12.2 Examples of traditional physiological experimental approaches in physiological ecology. (Top) The use of noninvasive techniques in a study of the mechanics of air breathing in fishes is demonstrated by sequential photographs of an unrestrained, undisturbed fish (*Hoplerythrinus unitaeniatus*) during a single air breath. (From Kramer, 1978.) (Bottom) The use of invasive techniques is exemplified by an anesthetized, restrained toad (*Bufo marinus*) with implanted instrumentation for measurement of blood flows and pressures. (Photograph courtesy of Dr. N. H. West.) Each experimental approach provides important data, but each approach also has its limitations (see text and Table 12.2).

TABLE 12.2 Some advantages and disadvantages of the two traditional methodologies in physiological ecology

Noninvasive		Invasive	
Advantages	Disadvantages	Advantages	Disadvantages
Intact, undisturbed animals with normal behaviors are used.	Animal usually remains a "black box."	Environmental variables are closely controlled. Thus, cause and effect are easily established.	Crucial ecological and environmental factors are often ignored.
Crucial ecological and environmental factors serve as experimental variables.	Mechanisms often remain undetermined or misinterpreted.	Morphological and physiological limitations manifested in field observations are defined.	Typically a small number of observations are made.
Typically a large number of observations are made.	Field observations (particularly negative data) may remain enigmatic upon close scrutiny.	Mechanisms behind whole-animal responses are revealed.	Relevance to field observations must be proven on case-by-case basis.
Populations as well as individuals are studied.	Relations of observed responses to environmental factors often are correlative, rather than causative.		Studies of populations, rather than individuals, are rare.
			Responses are likely to be altered by invasive experimental methods.

Note: see text for detailed comparisons.

considered an achievement. Whatever the underlying reasons involving data collection, noninvasive studies generally are based on larger numbers of observations than are typical of invasive studies. Acquiring higher *n* values not only adds statistical confidence to the conclusions, but also means that the scope of studies is more easily expanded to include differences between individuals from different populations.

Although the noninvasive physiological approach has enjoyed and will continue to enjoy great success in physiological ecology, it also has substantial and critical limitations (Table 12.2). Perhaps the most serious criticism is that in most instances the animal itself is treated as a "black box." Whereas the dynamics of interaction of the animal with its environment may be extensively categorized, the underlying internal physiological mechanisms supporting these interactions remain frustratingly beyond reach in most instances. For example, observations made in the 1960's that lizards heated faster than they cooled, and that dead lizards cooled more rapidly than living ones, clearly demonstrated that these reptiles could somehow regulate heat transfer between body and environment (see Bartholomew and Tucker, 1963). While the adaptive advantage of such a thermoregulatory capability was clear, the physiological adjustments and attendant costs and compromises remained uncertain until the actual mechanisms for physiological thermoregulation were thoroughly investigated by means of invasive methodologies, that is, by measurement of blood flow coupled with patterns of internal heat distribution (Heath, 1964; Weathers, 1970; Weathers and White, 1971). Clearly, interpretation and evaluation of environmental adaptations are most valid when both the *mechanism* and the ultimate *response* of an animal are equally well understood.

A second disadvantage of the noninvasive approach arises from the fact that interpretation of field observations in which an expected response or behavior fails to develop (i.e., a "negative finding") often remains largely enigmatic. Consider, for example, a hypothetical, noninvasive study centered on diving birds. Detailed field observations are made of two species of ducks, in both cases involving undisturbed, voluntarily diving birds intermittently foraging on the bottom of a pond. The collection of data includes timing the breath-holding intervals during many successive dives. One species of duck shows a mean dive interval of 180 seconds, whereas another phylogenetically closely related species "fails" to show long dives, typically remaining submerged for only thirty seconds. Short of actually measuring the physiological responses of each species during diving experiments in the laboratory (i.e., adopting an invasive approach), the observer is simply unable to conclude whether the species diving for briefer periods ultimately is physiological unequipped for longer dives (e.g., low oxygen stores, limited capacity for regional hypometabolism), or whether instead there are complex energetic (or behavioral) constraints tending to limit diving times. Knowing the underlying physiological mechanisms involved in regulating diving physiology – that is, opening the "black box" through invasive methodologies – is essential to an overall understanding of avian adaptations for diving.

Invasive Approaches

Like noninvasive approaches, invasive methodologies have their advantages and their limitations. Foremost among the advantages is the ability to discriminate the mechanisms that underlie responses observed in either laboratory or field settings. In the example of diving physiology referred to above, laboratory experiments characterized by invasive surgical procedures have allowed measurement of cardiovascular parameters, blood gases, and metabolite concentrations. Carefully controlled experiments using both voluntary and forced diving (the latter completely outside the domain of noninvasive methodologies) have revealed that in most diving animals the length of a voluntary dive generally is dictated by the available oxygen stores (Elsner and Gooden, 1983). Although many diving animals can greatly prolong dive times by switching to anaerobic metabolism after oxygen stores are depleted, aerobic dives are the norm, presumably because of the considerably greater metabolic efficiency of aerobic metabolism as compared to glycolysis (Hochachka, 1980). Hypotheses explaining the field observations of differing dive times between duck species in our hypothetical study could thus be tested by laboratory investigations of blood and tissue oxygen stores and/or mechanisms for oxygen conservation.

A second major advantage of invasive experimental techniques lies in the degree of experimental control of the environment (Table 12.2). In most natural habitats, environmental variables are constantly changing in a gradual, cyclic fashion. Most often there are multiple cycles of independent variables that may be either changing in phase or varying independently. This extremely complex situation contrasts markedly with typical laboratory experiments utilizing invasive techniques. In laboratory studies, all but one or a few environmental factors are maintained constant, while the variable(s) of interest is (are) manipulated according to some predefined, rigorous protocol. In essence, invasive approaches make a worthy sacrifice of environmental realism for the great benefit of analytical precision. Consider, for example, possible approaches for assessing environmental cues that stimulate the winter onset of hibernation in ground squirrels. Of course, one could take a noninvasive approach by attempting to quantify extensively the field conditions evident at the onset of hibernation. Even with the most sophisticated of measurements and analytical procedures, however, the capricious nature of natural habitats would probably dictate that many seasons of observations be required to establish nonequivocally cause and effect between possible environmental signals and the onset of hibernation. Because in an invasive approach the experimenter can *independently* regulate the environmental variables (e.g., temperature, photoperiod, food and water availability, etc.), then a first series of experiments involving different populations experiencing different laboratory conditions would more quickly and efficiently indicate the relative importance of the various environmental variables in inducing hibernation, especially when combined with measurements of oxygen consumption, brown fat metabolism, blood chemistry, etc.

Needless to say, the invasive approach also has its disadvantages (Table 12.2). A common feeling among many ecologists and field-oriented physiological ecologists is that laboratory measurements are made on animals that, at the very least, are denied normal behavior as an alternative mechanism for adjusting to environmental perturbations [Figure 12.2 (Bottom)]. At the worst, it is felt that physiological measurements are frequently made on experimental subjects that are physiologically compromised, if not actually moribund. Unfortunately, these criticisms are not easily refuted. Certainly, in any situation where an animal is confined, restrained, or anesthetized, the potentially important role of normal behavior has been obviated.

Examples abound which indicate that observations made under conditions of restraint may or may not apply to animals in the field. Several decades ago, the renowned physiologists Per Scholander and Laurence Irving documented the widespread occurrence of "diving bradycardia," a profound reduction in heart rate during water submersion. Even though Scholander (1940) observed that bradycardia sometimes failed to develop in freely diving seals, unlike restrained seals (which almost always showed the response), Scholander never clearly separated the effects of restraint, fright, and training from diving per se. In recent years, following the advent of sophisticated telemetry techniques (see below), it has been shown for a wide variety of animals that bradycardia during natural, voluntary diving is often muted as compared to during forced dives, if a bradycardia even develops at all (see Smith, Allison, and Crowder, 1974; Butler, 1982; Elsner and Gooden, 1983; Gallivan, Kanwisher, and Best, 1986).

If surgery is performed and postsurgical measurements are made, there is the additional question of whether the animal has actually recovered from the effects of anesthesia and surgery in the postoperative period. Often, advocates of the invasive approach simply report that they waited (arbitrarily?) for twenty-four hours postoperatively to begin measurements. Rather than collecting preliminary data to establish that twenty-four hours is the actual point of physiological stabilization after surgery, this point in time all too often is chosen on the basis of some ill-founded "industry standard." I suspect that twenty-four-hour measurements have become pervasive for the historical reason that, in the past, animals might not have survived for forty-eight or seventy-two hours after surgery!

The current state-of-the-art techniques for invasive experimentation, particularly involving miniaturization and postoperative care, now allow little or no excuse for experimentation on animals still compromised by surgery or anesthesia, but the stigma of working on moribund animals die hard both within and outside of the field. Ironically, there is an unfortunate consequence of increasing concern among physiologists about the postoperative condition of experimental animals, even when the utmost precautions have been taken. True "outliers" – for example, animals with an intrinsically "inferior" physiological phenotype – are usually quickly dismissed as animals that were unduly affected by the surgical procedures. Alternatively, a few outliers with superior performance can cause some experimenters erroneously to doubt the condition of the majority of their animals! As alluded to earlier, Chapter 7 has discussed the potential insights in physiological adaptation that can come from studying the physiology of outliers in addition to individuals more closely representing the "mean" of the species.

A second major disadvantage of invasive approaches to physiological ecology involves not so much the techniques as the individuals who wield them (Table 12.2). All too often, advocates of the invasive approach, who tend to come from more physiologically oriented backgrounds, tend to be unaware of the significant effects that subtle (and not so subtle) environmental factors can have. For example, consider Otto Loewi, who shared the 1936 Nobel Prize in Physiology and Medicine for his experiments on neurotransmitter actions at the synapse. His first definitive experiment showing neurotransmitter function was performed on a frog prepared at 3:00 A.M. one spring morning immediately following a "directive" in his dreams. Loewi was not only brilliant but also lucky. The same experiment performed at 3:00 P.M. in the autumn, a time and season when the heart of ranid frogs is least sensitive to neurotransmitters (Friedman, 1974), may have failed utterly!

Advocates of invasive techniques may actually be aware of the significance of environmental factors, but ultimately fail to treat them as possible major sources of variation in the data. These attitudes are not founded in any sort of physiologists' arrogance, but rather in human nature. When an experimenter is attempting to measure blood pressure and flow in three different arteries, measure oxygen and carbon dioxide partial pressures and pH of blood at these sites, all the time while varying in some complex fashion the fractional concentration of oxygen in inspired gas, it can perhaps be understood how he or she might not be minutely concerned with, for example, whether the animal is in a postabsorptive state, whether it has a defined thermal history, or even whether it is male or female. Of course, in many invasive experiments it may be the case that physiological adjustments produced by an experimental treatment are very large and completely obscure the effects caused by differences in blood glucose levels, thermal acclimation, or circulating levels of testosterone or estrogen. Yet, it is clear that, by failing to account for this type of variable (which, incidentally, could be the sole focus of ecologically oriented workers), experimenters using invasive technology are needlessly introducing additional variation into their data. All too often advocates of laboratory studies refer to this variation as "biological noise," which erroneously implies that it is indeterminable and uncontrollable, when in fact it results from variables that may actually be central to noninvasive studies.

Another disadvantage of invasive studies also stems from the frequently complex experimental methods and protocol. Because an individual preparation may consume the entire attention of an experimenter (or experimenters) for hours, days, or even weeks, it is often not realistic to pursue large sample sizes. Although the use of the rigorous statistical procedures for low numbers of observations has now become the rule rather than the exception (but see Chapter 10), the typically small numbers of experiments on comparatively few animals are not conducive to delineating subtle physiological effects. Moreover, because labor- and technology-intensive approaches realistically produce only small numbers of observations, important interpopulation differences are very rarely investigated. In the eyes of many advocates of invasive methodologies, such problems, if even recognized, are simply too time-consuming to explore.

In summary, invasive approaches, like noninvasive approaches, have made great contributions to physiological ecology, but also have important limitations and constraints that must be recognized.

Integration of invasive and noninvasive approaches

There are no conceptual or theoretical justifications for the methodological dichotomy in physiological ecology that has been outlined above. Rather, the different experimental approaches currently used exist largely because distinct schools of thought propagate only their own proponents by failing to encourage interaction outside clearly defined boundaries. Staunch proponents of either noninvasive or invasive physiological methodologies might each consider the other's approach as severely limiting, though historically the noninvasive school has been most vociferous in its criticisms. Is there a most effective methodology? Obviously not. Each experimental approach has made and will continue to make important contributions to ecological physiology, as presently defined. Most importantly, however, the advancement of our understanding of "adaptation" and "acclimation" *can no longer be fully served by either approach alone.*

Proponents of both approaches ultimately have a common goal in the form of understanding adaptive processes in physiological ecology (Table 12.1). Each approach contributes to a different element of that overall goal. Thus, it seems clear that the field as a whole is best served by achieving, if not a complete integration of invasive and noninvasive approaches, then at least a modification of how we carry out invasive and noninvasive physiological ecology. What can be done to facilitate methodological integration in the future?

There are at least three adjustments in the way we perceive and perform physiological ecology that could achieve a fruitful new coalition of invasive and noninvasive approaches to problem solving. These adjustments involve education, collaboration and the use of new technologies, and all three are closely interrelated.

Education (or reeducation) of physiological ecologists

Suggesting education as a solution to any problem is an overworked, even trite solution that frequently fails to bear fruit. Yet, education of young scientists newly entering the field of physiological ecology, as well as of established scientists who themselves have followed only the traditional paths, is at the heart of achieving methodological integration. The biases and stereotypic views that serve to enforce the dichotomy between invasive and noninvasive approaches to physiological ecology are strong ones, and their foundation rests firmly in the training, early experiences, and subsequent reinforcement that physiological ecologists receive.

Altering these attitudes is a formidable task. Consider briefly, for example, what I consider to be a typical attitude of "slash-and-burn" physiologists with regard to the effects of circadian or circannual rhythms in physiological processes. At some point in our biological training, every physiologist, however laboratory oriented, learns of circadian and circannual rhythms (though unfortunately often in the context of animal behavior rather than physiology). It is common knowledge that in numerous instances the time of day or season an experiment is performed can have a significant effect on the resulting data acquired. As but one example, the effects of time of day and season on metabolic rate (and the various processes that support metabolism) are notorious (see Aschoff, Daan, and Groos, 1982). Yet, very few physiologists actually test for the effects of rhythms, let alone adjust their experimental schedule if necessary. At the very least, a knowledge of circadian and circannual rhythms should argue for consistency in scheduling experiments. Yet, these biological variables (and many others very familiar to ecologically oriented workers) typically are not controlled for in invasive experiments and doubtlessly are the source of much of the so-called biological noise. Even when some change in protocol is made to take rhythms into account, it may still be misapplied. For example, the standard 12:12 photoperiod is definitely not "neutral," and actually may be physiologically debilitating to poikilotherms that are being maintained at low temperatures consistent with temperate fall or winter conditions (John Roberts, personal communication).

The need for education resides not solely with proponents of invasive physiology, as apparent from Table 12.2. Without taking the space to develop the many arguments parallel to those above which berate current modes of operation by many proponents of invasive physiology, it should be emphasized that ecologically oriented investigators frequently fail to recognize the crucial importance of understanding underlying mechanisms, and of the incomplete or even erroneous conclusions that can result from contemplation of an animal operating as a black box (Table 12.2).

Clearly, the need for (re)education exists in the field of physiological ecology, but what is the vehicle for this process? Journal editors and the reviewers that they depend upon can play a pivotal role in this process. Many investigators have good intentions in terms of improving the rigor of their experimental approach, but somehow never quite manage to find time either to read review articles or to attend workshops designed to heighten awareness of alternate experimental approaches or techniques. On the other hand, receiving stern (but appropriate) editorial comments on a submitted paper has an amazing way of focusing one's attention and motivating future efforts! As but one example, witness the ongoing revolution in statistical analyses of data in both invasive and noninvasive studies (Chapter 10). Less than ten years ago, the majority of investigators depended upon the *t*-test and linear regression for almost all of their statistical analyses, even when these approaches were inappropriate (as was frequently the case). The statistical sophistication of the physiological ecology community is now far greater. While there is no doubt that workshops and journal articles have served as useful educational sources, in many instances it has been the critical remarks of editors and reviewers that has initially stimulated reexamination of statistical approaches.

Thus, a careful examination of methodology by editors and reviewers ultimately can "reacquaint" even the most bench-bound of investigators with important ecological factors that they may have been failing to consider, or show ways in which enigmatic field observations might be clarified by laboratory investigations. Of course, once made aware of the advantages of alternate approaches, an investigator has a real obligation not only to take them into consideration in future experimental protocols where appropriate, but as importantly to ensure that persons under their tutelage are made aware of these approaches.

Expanded collaborative efforts

As physiological ecologists become more aware of the multiple biological facets of what used to be regarded as simple, straightforward problems, it becomes clear that single investigations will increasingly draw upon diverse approaches, drawing strength from the best that is offered by noninvasive and invasive approaches. Solution of multidisciplinary problems will increasingly require multidisciplinary skills. A thorough investigation of subspecies' thermal preferenda, for example, could easily employ use of field temperature recordings and associated animal distributions, laboratory measurements of temperature coefficients for metabolism, in vitro enzyme kinetics, and protein electrophoresis.

In numerous instances, single investigators have been able to garner the necessary resources and skills to undertake single-handedly a range of examinations extending broadly from field ecology to molecular biology. Such renaissance approaches have been facilitated by technological advances that simplify procedures formerly demanding extensive training.

There is, however, an inherent danger when a single investigator attempts to use a multitude of experimental approaches to test a complex hypothesis in physiological ecology. This danger, which might be termed "cumulative amateurism," results from errors introduced at many successive stages as a result of unfamiliarity with a particular technique or analytical procedure, and is somewhat analogous to serial propagation of mathematical errors.

Even when an investigator can successfully bring many different experimental approaches to bear on a research problem, in many instances it is probably most "cost-effective" in terms of time, quality, and resources to enter into a collaborative arrangement with an individual or individuals with the necessary skills and interests. This is certainly not an innovative idea. Ladd Prosser advocated collaborative ventures in comparative physiology in the late 1950's, and David Randall (in press) has predicted the increasing importance of collaboration in comparative physiological investigation in the next twenty-five years. It is clear that a satisfactory solution to many of the crucial unresolved problems discussed by other authors in this book will ultimately require a complex collaboration between ecologists, physiologists, systematists, statisticians, biochemists, and ethologists.

Unfortunately, many experienced investigators in physiological ecology have had mixed experiences with collaborative studies. One common complaint is that ". . . my collaborator was not as interested in my problem as I was." Clearly, the most fruitful collaborations will be ones in which *each* investigator has a strong interest in the outcome and will help to design and direct the experiments, rather than having one collaborator serving merely as a skilled technician. In this respect, collaborators should not be chosen simply on the basis of proximity or politics, but rather on the basis of genuine mutual interest and commitment. Initial establishment of the productive collaborations may necessitate considerable effort and time and, once established, may require considerable travel. Fruitful collaborations can be as important and long-lasting as a rental agreement or mortgage on a home, and at least as much effort should be devoted to setting them up.

Technology's role in methodological integration

Rapid advances in technology relevant to physiological ecology will play a crucial role in breaking down the boundaries between invasive and noninvasive physiology. In the past, for example, investigators primarily interested in studies with experimentally uncompromised animals behaving more or less normally in natural or simulated natural environments (i.e., proponents of the noninvasive approach) had to deal with animals as "black boxes" that were usually very reluctant to reveal their inner physiological secrets (see Table 12.1). However, recent technological advances in a number of areas point to a future in which astonishing amounts of morphological and physiological information that was formerly in the domain of traditional invasive physiology can be derived literally without touching the experimental animal.

Many of these new technologies are "spin-offs" from advances in medicine, and are being introduced into physiological ecology primarily by way of comparative physiologists. Among the most exotic contenders are computerized axial tomography (CAT) scanning, which allows imaging of internal structures and, in some instances, physiological processes. Positron emission transverse tomography (PETT) is a similar diagnostic procedure, but does require the introduction of a positron-emitting isotope at some point in the procedure. However, by use of a nuclear magnetic resonance (NMR) facility, regional intracellular pH, concentrations of ATP and other high-energy phosphate compounds, and a host of other variables can be determined in intact, undisturbed animals resting in a confinement vessel (Gadian, 1982; Axenrod and Ceccarelli, 1986).

Needless to say, CAT, PETT, and NMR require large centralized and shared facilities. A yearly maintenance contract on one of these facilities can easily exceed the average size of a typical National Science Foundation award in physiological ecology! While these technologies hold great promise in providing invasive information from noninvasive settings, such studies obviously for some time in the future will be carefully orchestrated collaborative efforts performed at a limited number of sites.

Not all technological innovations are so expensive or complex as to be beyond the reach of physiological ecologists. Many improvements, when coupled with an increasing awareness of the critical importance of minimizing surgical intervention, have allowed advocates of invasive approaches to move toward experimental conditions that minimize disruption and support normal behaviors. For example, there has been a radical miniaturization of instrumentation requiring surgical implantation. Figure 12.3 compares a conventional and bulky $400 electromagnetic blood flow transducer, designed to wrap completely around blood vessels, with a more recently developed and much smaller $25 Doppler ultrasound blood flow transducer, designed for localized attachment to the blood vessel wall. While Doppler transducers do not adequately serve in all instances where electromagnetic transducers have been used, miniaturization as typified in Figure 12.3 has in many instances greatly reduced postsurgical stress and interference, with lowered cost being a highly welcome accompaniment! As another example,

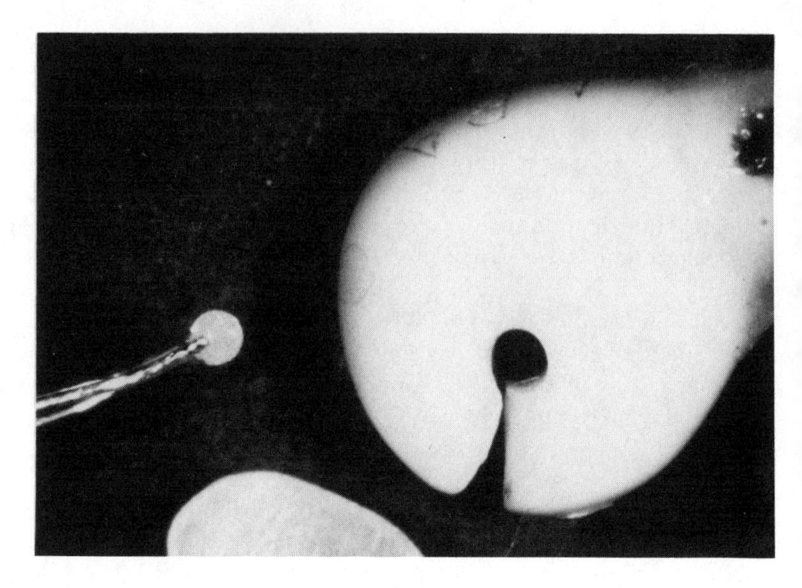

FIGURE 12.3 Alternative techniques for monitoring arterial or venous blood flow. The large, circular probe is an electromagnetic flow transducer designed to encircle the blood vessel. The smaller, disk-shaped plate is an ultrasonic Doppler flow probe, which is designed to be attached directly to the vessel wall. For a scale reference, the lumen of the electromagnetic transducer is 1.0 mm in diameter. (Photograph courtesy of Dr. A. W. Pinder.)

detailed physiological information can now be gleaned from postmortem analysis following injection of radioactive microspheres designed to be trapped preferentially in specific vascular beds. Such experiments still require implantation of indwelling catheters under anesthesia, but even in the areas of anesthesia and catheter implantation there is ongoing development of new materials and techniques designed to hasten full recovery prior to actual data collection.

One of the more exciting areas of development of experiments formerly in the domain of invasive physiology is the movement of animals from the laboratory back into the field. As a partial alternative to measuring metabolic rates in confined animals in respirometers, it has been possible for several years to measure CO_2 production and water fluxes in animals released in the field following a single injection of doubly labeled water (Nagy, 1975, 1983). Telemetry also shows great promise in breaking down the boundaries between invasive and noninvasive approaches to physiological ecology. Measurement of body temperature and location, which has been used in field studies for many years, is now being used in innovative new ways to provide

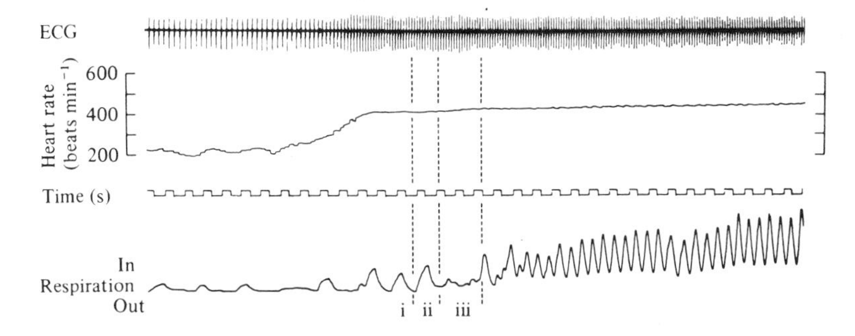

FIGURE 12.4 Data on heart and respiratory performance telemetered from a barnacle goose (*Branta leucopsis*) before and during voluntary, unrestrained flight. At hatching the goose was imprinted on one of the experimenters. The experimenter, along with the receiving electronics and recorder, sat in the back of a truck while the goose remained on the ground. When the truck began to drive away, the goose began to fly to stay within sight of the foster parent. The vertical dashed lines indicate when the truck began to move (i), when the goose began to run and flap its wings (ii), and when the bird became airborne (iii). (From Butler and Woakes, 1980.)

information on fields as diverse as genetics and systematics (see Chapter 9). More complex information, including heart rate, blood pressure, blood flow, electromyograms, etc., are also now being telemetered in animals in the field (see Figure 12.4; also Cheeseman and Mitson, 1982).

Clearly, technological advances will be at the forefront of changes in physiological ecology over the coming years. Ultimately, however, the full benefits of this burgeoning technology will be realized only if it is successfully coupled with a heightened awareness that the existing artificial dichotomy between invasive and noninvasive approaches is not in the best interests of advancing physiological ecology. In other words, *we must avoid using technological advances simply to fortify traditional thinking.*

The dangers of technology

The reductionistic approach, in which the whole (in this context, the animal in its environment) is reduced to its constituent parts (e.g., reflexes, enzymes, amino acids, genes) is firmly entrenched in comparative and ecological physiology (Prosser, 1975; Waterman, 1975; Ross, 1981; Greenberg, 1985). As every reader will appreciate, this approach has been validated repeatedly. We simply must know the building blocks upon which the whole animal's responses (or even those of populations) are based, as has been articulated in Chapter 5 of this volume. Certainly, extreme concentration on a particular problem, sometimes demanding years of effort by an entire laboratory group, has ultimately yielded major breakthroughs and this is, after all, the stuff that Nobel Prizes are composed of.

Technology also has a "dark" side which is already infiltrating comparative physiology and biochemistry. One of the major problems spawned by technology is what might be termed "excessive reductionism." While the previous section has argued that technological advances hold great promise in fostering an integration of diverse physiological approaches, our fascination with technology over the past few decades has inadvertently contributed to the rampant "encyclopedism" and "matrix filling," alluded to by Ross (1981) and criticized in Chapter 3. Sophisticated new technologies may very well require a rather large investment of time and money to realize their full potential. In the prevailing climate of stiff competition for limited research resources, an investigator who has made a major commitment to a new technology faces considerable pressure to produce a large flow of data. One way in which some investigators have succumbed to the pressure to produce data immediately is to apply the new technology repeatedly to an extremely narrow research problem. Thus, as a hypothetical example, an investigator with new facilities for rapid, accurate (and expensive) measurement of blood respiratory properties might be tempted to perform the same analyses on a great multitude of species (e.g., "Respiratory Properties of Hemocyanin from Brachyuran Crabs: XXIV – the Shore Crab, *Carcinus*"). Such studies essentially constitute surveys without hypotheses and represent a form of creative poverty disguised by high technology.

Technology now allows us to telemeter electroencephalograms from spawning salmon, and to measure oxygen-hemoglobin binding within a single red blood cell! While it is important, even essential, to have the detailed data base that derives from a reductionist approach supported by technological innovation, it is equally important that we frequently take stock to see how (and whether) our experimental approaches fit into a broader perspective and help to achieve the goals of physiological ecology as a whole.

Conclusions

These are exciting times as physiological ecologists reassess their goals and attempt to predict, if not actively alter, future directions in this well-established field of research. As I have attempted to convince the reader in this essay, a critical reexamination of the physiological methodologies that physiological ecologists use must be considered as essential in establishing "new directions" and pursuing new goals that contribute to our understanding of physiological evolution and adaptation. Failure of physiological ecologists to integrate divergent methodological approaches, either along the lines I have suggested or in many additional ways that I have not envisaged, will not in any way detract from the current strength and importance of physiological ecology in modern biology. However, an elimination of the current dichotomous trend in approaches to physiological ecology might propel us more rapidly toward our fundamental goal of understanding animal evolution.

Acknowledgments

Martin Feder, Juan Markin, John Roberts, and Alan Smits contributed many comments and ideas of both a philosophical and literary nature. The author received financial support from NSF Grant DCB 86-08658 during the preparation of this manuscript.

References

Aschoff, J., Daan, S., and Groos, G. A. (1982) *Vertebrate Circadian Systems: Structures and Physiology.* Berlin: Springer-Verlag.

Axenrod, T., and Ceccarelli, G. (1986) *NMR in Living Systems.* Dordrecht: Reidel.

Bartholomew, G. A., and Tucker, V. A. (1963) Control of changes in body temperature, metabolism and circulation in the agamid lizard, *Amphibolurus barbatus. Physiol. Zool.* 36:199–218.

Butler, P. J. (1982) Respiratory and cardiovascular control during diving in birds and mammals. *J. Exp. Biol.* 100:195–221.

Butler, P. J., and Woakes, A. J. (1980) Heart rate, respiratory frequency and wing beat frequency of free flying barnacle geese *Branta leucopsis. J. Exp. Biol.* 85:213–226.

Cheeseman, C. L., and Mitson, R. B. (1982) *Telemetric Studies of Vertebrates. Zool. Soc. Lond. Symp.,* no. 49. New York: Academic Press.

Elsner, R., and Gooden, B. (1983) *Diving and Asphyxia.* Cambridge University Press.

Friedman, A. H. (1974) Serendipity and chronobiology in pharmacology. In *Chronology,* ed. L. E. Scheving, F. Halberg and J. E. Pauly, pp. 163–167. Tokyo: Igahu Shoin Ltd.

Gadian, D. G. (1982) *Nuclear Magnetic Resonance and Its Application to Living Systems.* Oxford: Oxford University Press (Clarendon Press).

Gallivan, G. J., Kanwisher, J. W., and Best, R. C. (1986) Heart rates and gas exchange in the Amazonian manatee (*Trichechus inunguis*) in relation to diving. *J. Comp. Physiol.* [*B*]. 156:415–423.

Greenberg, M. J. (1985) Ex bouillabaisse lux; the charm of comparative physiology and biochemistry. *Am. Zool.* 25:737–749.

Hasler, A. D., and Scholz, A. T. (1983) *Olfactory Imprinting and Homing in Salmon.* Berlin: Springer-Verlag.

Heath, J. E. (1964) Head-body temperature differences in horned lizards. *Physiol. Zool.* 37:273–279.

Hochachka, P. W. (1980) *Living Without Oxygen.* Cambridge, Mass.: Harvard University Press.

Kramer, D. L. (1978) Ventilation of the respiratory gas bladder in *Hoplerythrinus unitaeniatus* (Pisces, Charcoidei, Erythrinidae). *Can. J. Zool.* 56:931–938.

Nagy, K. A. (1975) Water and energy budgets of free-living animals: measurement using isotopically labeled water. In *Environmental Physiology of Desert Organisms*, ed. N. F. Hadley. Stroudsburg, Pa.: Dowden, Hutchinson and Ross.
Nagy, K. A. (1983) The doubly labeled water ($^{3}HH^{18}O$) method: a guide to its use. Publ. no. 12–1417. Los Angeles: University of California.
Prosser, C. L. (1975) Prospects for comparative physiology and biochemistry. *J. Exp. Zool.* 194:345–347.
Randall, D. J. (in press) Comparative physiology: the next 25 years. *Can. J. Zool.*
Ross, D. M. (1981) Illusion and reality in comparative physiology. *Can. J. Zool.* 59:2151–2158.
Scholander, P. F. (1940) Experimental investigations on the respiratory function in diving mammals and birds. *Hvalradets Skrifter* 22:1–131.
Smith, E. N., Allison, R., and Crowder, E. (1974) Bradycardia in a free ranging alligator. *Copeia* 1974:770–772.
Waterman, T. H. (1975) Expectation and achievement in comparative physiology. *J. Exp. Zool.* 199:309–344.
Weathers, W. W. (1970) Physiological thermoregulation in the lizard *Dipsosaurus dorsalis*. *Copeia* 1970:549–557.
Weathers, W. W., and White, F. N. (1971) Physiological thermoregulation in turtles. *Am. J. Physiol.* 221:704–710.

Discussion

DAWSON: Some physiological ecologists are doing fairly faithful replications of what used to be called tolerance physiology. There are certainly invasive, or slash-and-burn-type physiologists who ignore the combination of the animal and its environment that Bartholomew mentioned. One happy development, however, has been a resurgence on the part of physiological ecologists of activity that is mechanistically oriented. This brings them closer to being comparative physiologists, who, in addition to having a concern for environmental problems, realize that the use of animal models has very powerful implications for the characterization of mechanisms. The archetypal dichotomy into invasive and noninvasive approaches is not one that I am comfortable with. The physiological ecology that I aspire to and I hope the field will come to represent more and more, is one in which a continuum exists between what I might regard as an initial phase, which has a high degree of environmental orientation, and a subsequent one that deals with mechanisms. Examination of problems in relationship to environmental situations is enriched by a concern with mechanisms. We should not simply collaborate. We should all expand our thinking to include a broad range of physiological and ecological approaches. I feel very strongly about this because recent years have witnessed not just the development of noninvasive techniques, but also an immense number of developments that allow us to investigate both mechanisms and patterns in evolution with what, at one time, seemed exclusively biochemical or physiological approaches.

One of the important issues that we should deal with here is where organismal biology is going. We should not use the term "organismal" except in reference to the philosophical arena in which we operate. Organismal biology should be aggressively seeking to move to the population level; it should be seeking to move, as appropriate, to the molecular or other intermediate levels. One thing that will distinguish this new organismal biology from other fields is the obligation to fit its findings at whatever level back into an organismal framework and having that interface with the environment.

FEDER: I agree, especially with regard to the future directions of organismal biology and physiological ecology. Burggren has accurately characterized a dichotomy in one particular respect. A major distinction between the two approaches is that by its very nature, the invasive approach eliminates behavioral responses to an environmental variable. If it does not eliminate behavior, it potentially may frustrate or confound behavior. The less invasive approach typically admits both behavior and the hard-wired physiological responses. These approaches are different, not one better than the other, but complementary.

DAWSON: But they should be carried out, when it is appropriate, within the same laboratory. I am calling for invasiveness when it is appropriate to be invasive. My own experience ranges from behavioral and microclimatic observations, to enzymatic observations. If you define your problems clearly, you should take them where they lead. I would submit that that is one of the challenges and rewards in this field: if you have a lively curiosity, you can satisfy it in more ways than perhaps the more stereotyped.

BURGGREN: The division of studies into invasive and noninvasive schools is an anachronism. I would like to see these boundaries break down. Dawson also used the word "continuum," which is appropriate to describe the methodological approaches that are available to us. We should use the appropriate experimental approach and technology as called for.

I worry in that I seem to hear Dawson argue against collaboration. As the fields get more complex and as new methodologies become available, to maintain a first-rate approach you either must retool and learn the new methodology or you have to entrust your project to collaborators. Admittedly, one problem with extensive collaboration, even if you can find someone who is interested in your problems, is that there can be too many chiefs and not enough Indians, resulting in no central person having an overview of the project.

DAWSON: I was not eschewing collaboration. The point is the difference between collaboration in which you are truly at an interface, and collaboration in which you are just subcontracting. I like to avoid subcontracting.

BURGGREN: Perhaps the best of both worlds is to be a general contractor in building the house. You deal with the people who are making the blocks and the people who are doing the wiring and the plumbing, but you do have the overview and can envisage from the blueprints what the house will be like.

POWERS: There are some advantages to being a general contractor, but one has to be careful to whom one contracts. Some of the best kinds of collaborations are those that occur between students and faculty members. It is important, however, that you have a basic working knowledge of the other field. You can not be an expert in all fields: you might end up being a jack-of-all-trades and a master of none. It is important that you are competent in your field and that you are knowledgeable in other fields and that you subcontract to the most economically feasible source, that is, students, or a postdoc in some cases, with whom you interact on a routine basis. If you do that, you have a much greater chance of success.

JACKSON: The dichotomy that I am more sensitive to is between studying a mechanism for the sake of the mechanism and studying it in order to understand an ecological problem. Physiologists justify their study to colleagues and granting agencies on the basis of understanding mechanisms rather than in terms of any broader biological or environmental problem.

HUEY: Many of the techniques used in mechanistic physiology could be applied in an ecological context as manipulations. Physiologists have developed many techniques to control maximum oxygen consumption and other physiological activities, and we can test those in an ecological context.

BARTHOLOMEW: With the development of microelectronics, there is the possibility of the invasive and noninvasive approaches finding a common path that is the best of both worlds. I personally have moved back and forth, invasive to noninvasive, depending on the problem, always reluctantly to the invasive. Now it is becoming more probable that the two approaches will end up using the same technique because it is the best technique.

BURGGREN: And, perhaps, the least expensive technique! While some technological advances, like NMR, have big price tags, many of the technological advances are inexpensive. For example, in comparing the new versus the traditional systems for measuring blood flow, the traditional electromagnetic transducers were about $400; the new pulsed Doppler ultrasound transducers are about $25 and we throw them away when they get blood clots on them. Technology can also mean a reduction in expenses.

13 The use of models in physiological studies

PETER SCHEID

Introduction

The term "model" is widely used in modern scientific literature although with very different meaning. The physicochemist constructs a *model* to clarify the spatial pattern of a chemical compound; the physiologist applies the hairpin, countercurrent *model* to explain the concentrating mechanism of the kidney; the experimental biologist uses a particular animal *model*; and these are only a few examples of the use of the term. Dictionaries, like *Webster's New World* (Guralnik, 1976) or the *American Heritage Dictionary* (Morris, 1982), list several definitions of the term, but this list does not appear to cover comprehensively most of the cases in which it is used in science.

The following tentative definition may describe the typical application of the model as used in physiological studies:

A model is an image of part of the physical or mental world apt to explain or predict observations.

In this sense, normally two types of models, which I refer to as the "interacting model" and the "structure-function model," are used in physiological studies.

Interacting model

Several responses of the organism come about as the sum of a great number of single effects. This is true for respiration, body temperature, and limb movement, to name a few. To understand qualitatively and describe quantitatively this response, the systems engineer's approach of modeling is often applied, in which the complex interactions between parts in the system are described in terms of control circuits, (e.g., feedback or feedforward loops). The properties of the single elements in the control circuit are usually obtained by studying them in isolation (e.g., by experimentally opening loops).

These models are thus mainly heuristic and empirical in nature. Their elements are defined as functional subunits. Structure-function relationships are not typically of major concern in their construction. The organism is viewed as a concert of several black boxes, each of which is characterized by a particular input-output relationship, but whose internal working remains unclear. It is the knowledge of each of these elements and of the peculiar mode in which they are tied together that enables the interacting model to make predictions about the behavior of the system.

Structure-function model

Definition

Unlike the interacting model, the structure-function model is not composed of functional black boxes, but is built with distinct constraints from the anatomical or structural reality. Structure-function relationships are of significance at all levels of the organism, from the subcellular to the whole animal, and structure-function modeling is important in investigations at all these levels.

With the increasing power of resolution of modern microscopy has come the hope of deriving an understanding of cellular function from the structural image alone. So far this has not been realized, and the gap between structure and function may never be bridged completely. Physiological studies of cell function thus have to consider both the structural appearance and the functional behavior, namely by building structure-function models. Of particular importance is modeling, however, on the supracellular level. It is widely accepted today that several functional elements (e.g., buildup and maintenance of concentration gradients across the plasma membrane, response to hormone substances, release of chemical agents) are very similar in corresponding cells of many animals. This is certainly an advantage for the physiologist who may use "lower" animals for studies of general cell function.

In striking opposition to these similarities in the functional elements, there are tremendous differences among animals in the function of organs, manifested in the central nervous system, and the respiratory, cardiovascular, and other systems. These functional differences can only to a small degree be explained as differences in the elements of which they are built. They are rather brought about by peculiarities in which these elements are arranged to form the particular organ. As an example, ion pumping and restricted water permeability are necessary requirements for a kidney to excrete urine of varying osmolarity. The degree of this concentrating ability is, however, closely linked to the structural arrangement of the tubular system in the kidney medulla to enable the countercurrent flow of tubular fluid in the loop of Henle and the collecting ducts. The structural arrangement thus becomes a crucial parameter for a quantitative description of organ function, and the structure-function model serves this purpose.

Model and real structure: the "best" model

In constructing a structure-function model, a question of relevance is whether the model should represent an exact image of the underlying structure. Assuming you model the real structure as closely as possible, two possibilities can occur. First, the model may not function like the real object; in that case, the model is wrong and must be revised. Second, the model may behave as expected; in this case, it cannot tell you more than the biological object from which it is built, and the model is again useless. Hence, the model must not be a faithful image of reality.

If the model thus is not to represent a faithful image of reality, what are the criteria upon which its construction rests? Constructing a model is never a task by itself. Rather, the model is built when it is needed to *answer a specific question*, and, since many physiological questions can be asked about a system, many different structure-function models about the same object are possible. To serve this purpose, the model must consider all those elements of the underlying structure that are of qualitative and quantitative relevance to the particular question, but must abstain from including all the other elements.

There are several important consequences. First, the structure-function model typically constitutes an abstraction of the underlying real object, which neglects and has to neglect many structural elements, namely those that are irrelevant to the question. The structure-function model does not necessarily resemble the underlying object. Second, this abstraction results in simplifications of the complex biological structure. Models may thus be simple and should be simple to allow a comprehensive mathematical treatment of the question and hopefully a general solution. Third, each question requires a different model. Thus, there is not *the* best model of an organ or any biological system, but there is an appropriate model for any given question into the function of this system.

Constructing a model constitutes an analytical task in that the object must be analyzed (i.e., unraveled or disentangled) to find the elements of significance. This task is often neither easy nor trivial. In the following text, I will give examples for the use of structure-function models in physiological studies. These examples shall also serve to demonstrate how models are built, modified, and tested for validity.

Application: external respiration

Structure-function models have widely been used in the analysis of lung function. The mammalian alveolar lung, with its several hundred million alveoli and air passages leading into them, is obviously a highly complex design, and

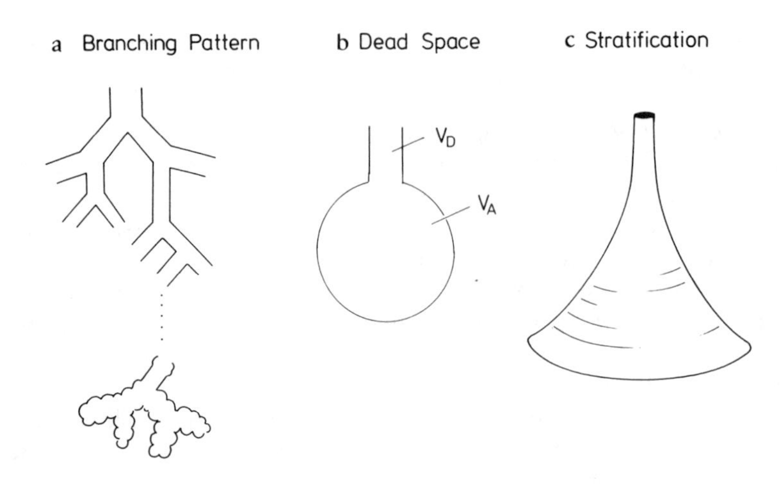

FIGURE 13.1 There are different lung models for different questions regarding the function of the lung: (a), model of regular branching pattern of the airways; (b), model for gas exchange, dividing air spaces into those without (dead space) and with gas exchange (alveolar space); (c), trumpet model for the study of intrapulmonary gas mixing or stratification.

significant simplification is needed to construct lung models. This simplification and the resulting model depend on the physiological question.

Figure 13.1 shows three conspicuously different models of the same biological structure, the alveolar lung. Figure 13.1a schematically depicts a model of the average airway geometry, which has quantitatively been presented as Model A by Weibel (1963). This model has been extremely successful in many studies in which quantitative data on the bronchial tree were needed. The simple dead space model of Figure 13.1b lumps the conducting airways into a single dead-space compartment (V_D), and combines the alveoli into one alveolar space (V_A). This model is helpful for understanding alveolar gas exchange, and it may be slightly modified (e.g., by an alveolar dead-space compartment) to introduce functional lung heterogeneity (cf. Piiper and Scheid, 1980). Figure 13.1c shows the trumpet or thumbtack model which relates the longitudinal distance along the airways to the combined airway volume. This model has successfully been used to evaluate gas mixing in the alveolar region, often termed "stratification" (Scheid and Piiper, 1980a). The fact that earlier models have failed to arrive at correct conclusions about gas mixing in the alveoli is not due to their being too simple as compared to the real lung structure, but to their being inappropriate with respect to the questions posed. These examples of an undoubtedly much longer list exemplify the important, yet often neglected, fact that there is not one correct model but that the appropriate model depends on the question being asked. There are distinct similarities with the arts: whereas the most accurate model of an object is a photograph, the message of the artist painter may often be beyond this trivial reality. The phrase "Art starts where nature ends" has its counterpart in science, and particularly in physiology.

Interpreting model and normalizing model

The structure-function model is used in two ways: to explain unexpected observations and to analyze measurements quantitatively. This distinction does not define different models but rather defines different ways in which one and the same model is applied. Thus, in the first case, the model is used for qualitative interpretation and may be termed an "interpreting model." In the second application, the model is used in a quantitative way, particularly to eliminate, and thus account for, experimental variability. It may be referred to as a "normalizing model."

Interpreting model

The starting point for the development of an interpreting model is typically an experimental observation that does not fit expectation. The task is to define those structural or functional elements that are crucial for explaining the observation. For this, one must construct the simplest model that allows explanation, and any unnecessary details must be avoided.

Gas exchange in avian lungs may serve as an example. Zeuthen (1942) was the first to report that the partial pressure of CO_2 in expired air in birds (P_{ECO_2}) may exceed the value in arterial blood P_{aCO_2}, an observation that is not easy to explain on the basis of alveolar gas exchange. Zeuthen invoked the structural arrangement in the gas-exchanging part of the lung to explain this observation qualitatively; and Scheid and Piiper (1970), on the basis of the anatomical arrangement, proposed the crosscurrent system as an appropriate model for gas exchange in the avian parabronchial lung (cf. Scheid, 1979). As is shown in Figure 13.2, this model accounts for the structural peculiarity of avian parabronchi that a given blood capillary has contact for gas exchange with only a short stretch along the ventilated parabronchial tube. The fact that blood flow to the segments of the parabronchus is in parallel, while air flow is in series, can explain the observation that P_{aO_2} (the P_{O_2} of arterial blood) exceeds P_{EO_2} (the P_{O_2} of expired air) while blood P_{O_2} is below gas P_{O_2} in each segment.

There was, however, another model that also appeared well suited to explaining the experimental observation of P_{ECO_2} exceeding P_{aCO_2}: the countercurrent model (Schmidt-Nielsen, 1971). To determine which of the two models is the more appropriate for parabronchial gas exchange, more information, structural or functional, was needed. Structure appears to favor the

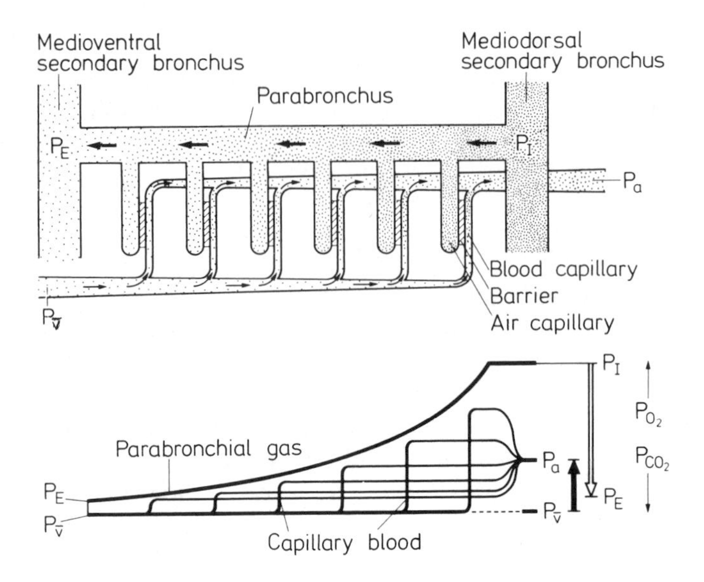

FIGURE 13.2 Crosscurrent model of avian parabronchial gas exchange. The parabronchial tube extends between the mediodorsal and medioventral bronchi. Blood capillaries, originating from a branch of the pulmonary vein, contact radial extensions of the parabronchus, the air capillaries. Air flows from the mediodorsal (partial pressure, P_I) to the medioventral secondary bronchus (P_E) through the parabronchus, whereby gas is exchanged with the blood across the blood-gas tissue barrier. (Density of stippling indicates in qualitative terms the level of P_{O_2}.) The P_{O_2} profile below shows that blood in the first (with respect to air flow) blood capillaries assumes a high degree of arterialization whereas the reverse is true for the last blood capillaries. While P_E reflects the gas with lower P_{O_2}, P_a constitutes an average of end capillary blood values. (From Scheid, 1979.)

crosscurrent system. However, exclusion of one of these competing models was accomplished by a crucial experimental manipulation (Scheid and Piiper, 1972), which would yield basically different results for the two models. This experiment, in which parabronchial air flow was experimentally reversed, favored the hypothesis of the crosscurrent and ruled out the countercurrent model; it is now accepted that diffusion and convection in connection with the structural arrangement of the crosscurrent system can explain avian parabronchial gas exchange. This discussion illustrates the well-known fact that a model can never be proven by experimental results that fit predictions, but only disproven.

Another unexpected observation was later reported by Davies and Dutton (1975), who found that P_{ECO_2} can exceed even mixed venous P_{CO_2} ($P_{\bar{v}CO_2}$) in the chicken. Because this was not simply explained by the parabronchial model, the authors proposed the charged membrane hypothesis to explain the observation. This theory, which had earlier been applied to the alveolar lung (Gurtner, Song, and Farhi, 1969), was not undisputed on experimental and theoretical grounds (Scheid and Piiper, 1980b; Piiper and Scheid, 1980). Meyer, Worth, and Scheid (1976) confirmed the observation of P_{ECO_2} exceeding $P_{\bar{v}CO_2}$ in the chicken, but offered a different explanation, namely the interaction of O_2 and CO_2 in binding to hemoglobin, known as the Haldane effect. Although the Haldane effect is in principle the same in bird blood and in mammalian blood, its effect on CO_2 removal is different in the alveolar and parabronchial lung, and can in particular lead to P_{ECO_2} being greater than $P_{\bar{v}CO_2}$ in the crosscurrent parabronchial lung. The crucial experiment that Meyer et al. (1976) conducted to discriminate between their explanation and that of Davies and Dutton, was rebreathing with blood-gas equilibrium for both O_2 and CO_2, for which the charged membrane hypothesis would predict gas P_{CO_2} to exceed blood P_{CO_2}. The fact that gas and blood P_{CO_2} were identical was proof against the charged-membrane hypothesis and corroborated the crosscurrent system with diffusion and convection, indicating that for particular studies the Haldane effect would have to be considered.

Unexpected observations and the need to explain them had thus shaped a model which proved useful for a qualitative understanding of gas exchange in avian lungs, and could hence be used in quantitative studies as well. Piiper and Scheid (1975, 1984) have analyzed the main gas-exchange organs of vertebrates which they found could be represented by four different models (Figure 13.3). These models differ only in the structural arrangement of blood and medium (gas or water) conducting structures at the site of gas exchange.

Normalizing model

This kind of model can be helpful in several ways in experimental and theoretical studies. Suppose the lung gas-exchange function is to be assessed in a given animal and is to be compared with that of a second animal, a specimen of the same or a different species. For this, arterial oxygenation (e.g., arterial P_{O_2}, P_{aO_2}) may be measured. Differences in P_{aO_2} are, however, not unequivocally indicative of differences in lung function, because P_{aO_2} depends on a number of factors not all of which are related to lung function. These additional factors include the anatomy of the lung (e.g., thickness of gas/blood separating membrane or its area), physical parameter (e.g., blood flow, ventilation, hematocrit, etc.), and the structural arrangement. Because P_{aO_2} is thus a function of a great number of parameters – some known, some unknown but measurable, others not even measurable – direct comparison between measurement of P_{aO_2} between animals, or even at different situations in one animal, are of little value for the assessment of lung function, because any of several parameters may have changed.

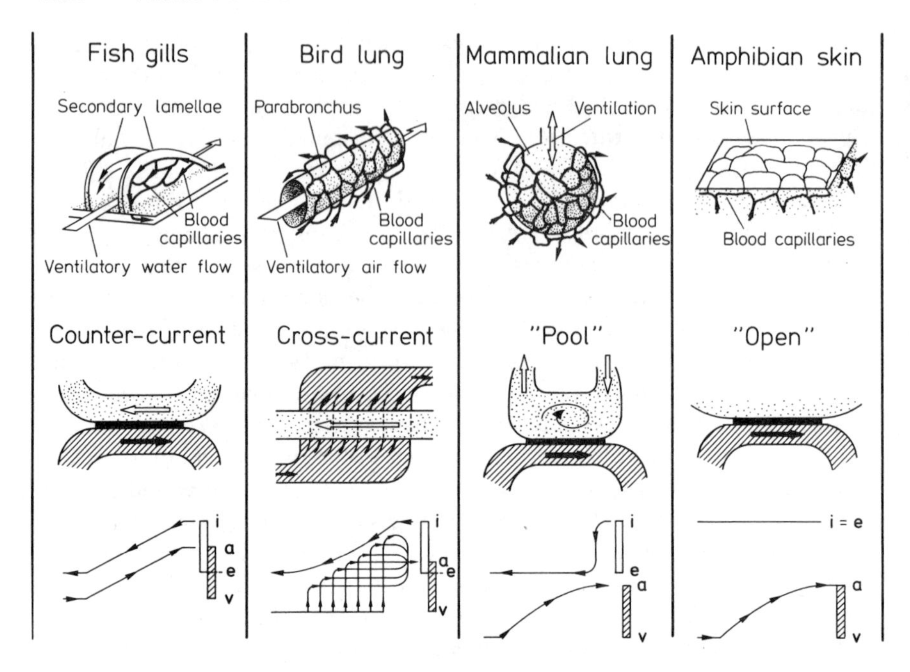

FIGURE 13.3 Four models proposed by Piiper and Scheid (1975) for external gas exchange in vertebrates. Structural elements are schematically represented in the upper panels; the model, in the middle; and the partial pressure profiles in the medium (air or water) and blood, in the lower panels.

The structure-function model, however, describes the quantitative relationship among these parameters and represents the way in which they impinge on P_{aO_2}. Hence, if some of these parameters are experimentally determined, they may be inserted into the formulas describing the model to calculate a model parameter that constitutes an estimate of gas exchange normalized for differences in the measurable data. Such a model parameter is the lung diffusing capacity, D_L, which is an estimate of structural, physical, and functional factors. D_L is, however, largely independent of ventilation, blood flow, O_2 binding properties of blood and other parameters, which affect P_{aO_2} without changing the lung gas-exchange function in its strict sense.

In this example, the structure-function model has been used to calculate a characteristic parameter which is normalized to differences in measurable factors, that are thus accounted for. In such applications, the structure-function model may be termed a "normalizing model." Obviously, the correct structure-function model must be found before a normalizing model can be applied, so that interpretation and normalization may be considered as sequential steps in the application of models.

Model assumptions

Because structure-function models constitute simplifications of real objects, their application is burdened by assumptions. In fact, assumptions may be viewed as the quantitative counterpart of the model simplifications, which were suggested to be of vital importance for the appropriateness of the structure-function model.

The significance of these assumptions for the results obtained by the model depend on the problem under investigation. There are three types of assumptions in respect of their significance in a particular study: (1) critical under all conditions; (2) critical only under certain conditions; (3) not critical under any condition. A method for determining the significance is to use the model with and without the assumption, all other parameters being identical, and compare the results. If the results do not differ significantly, then the assumption does not constitute a significant constraint to the validity of the model. If the assumption proves to be significant, either the model can be expanded by dropping the assumption; or, if this proves too complicated, the influence of the assumption can be estimated in a separate investigation, and the model result can be corrected accordingly.

An example is the influence of the curvature of the O_2 dissociation curve in the analysis of O_2 exchange in lungs. Its inclusion would render the models of the vertebrate gas exchange organs (Piiper and Scheid, 1975) very complicated and would preclude a number of general conclusions. Fortunately, calculations and graphical analysis, known as the Bohr integration, show that the assumption of linear dissociation curves is not crucial for the analysis of the lung O_2 diffusing capacity, D_L, provided that measurements are made in hypoxia. This is one reason why D_L should be measured in hypoxia. The calculations show further that the curvature is of advantage to O_2 uptake, in normoxia, and that D_L would be overestimated when calculated from measurements in normoxia that use the alveolar model with a linear dissociation curve.

Conclusions

Models are not only helpful but often indispensable in quantitative biology. The proper model must be chosen carefully, and its validity in and applicability to the problem under study must be checked carefully. Models must be simple, and the degree of simplicity is exclusively determined by the problem itself. Assumptions resulting from this simplification must be checked carefully, and an appropriate compromise is needed between complexity of the model and general validity of the model results. In modeling, mathematics must be used as a tool, and the biologist must refrain from complicating the model beyond the absolute necessity. The simplest model that serves the purpose of explaining or analyzing data is the best.

It is often asked why quantitative modeling is not more often used in physiological studies and why it has so often failed to yield general answers to important questions. There may be many responses to this question, among them the following. Quantitative modeling necessitates a certain amount of mathematical expertise. Biologists, mostly not trained in the mathematics needed, have often sought help from mathematicians. It appears to be extremely difficult for the biologist, unfamiliar with mathematics, to explain to the mathematician the biological frame into which to build the model, and for the mathematician, unfamiliar with biological problems, to translate the structure-function model into mathematical terms. It seems, therefore, mandatory for biologists to learn the basics of mathematics at least to the point where they can confront mathematicians with clearly posed questions and can control their results, at least intuitively, from their experience with biology.

References

Davies, D. G., and Dutton, R. E. (1975) Gas-blood P_{CO_2} gradients during avian gas exchange. *J. Appl. Physiol.* 39:405–410.

Guralnik, D. B. (1976) *Webster's New World Dictionary of the American Language*, 2nd college ed. Cleveland, Ohio: World Publishing.

Gurtner, G. H., Song, S. H., and Farhi, L. E. (1969) Alveolar to mixed venous P_{CO_2} difference under conditions of no gas exchange. *Respir. Physiol.* 7:173–187.

Meyer, M., Worth, H., and Scheid, P. (1976) Gas-blood CO_2 equilibration in parabronchial lungs of birds. *J. Appl. Physiol.* 41:302–309.

Morris, W. (1982) *The American Heritage Dictionary of the English Language.* Boston: Houghton Mifflin.

Piiper, J., and Scheid, P. (1975) Gas transport efficacy of gills, lungs and skin: theory and experimental data. *Respir. Physiol.* 23:209–221.

Piiper, J., and Scheid, P. (1980) Blood-gas equilibration in lungs. In *Pulmonary Gas Exchange*, vol. 1, ed. J. B. West, pp. 131–171. London: Academic Press.

Piiper, J., and Scheid, P. (1984) Model analysis of gas transfer in fish gills. In *Fish Physiology*, vol. 10, part A, ed. W. S. Hoar and D. J. Randall, pp. 229–262. London: Academic Press.

Scheid, P. (1979) Mechanisms of gas exchange in bird lungs. *Rev. Physiol. Biochem. Pharmacol.* 86:137–186.

Scheid, P., and Piiper, J. (1970) Analysis of gas exchange in the avian lung: theory and experiments in the domestic fowl. *Respir. Physiol.* 9:246–262.

Scheid, P., and Piiper, J. (1972) Cross-current gas exchange in avian lungs: effects of reversed parabronchial air flow in ducks. *Respir. Physiol.* 16:304–312.

Scheid, P., and Piiper, J. (1980a) Intrapulmonary gas mixing and stratification. In *Pulmonary Gas Exchange*, vol. 1, ed. J. B. West, pp. 87–130. London: Academic Press.

Scheid, P., and Piiper, J. (1980b) Blood-gas equilibrium of carbon dioxide in lungs. A review. *Respir. Physiol.* 39:1–31.
Schmidt-Nielsen, K. (1971) How birds breathe. *Sci. Am.* 225(6):72–79.
Weibel, E. R. (1963) *Morphometry of the Human Lung.* Berlin: Springer-Verlag.
Zeuthen, E. (1942) The ventilation of the respiratory tract in birds. *Kgl. Danske Videnskab. Selskab Biol. Medd.* 17:1–50.

Discussion

FEDER: I wish that Scheid's talk had been first in the workshop because it speaks to a central dilemma. We are talking here about adding on to the scope of the field: expanding into genetics, molecular biology, and statistical methodologies that are not widely used. On one hand, while additions and expansion into each one of these fields provides the potential for reaching a more general understanding of ecophysiological interactions, they also introduce additional variables that need to be studied. This complicates the analysis. On the other hand, if one is interested in a specific analytical solution, obviously one should try to reduce the number of variables. That has its own costs in that it potentially confounds the generality of the conclusions at which one arrives.

POWERS: One of the dangers in modeling is that models may be produced that are independent of real data. I like your emphasis on the importance of taking data and seeing whether a model is a valid fit to the data and then refining the model, depending upon whether or not it explains all of the variance. That is very important because essentially you are using the model as a testable hypothesis. Also, this approach can be very useful in deciding what sort of data you need to build a better model or to ask the question differently. The approach is the same whether the model concerns molecular interactions or ecological changes.

ARNOLD: Scheid made a point of addressing the issue as to why modeling is not used more often in physiological studies, indicating that, in some cases, a fear of mathematical treatments scares investigators away. Another factor, however, is that people think that you can prove anything that you want with a model, that models are infinitely malleable and can produce any result at will. As a consequence, some people put no more stock in the results of a formal model than they would in the most casual verbal rendition of a phenomenology. I wonder if you might comment on that attitude. Have you ever had one of your models give you a result that did not reinforce your intuition?

SCHEID: This is an important point. It is said that with five variables you can construct an elephant and with six he will wave his trunk. This is the risk when you get too far from the data. The problem is really that you must be extremely disciplined in testing your assumptions and incorporating only important parameters. I can think of a number of papers that have violated this point. When you try to build the models the way that I proposed, keeping them simple and adding only tested values, then I think that you are not in so much danger of falling into this trap.

POWERS: I would like to add a couple of cautionary notes. What I have found, especially among students and unfortunately also among colleagues and myself to a certain extent, is ignorance about criteria for a good fit of data to models. Especially with canned programs, there is a real danger that you plug the data in, you get a number out, you publish it, and you are done with it.

FUTUYMA: I breathed a sigh of relief when I heard the presentations of Burggren, Scheid, Lindstedt, and Jackson because it had seemed to me that this whole field was an empirical one. It was greatly reassuring to see that model building and general conceptual approaches are also part of the field. I would like to take exception to the claim that models should be at all times tied to data. There are many different kinds of models, and in evolutionary biology the most important models are those that existed before any data existed with which to test them. The entire structure of population genetics, which is probably the most rigorous quantitative theory in biology, existed by the late 1920's and 1930's, before we had any data whatsoever to fit into selection theory, to fit into concepts of effective population size, gene flow, or anything else. Rather, what had been constructed was a complex theoretical superstructure, which then served as an organizing force for asking questions. It is important to distinguish between models that are built specifically to interpret the data on very specific systems versus models that are much more general in nature and serve to organize thinking and identify theoretical impossibilities.

PACKARD: We are discussing two different kinds of models. One is a mechanistic, predictive model, whereas the other is a statistical, descriptive, analytical model.

FUTUYMA: I disagree. I am talking about a third class of model altogether.

PACKARD: I think that Scheid discussed essentially two types of models. Most of his remarks were directed toward his interpretive or predictive model and relatively few to the second model. Could I direct a question to Futuyma? Is there any difference between this model and the mechanistic, predictive model? I think that they are just different degrees of refinement of the same thing.

FUTUYMA: This comes fairly close to what I am talking about; still, I think there is a difference.

DAWSON: If we have a refined model, sometimes we can get to things that are not convenient to deal with experimentally. It is possible by modeling to assess the reasonableness of a conclusion. It is not inevitable that one always confirms intuition. A good model, I think, should help in dealing with things that are hard to manipulate in the real world.

SCHEID: Are the models that I discussed the correct ones? My terminology for the regressing model is inadequate for a simple reason. Regression analysis is a phenomenological tool: you have data, look at two variables, and test for a correlation. My approach in this regressive model is different. I erect the model on theoretical grounds and make one or two measurements to test the model. What I want to do is take out the variability, but I do it on a mechanistic model. So what I do is really standardize and normalize or reduce the data. So would not a reducing model or normalization model be a more appropriate term than a regressing model, which implies phenomenology? With respect to Futuyma's notion of more general models that precede data, I am inclined to think not.

FUTUYMA: It never precedes all data. We do not come up with biological theorems by pure a priori knowledge. . .

SCHEID: You must have observations.

FUTUYMA: You have some observations. But, in the case, for example, of population genetic theory, the observations were Mendelian inheritance. When Fisher and Haldane and Wright erected the theory of population genetics, which still serves as the fundamental theory of evolution, there were no measurements of selection coefficients. No one had ever measured natural selection. No one had ever measured gene flow. We knew only that organisms moved around, and carried their genes with them. We had neither the concept of what effective population sizes were nor even the concept of effective population size in comparison to census population size. Just from the rules of transmission genetics and the concepts of natural selection and finite populations, it was possible to erect an a priori theory of the factors of evolution, which has held up practically without change for fifty years.

SCHEID: I had two models. One was to look at control regulation of homeostasis. This is a system engineer's approach to modeling. The other is the mechanistic model. Did I miss one other important model? Can you think of any other type of model for this function?

FUTUYMA: I am not the right person to answer that, because I do not know if there are other models in physiology. I can describe other models in ecology and in evolutionary biology that I think differ from those in that you have a notion of a phenomenon and then you try to make a priori predictions about the nature of that phenomenon.

BARTHOLOMEW: I think that I can give a possible linkage here. You may be understating the magnitude of information within population ecology. There was a very well documented case for natural selection, and knowledge about natural history and populations of animals. It was against this background of very substantial observational and nonquantitative information that the models were made. It is like saying that Scheid's first model could explain respiratory rhythms, breathing in and out. You watch vertebrates do this, and there is no doubt that they are breathing in and out. You know roughly that this involves their ribs and lungs and movement of air, and then you build a model. It is all black boxes, but it has a central control and it has various time units in it. I think that they are really the same.

POWERS: I think that you are all talking about the same thing. Your models differ in levels of refinement. It reminds me of Plato's allegory of the cave. We are all looking at shadows on the wall, describing reality as we see it, and somebody gets out of the cave and sees reality in a different light. He comes back and nobody believes him. There is a continuum of ideas here, and I think that they are not at all inconsistent. Depending on the particular model or particular theory you are generating, it depends on how much data are available or need to be available to test the model.

RANDALL: I can not think of any other models but interactive models or structural models. I do not think that it matters whether your model is based on a large amount or a minimum amount of data; that is almost immaterial, except that obviously the reference to data determines the value of the model.

PACKARD: There still is an important distinction. The so-called mechanistic model is relating the response causally to the variables in the equation. The second is relating them in a correlative manner to the variables in the equation. It makes a big difference then as to how you proceed from the equations to testing further hypotheses. The second model is essentially a multiple linear regression model, a correlative model. No causal relationship is implied between the dependent and the independent variable in the model. There may be, but you cannot make that inference. The only way that you can really find out whether there is a causal relationship is to proceed to the first of the two models and make a set of mechanistic predictions, and then find out if the predictions are satisfied.

14 Symmorphosis: the concept of optimal design

STAN L. LINDSTEDT AND
JAMES H. JONES

Optimality: history and definition

Optimality is a term that is frequently used in biology without a clear understanding of its meaning and ramifications. The concept of optimization was first formalized as an offshoot of maximization theory, applying calculus to find inflection points in functional relationships of pairs of variables. In turn, the "principle of optimality," as it was first introduced by Bellman (1957), developed from dynamic programing and relied on the calculus of variations to find integrated optima to multiple simultaneous equations. Optimality theory was soon incorporated into economics (e.g., Lipsey and Steiner, 1975) and biology (Rosen, 1967; Schoener, 1971; Hämäläinen, 1978; Alexander, 1982) as a quantitative approach to cost-benefit analysis. Although optimization as a guiding force or goal in shaping markets or in governing living systems is intriguing as a working hypothesis, this idea has proven to be difficult to test experimentally and is not universally accepted.

Optimality theory is now a formal mathematical construct whereby a maximizing solution to a specific function is found while the value of defined cost functionals is minimized (Schoemaker, 1984). Optimal solutions always represent choices among a discrete number of defined alternatives or possibilities. A result is optimal only within the context of the alternative solutions being considered and the cost functional utilized in the process. Despite these limitations, optimality theory is frequently applied to empirical problems in an attempt to understand the "economy of nature" in designing biological systems (Rosen, 1967; Hämäläinen, 1978; Alexander, 1982).

This discussion considers the application of optimality to the relationship between structure and function in the design of animals. We present evidence that both supports and refutes the optimization of structure-function integration in animal design and we discuss its implications for comparative physiology, morphology, and evolutionary biology. Finally, we consider the utility of identifying optimizing principles: how might such a mechanism affect our understanding of physiological ecology?

Optimization of form and function: symmorphosis

The heuristics of optimality in animal design dates back to the earliest recorded works in the fields of morphology, physiology, and the study of structure-function relationships in animals. Aristotle, in *De Partibus Animalium*, elaborated on functional balance between structures in an animal, stating that "nature invariably gives to one part what she subtracts from another." Galileo (1638) developed the quantitative concept of matching structure and function with his discourse on the scaling of bone size to body size, the first consideration of similitude in biology. Cuvier (1800) further expounded on this idea when he stated:

> It is on this mutual dependence of the functions and the assistance which they lend one to another that are founded the laws that determine the relations of their organs; these laws are as inevitable as the laws of metaphysics and mathematics, for it is evident that a proper harmony between organs that act one upon another is a necessary condition of the existence of the being to which they belong.

D'Arcy Thompson's *On Growth and Form* (1917) represented a major development of this approach by introducing the direct quantitative analysis of structural components in the context of the dynamic functional demands to which they are subjected. All these historical considerations of the quantitative relation of structure to function in animals contributed to the development of a postulate of optimality in animal design. In 1981, Taylor and Weibel introduced the principle of "symmorphosis," which they defined as

> a state of structural design . . . whereby the formation of structural elements is regulated to satisfy but not exceed the requirements of the functional system. . . . No more structure is formed and maintained than is required to satisfy functional needs.

Simply stated, symmorphosis considers structures to be optimally designed to satisfy functional requirements.

Restrictions and assumptions

Implicit (and explicit) restrictions govern the application of optimality theory. We must consider these before attempting to evaluate whether or not structures in animals are designed optimally. Three fundamental criteria must be identified before we can utilize optimality theory:

1. What functions (physiological, biochemical, ecological, or behavioral, but expressible in mathematical terms) or goal of the animal is being maximized? Although the "goals" and "costs" may be trivial to identify in economics, they are often both ambiguous and variable in biology. Optimal foraging theory states that animals are expected to behave to maximize energy intake per unit time (Schoener, 1971). However, foraging behavior may

sometimes be directed by a need for trace elements, avoidance of toxicants (see Rand, 1985), or minimization of intake of metabolically- or water-costly by-products such as salt. Any one, all, or none of these may be "optimized" under the appropriate unique suite of environmental conditions.

2. Identification of the cost functional to be minimized in a biological optimization problem may be equally difficult. The cost of greatest importance (in terms of optimization) may simply be the energy required to construct and maintain additional structure, or it may be in the form of compromises to other structures or functions. *No* adaptation is possible without cost (Mayr, 1982).
3. Optimization can operate only within a well-defined set of constraints. Structural constraints include allometric (Calder, 1984; Lindstedt and Swain, in press) and geometric (Alexander, 1982; McMahon, 1984) physical principles, among others. Most organs serve multiple functions. Optimization for one function may affect others. Mayr (1976) noted that an animal's phenotype is a compromise among various conflicting pressures; hence "every specialization is bought at a price." We sometimes overlook this fact in physiological research because we have been taught to isolate single dependent and independent variables while controlling others. In reality, every isolated feature must operate within constraints set in part by multiple demands.

Optimality in evolution

The principle of symmorphosis assumes that animals incur a selective penalty for maintaining structures in "excess" of the immediate demand. This idea, implied by Aristotle and Cuvier, was explicitly stated by Darwin in *The Origin of Species* (1859): ". . . natural selection will tend in the long run to reduce any part of the organism, as soon as, through changed habits, it becomes superfluous."

The quantity of structure formed and maintained and therefore the ultimate functional limitation to performance, is set by an animal's genotype, which selection can alter. At least since the evolutionary synthesis of the 1930's and 1940's, we have appreciated the actions that natural selection has on phenotypes. Traits that become prevalent in a species are those conferring a selective advantage in its environment. Physiological components that interact with the environment must play an essential role in selection. Thus the *genotype* determines (either proximately or ultimately) the quantity of *structure* that is built; the quantity of structure sets limits to the *function* that the animal can achieve; and this function is acted upon by *natural selection* to select the genotype that is promulgated (Figure 14.1).

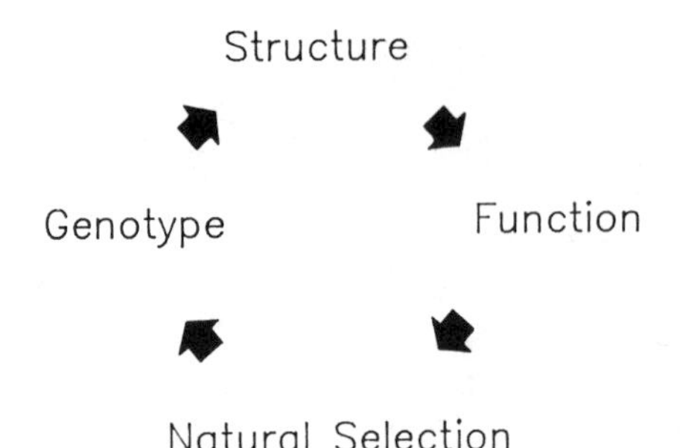

FIGURE 14.1 Schematic of the interrelationship between form, fitness, and function in the evolution of animals. Natural selection acts upon the phenotype; the functional properties of the phenotype are to some extent limited by the structural attributes of the animal; the structural elements are largely determined by the genotype (although plastic structures, e.g., skeletal muscle, may respond to functional demands); and the genotype is largely a result of natural selection acting on previous phenotypes in the lineage.

Could natural selection produce symmorphosis?

Is symmorphosis a truism? After all, isn't any structure designed to do what it does? Are there reasons to question the validity of the concept of symmorphosis a priori?

1. What if functional demands on a given structure conflict? Can such a structure be optimally designed? During the course of evolution, subcellular organelles, cells, tissues, organs, and even organ systems have been "exapted" (Gould and Vrba, 1982) to perform novel functions. Most structures continue to serve multiple functions. An animal is obviously more than a compilation of isolated structures. There must be anatomical and functional compromises as structures are called upon to perform multiple tasks or are constrained, for example, geometrically, within the organism. For instance, should skin be designed optimally for temperature regulation, exteroreception, osmoregulation, encryption, or as a barrier to infection? The answer must depend on both the environment and the animal's niche within it, which together define the selective forces acting on a population. For a 4-g pocket mouse in the dry plains of western North America, the skin must form an effective vapor barrier to prevent excessive water loss. Indeed, their skin has an exceptionally low water permeability (Lindstedt, unpublished). In contrast, a muskrat's pelage can be virtually porous to water vapor but must provide adequate insulation during dives in cold mountain ponds (Harlow, 1984). Is the skin optimally designed to solve these obvious demands in either of these species? All tissues, many

organs, and certainly some organ systems must accomplish multiple tasks; it seems an unlikely engineering feat that their design is optimized for each one.

2. Why should animals be designed optimally rather than just adequately? For natural selection to favor the optimal over the adequate, some associated evolutionary cost or penalty must be levied on mere adequacy. Extinction is an obvious penalty for maintaining inadequate structure to perform a given task. Much less apparent is what the penalty might be for too much structure. For example, the cross-sectional area of the trachea must be sufficient to allow adequate ventilation to support an animal's O_2 demand. What might be the selective pressure that prevents the trachea from being somewhat larger than just necessary, as symmorphosis would dictate? If we consider an animal as a series of interconnected physiological, ecological, and behavioral networks, parsimony alone argues for adequate over optimal design as a result of the net selective pressure.
3. Is symmorphosis testable? One serious criticism of optimality theory in biology is that it is often untestable (see Lewontin, 1977; Maynard Smith, 1978). Apparent "tests" of optimal design are often forgone confirmations. In this approach, a system of progressive approximation is used in which an extant system is compared to an optimality hypothesis. Any discrepancies simply call for a new optimality argument to be superimposed on the original hypothesis. Examples of the abuse of this approach and entertaining reading are both plentiful in Lewontin (1977) and Gould and Lewontin (1979). The invocation of additional selective pressures to explain an observed structural component can only lead to making "adaptation a metaphysical postulate, not only incapable of refutation, but necessarily confirmed by every observation" (Lewontin, 1977). Unless we are able to identify and quantify (in relative terms) the selective pressures that are acting on an animal, we cannot determine whether its structures have been optimized in response to those factors.

An alternative approach to optimization testing proceeds with the assumption of what feature or function is being optimized (i.e., the nature of the selective pressure in terms of maximization or minimization of functions) and evaluates the effectiveness of adaptation. The former approach may never be amenable to unbiased testing. The latter approach is testable. In fact it is the experiment of domestic breeding. That given traits may be isolated and maximized is a testimony to the effectiveness of goal-directed selection. How often, if ever, does this occur in nature?

Are some structures built optimally?

In the case of the mammalian respiratory system, Taylor and Weibel (1981) proposed that the function being maximized was O_2 flux through the system.

The penalty assessed for maintaining "excess" structure is presumably either the energetic cost of building and maintaining unneeded structures themselves, a concept ultimately dating back to Boltzmann (1886, quoted in Thompson, 1917), and/or the functional loss of space occupied by "excess" structure displacing other structures.

Two fundamental assumptions are inherent in this interpretation of structure-function relationships. First, according to this model, an organ system is assumed to serve only a single or primary purpose, and that purpose is the function that is being optimized. Indeed, the statement that gas exchange is the primary or cardinal function of the lung is found in nearly all treatises on respiratory physiology (Burri, 1985; Weibel, 1985; West, 1985). This premise ignores the fact that a mammal in which the other functions of the lung (e.g., temperature regulation, angiotensin activation) fail, dies as readily as one with impaired gas exchange.

A second assumption is that the phenotype is subject to stable and uniform selection. Phenotypes that exist now are products of selection that acted on preceding generations. Gould (1983) put it, "In many cases, evolutionary pathways reflect inherited patterns more than current environmental demands." Therefore, an extant species may not be optimized unless the rate of environmental change is slower than the rate of evolutionary change (see Van Valen, 1973; Wright, 1982). Only in a stable environment might optimization occur.

Design of the mammalian respiratory system

To test the validity of symmorphosis, one must quantify both functional demand as well as structural capacity. The mammalian respiratory system is well suited to a test of symmorphosis. Taylor and Weibel (1981) defined the functional demand of the respiratory system as the maximum rate of O_2 flow through the system, $\dot{V}_{O_2max}$. They then correlated $\dot{V}_{O_2max}$ with structural features associated with the five steps that make up the O_2-transport cascade (Dejours, 1975): ventilatory convection, pulmonary diffusion, cardiovascular convection, peripheral tissue (primarily skeletal muscle) diffusion, and intramyocyte diffusion into the mitochondria, where O_2 is reduced by the enzymes of the electron transport system (Figure 14.2). Because they studied mammals varying widely in body mass (M_b) and $\dot{V}_{O_2max}$, they were able to compare allometric changes in structural features with allometric changes in functional demand, $\dot{V}_{O_2max}$.

How did the results of Weibel and Taylor's analysis of the mammalian respiratory system fit with their working hypothesis that both sets of features (structural and functional) would vary identically with changes in M_b? They compared the allometric scaling of $\dot{V}_{O_2max}$ to measures of three structural components of the O_2 transport cascade: volumes of mitochondria (V_{mito}) in two locomotor muscles and the diaphragm (indicative of the total oxidative

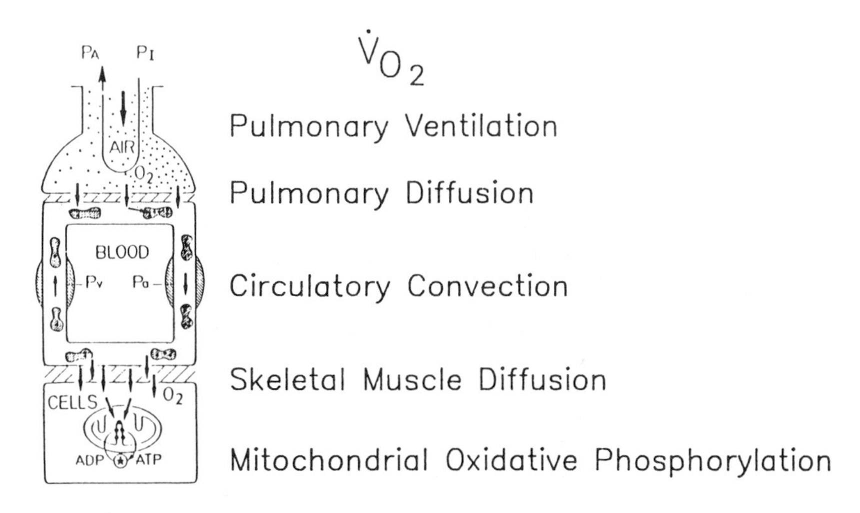

FIGURE 14.2 Schematic representation of the cascade of structural elements through which O_2 flows in the mammalian respiratory system from ambient air to its site of reduction in the mitochondria. The functional capacity at each step of the cascade is taken to be the maximum rate of O_2 consumption of the animal ($\dot{V}_{O_2max}$), and some of the structural elements responsible for the flow of O_2 through the system include the pulmonary diffusing capacity (D_{LO_2}), aerobic capacity of the heart (related to its total mitochondrial volume, V_{mito}), skeletal muscle diffusing capacity for O_2 (D_{smO_2}), and the capacity of the skeletal muscles to phosphorylate oxidatively ATP (related to the total skeletal muscle mitochondrial volume, $V_{mito(sm)}$).

capacity for ATP phosphorylation within the working muscle), total numbers of capillaries (N_{cap}) in the same muscles, which provide an initial estimate of the diffusing capacity for O_2 in the skeletal muscle (D_{smO_2}); and a morphometric estimate of the pulmonary diffusing capacity for O_2 (D_{LO_2}). They found that $\dot{V}_{O_2max}$, V_{mito}, and N_{cap} all scaled with nearly identical body mass exponents (0.81); however, D_{LO_2} scaled with a significantly greater exponent (1.00). (For a discussion of allometry and the interpretation of exponents, see Chapter 10.) This last finding contradicted the hypothesis of symmorphosis.

Since the original studies, several related findings have been published that expand the data base for understanding structure-function relationships in the mammalian respiratory system. The total (whole animal) volume of skeletal muscle mitochondria bears a constant and fixed relationship to $\dot{V}_{O_2max}$ in mammals of greatly differing M_b and mass-specific $\dot{V}_{O_2max}$ ($\dot{V}_{O_2max}/M_b$) (Hoppeler et al., 1984b, 1986a; Hoppeler and Lindstedt, 1985; Taylor and Jones, unpublished data). Several structural indices of the diffusing capacity of capillary beds in the peripheral skeletal musculature [D_{smO_2}, N_{cap}, capillary length

(J_{cap}), capillary volume (V_{cap}), and capillary surface area (A_{cap})] are each present in constant proportions to V_{mito} and thus to $\dot{V}_{O_2max}$, again in conformance with the predictions of symmorphosis (Hoppeler, Roesler, and Conley, 1986). These results are now confirmed beyond the cascade of O_2 resistances: V_{mito} and $\dot{V}_{O_2max}$ are matched in mammals not only to each other, but to specific activity of the skeletal muscle myosin ATPase and apparent rates of muscle shortening (Lindstedt, Hoppeler, Bard, and Thronson, 1985).

Hoppeler et al. (1984a) examined the relation between total V_{mito} in cardiac muscle and M_b and found that this structural variable scaled differently (as $M_b^{0.93}$) than the reported scaling of $\dot{V}_{O_2max}$ (as $M_b^{0.81}$). This finding suggested that the power output of the heart does not scale directly with $\dot{V}_{O_2max}$, and, to the extent that power output of the heart is an index of circulatory capacity, therefore might not be designed in conformance with symmorphosis. In addition to these morphometric studies, Jones et al. (1985) and Karas, Taylor, Jones, Reeves, and Weibel (1987) confirmed "excess" D_{LO_2} (relative to $\dot{V}_{O_2max}$) and cardiac output potential in larger mammals. They did this by running goats at $\dot{V}_{O_2max}$ while they were breathing a gas mixture containing 21% O_2, then running the same goats at $\dot{V}_{O_2max}$ while breathing 15% O_2. The goats achieved the same $\dot{V}_{O_2max}$ in both sets of experiments, demonstrating that they had greater O_2 transport capacity in their pulmonary and cardiovascular systems than was necessary to meet their functional needs at $\dot{V}_{O_2max}$ under control (21% O_2) conditions.

The results of these studies of the mammalian respiratory system show that structural variables sometimes do (total skeletal-muscle V_{mito} and D_{smO_2}) and do not (total cardiac muscle V_{mito} and D_{LO_2}) scale identically with $\dot{V}_{O_2max}$, the chosen index of the functional capacity of the system. Does this lack of complete structure-function matching prompt rejection of symmorphosis and indicate that the mammalian respiratory system is designed suboptimally?

In a recent reappraisal of symmorphosis, Jones (unpublished data) found that the distribution of $\dot{V}_{O_2max}$ as a function of M_b in the complete data set showed increasing variance with increasing M_b, but the distribution was bounded on the upper side by a line proportional to $M_b^{0.94}$. This exponent was significantly different from the 0.81 reported by Seeherman, Taylor, Maloiy, and Armstrong (1981), which was for the "average" $\dot{V}_{O_2max}$ of the arbitrarily chosen group of animals in that study. Jones identified the ten species of mammals that formed the upper boundary defining the 0.94 exponent as being those in which natural or artificial selection increased aerobic (or athletic) capacity: horses, canids, and small wild rodents (in which high aerobic capacities may be simply an allometric consequence of being small).

When Jones reassessed the discrepancy between the scaling of total cardiac V_{mito} and $\dot{V}_{O_2max}$, he found that seven of the eleven species analyzed by Hoppeler's group were species in the athletic group and these appeared to deter-

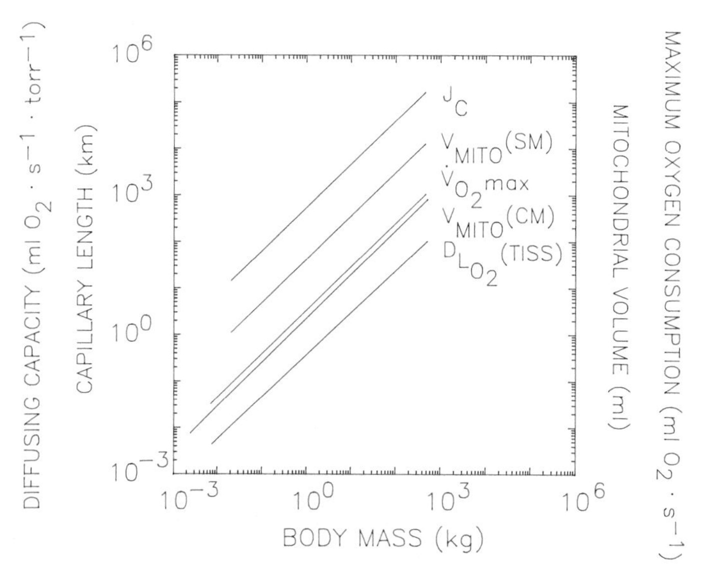

FIGURE 14.3 Relation of structural variables in the O_2 transport system and $\dot{V}_{O_2max}$ as functions of body size. The mass exponents (slopes of the regression lines on log-log coordinates) range between 0.91 and 0.94; none is statistically distinguishable from the others. J_{cap} is total skeletal muscle capillary length; $V_{mito(sm)}$ is total skeletal muscle mitochondrial volume; $\dot{V}_{O_2max}$ is the maximal rate of O_2 consumption for an animal running on a treadmill; $V_{mito(cm)}$ is total cardiac muscle mitochondrial volume; D_{LO_2} is the diffusing capacity of the lung for O_2.

mine the slope of the regression line. When the scaling of cardiac muscle V_{mito} was compared to $\dot{V}_{O_2max}$ in athletic species only, their exponents scaled identically (0.93 vs. 0.94). Similarly, the regressions of D_{LO_2} and $\dot{V}_{O_2max}$ in athletic species had nearly identical exponents (0.91 vs. 0.94) (Figure 14.3). Do these findings mean that the respiratory systems of some, but not all, mammals are designed according to symmorphosis? We regard the disparate scaling of some structures with $\dot{V}_{O_2max}$ in the original study to indicate that symmorphosis (at least as originally defined) is not generally applicable. One limitation of the original definition is that V_{O_2max} was considered the function that was maximized and to which the structural features of the system should be matched. In reality, however, very few species of mammals, including the majority of those in the original study (Taylor and Weibel, 1981), are strongly selected in nature or domestication for high aerobic performance. Clearly, evaluating a structure-function match with optimality theory that is based a priori on an invalid maximizing function (most species of mammals not being selected for $\dot{V}_{O_2max}$) must lead to a finding that symmorphosis does not hold. This is not to say that no mammalian respiratory systems are built in accordance with symmorphosis; however, if the functional demands to which their structures are optimally designed to match are diverse, attempting to demonstrate a match inevitably leads into Lewontin's (1977) trap of metaphysical postulation and tautological confirmation of the original hypothesis.

The species in which structures of the respiratory system scale identically with $\dot{V}_{O_2max}$ may be viewed differently. The observation that structure and function scale similarly in these species is in conformance with the predictions of symmorphosis, although it does not prove its validity. To demonstrate that the principle holds, each and every one of the structures in the system must be limiting to O_2 flux. Otherwise the identical scaling of structure and function may simply indicate that all sizes of athletic mammals have the same relative excess quantity of a given structure.

Notably, all species that define the high aerobic capacity line (exponent of 0.94) in the scaling of $\dot{V}_{O_2max}$ to M_b are species that because of their size (e.g., small wild rodents), their activity patterns in the wild (canids), or their selection by man (horses) would seem (ex post) to have experienced strong selection for aerobic performance for numerous generations. These species meet both criteria for the application of optimization analysis: they have experienced a strong selective pressure for a single functional trait (perhaps most obvious in race horses, where performance on the track, a function of $\dot{V}_{O_2max}$, has been the single phenotypic trait for which they have been selected), and the selection pressure for that trait has been historically stable and strong (again clearly depicted by horses, in which man has maintained selective pressure at a maximum level even as performance improved through the genotype).

To summarize, most mammalian species (at least as represented by the Taylor/Weibel and subsequent studies) show a direct quantitative relation between the respiratory system's maximal functional capacity ($\dot{V}_{O_2max}$) and some structural elements of the system (e.g., skeletal muscle V_{mito} and D_{smO_2}) but not others (e.g., D_{LO_2} and possibly cardiac-muscle V_{mito}). Species under strong selection for high aerobic performance show direct quantitative relations between $\dot{V}_{O_2max}$ and all structural features that have been measured. (We note the unavoidable tautology in identifying the athletically selected species as those that have high aerobic capacities. To do otherwise, we would have to breed "high performance" goats or sloths.)

The "typical" animals pose intriguing questions: What constraints set by functional demand necessitate that their "excess" structure be built and retained? Are the components of the system that are built in excess in the "typical" mammal genetically determined and fixed to be the same size as in athletic species, and is the maximum functional capacity of the system ($\dot{V}_{O_2max}$) determined simply by the more plastic components (e.g., V_{mito})? If so, what limits the development of the plastic components to prevent "typical" species from attaining the same level of performance as athletic species? Can our quantitative understanding of these relationships and contrasts between structure and function give us insights to the selective forces that determined them? Questions such as these exemplify the potential utility of analyzing structure-function relations in terms of the concept of symmorphosis. If symmorphosis is applicable in situations where appropriate criteria exist (single or dominant selective pressure acting on a system, continuous history of that selective pressure), it can be used to minimize variability that would otherwise obscure the patterns we seek to detect.

Symmorphosis vastly enhances the potential for our understanding the interrelationships of the alliterative trio: form, function, and fitness. By applying our quantitatively based understanding of structure-function relations, we have a powerful tool to assess mechanisms of natural selection. By considering how natural selection acts on the functional component of the phenotype, we can evaluate the resulting changes in structure.

Design of the mammalian kidney

Among mammals, few functions must be as closely regulated to support life as the maintenance of a constant internal environment. Chemical, pH, and volume balance are critical responsibilities of the kidneys. Despite this singular functional importance, interspecific variation in kidney structure is great. Might this variance be a consequence of optimal design?

The smallest mammals are apparently the most water stressed. Because of their high ratio of surface area to volume, coupled with inherently high rates of respiratory water loss, specific rates of water turnover are orders of magnitude higher in mice than bison, for example. In addition, many of the smallest desert rodents are granivorous, surviving without access to free water. The kidneys of these animals show remarkable structural and functional adjustments resulting in water conservation by production of very hypertonic urine (approaching 6000 mOsm/kg among the mice of the genus *Perognathus;* see Bard, 1984). Does this represent an optimally designed kidney for desert survival?

What are the identifiable constraints of design in the mammalian kidney? Body size profoundly affects kidney design (Calder and Braun, 1983). This is not just an inevitable correlation with body size (see Lewontin, 1977; Lindstedt and Swain, in press). The filtration pressure in the glomerulus does not change with increasing body size. However, the filtration demand does increase with body size (in direct proportion to metabolism). As nephron length is limited by a constant filtration driving force, the number of nephrons must increase as M_b increases. The consequence of set nephron length and increasing nephron number is that the geometry of the kidney changes and the relative size of the renal medulla decreases. Because urine-concentrating ability is a direct and linear function of relative medullary thickness

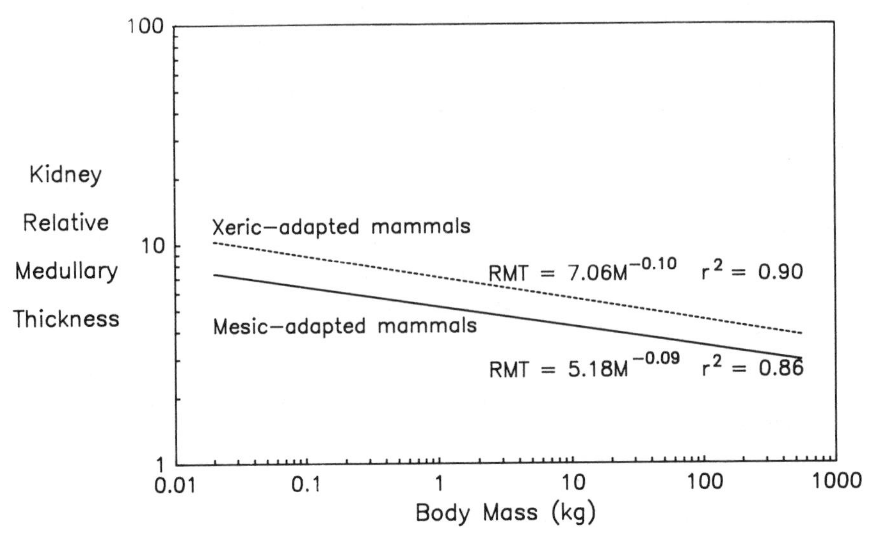

FIGURE 14.4 Body size acts to constrain the dimensions of mammalian kidneys. Hence, across a broad range of sizes there is a regular relationship between the relative thickness of the renal medulla and body mass. The result is a parallel scaling of maximum urine-concentrating ability. However, within a size range, those animals adapted to dry conditions have a greater medullary thickness and thus produce a more hypertonic urine. In this case, body size seems to set the standard with adaptive deviation (Calder, 1984) possible within limits. (Modified from Greegor, 1975.)

(Lindstedt, 1980), all smaller mammals are inevitably capable of producing a more hypertonic urine than all larger mammals. Despite the major influence of body size on renal structure, variation in structure is still present within a size range. This variation may represent optimization for a specific function. Within a species, kidney structure responds to antidiuretic hormone (ADH) (Trinh-Trang-Tan, Diaz, Grunfeld, and Bankir, 1981), and between species, to apparent water stress (Greegor, 1975) such that adaptation is apparently superimposed on the basic body size-dependent pattern (Figure 14.4).

Similarly, the "adaptation" of countercurrent respiratory water conservation, originally proposed as a specific adaptation for desert survival (Schmidt-Nielsen, Hainsworth, and Murrish, 1970), was subsequently demonstrated to be an inevitable consequence of small size, present in the most mesic-adapted of mammals to the same extent as the most xeric-adapted species (Schmid, 1976). The mere existence of a structure is not sufficient evidence to argue for its being optimally designed, as it may well be a consequence of other, though less obvious, constraints or demands.

The mammalian kidney provides another structural example that is relevant to consideration of symmorphosis. An evolutionary structural "theme" of sorts is the presence of some factor of safety. Do safety factors represent excess structure or do they represent a potentially optimal compromise of cost and risk? Alexander (1982) provides quantitative evidence in support of the latter possibility, especially in the structure of mammalian long bones. Any structure may be a consequence of the cost of producing and maintaining structure relative to the cost of failure. Though the costs involved may be less obvious than in the example of mammalian bone structure, mammalian kidneys are built with a factor of safety. A large fraction of nephrons may be lost with no apparent compromise in renal function; yet all species have about the same margin of error (Calder and Braun, 1983). Are nephrons present in excess, or is what appears to be an excess an optimal solution to the inevitable loss of functional nephrons that occurs during an animal's lifetime? In this case, might the structure-function relationship that is potentially optimized be ubiquitous among mammals, rather than evident in only those species that have been selected for very specific traits, as in the mammalian respiratory system.

Conclusions

In considering whether structures are optimized to their functions, we must be able to identify a function that is being maximized by the optimization procedure and a cost that is being minimized. In this paper we have considered how optimization might play a role in determining the structure-function relationship of two mammalian organ systems: respiratory and renal. Three patterns emerge that might place constraints on these structures being optimized for the functions they serve:

1. When a structure has been strongly selected for a single dominant function over a long period, its design may be optimized (i.e., built symmorphotically) for the performance of that selected function. If the function being selected has changed in the relatively recent past, the structural design may be evolving toward a new optimum but will not currently be optimized. If the structure has been selected for the performance of multiple functions, its design may be optimized to their integrated selective pressure. However, it may not be possible to demonstrate such optimization if it exists.
2. The ubiquity of safety factors designed into structures may represent optimal compromises between the costs (e.g., energetic, volumetric) of building and maintaining excess structures and the penalties associated with failure of those structures to function. Such optimization cannot be demonstrated because of the impossibility of ascribing appropriate mathematical functions to these variables.

3. Structures that exhibit great plasticity and readily adapt to meet changing functional demands may inherently be designed according to the principle of symmorphosis. Such structures will always be built just to meet, but not exceed, functional demand.

Great debates rage in the evolutionary literature as to the existence, nature of, and advantages of adaptation in structure and function (Gould and Lewontin, 1979; Grant and Grant, 1982; Conrad, 1983; Mayr, 1983; Hartl, 1985). The application of structure-function evaluations, if applied with a judicious consideration of the possibilities of the existence of symmorphotic relations, could potentially provide powerful quantitative tools to apply to these problems. Both nature and mankind have performed numerous experiments in selection, whether in terms of extreme environments, domestic breeding, or the allometry of changing body size. Both have created unique, strong, and potentially identifiable selective forces through evolutionary history.

In applying optimality criteria to biological systems, it is imperative to keep in mind the limitations of the approach and the potential abuse of information derived from it. As Schoemaker (1984) has pointed out, "When scientific data fit some optimality model (ex post), it does not follow that nature therefore optimizes. Since optimality is in the eye of the beholder, it is the latter who optimizes rather than nature." Darwin realized that nature pays no premium for perfection and that the key to survival is adaptability rather than optimality. However, in applying the optimality heuristic to our understanding of the continuum of form, function, and fitness, Maynard Smith (1978) aptly summarized the underlying principle, "The role of optimization theories in biology is not to demonstrate that organisms optimize. Rather, they are an attempt to understand the diversity of life."

References

Alexander, R. McN. (1982) *Optima for Animals.* London: Edward Arnold.

Bard, K. M. (1984) Structural and functional nephron heterogeneity in the Wyoming silky pocket mouse. Master's thesis, University of Wyoming, Laramie.

Bellman, R. (1957) *Dynamic Programming.* Princeton, N.J.: Princeton University Press.

Burri, P. H. (1985) Development and growth of the human lung. In A. P. Fishman and A. B. Fisher, *Handbook of Physiology,* sec. 3, vol. 1, *Respiration.* pp. 1–46. Bethesda, Md.: American Physiological Society.

Calder, W. A., III (1984) *Size, Function, and Life History.* Cambridge, Mass.: Harvard University Press.

Calder, W. A., III, and Braun, E. J. (1983) Scaling of osmotic regulation in mammals and birds. *Am. J. Physiol.* 244:R601–R606.

Conrad, M. (1983) *Adaptability: The Significance of Variability from Molecules to Ecosystems.* New York: Plenum Press.

Cuvier, G. (1800) *Leçons d'Anatomie Comparée.* Quoted in E. S. Russell, *Form and Function.* London: Murray, 1916.

Darwin, C. (1859) *The Origin of Species by Means of Natural Selection.* New York: D. Appleton, 1898.

Dejours, P. (1975) *Principles of Comparative Respiratory Physiology.* Amsterdam: North Holland/Elsevier.

Galileo, G. (1638) *Discorsi e dimostrazioni matematiche intorno a due nuove scienze attenenti alla Mecanica ed ai Muovimenti Locali: appresso gli Elzevirii.* Trans. H. Crew and A. de Salvioi, 1914.

Gould, S. J. (1983) *Hens' Teeth and Horses' Toes.* New York: W. W. Norton.

Gould, S. J., and Lewontin, R. C. (1979) The spandrels of San Marco and the Panglossian paradigm: a critique of the adaptationist programme. *Proc. R. Soc. Lond.* [*B*] 205:581–598.

Gould, S. J., and Vrba, E. S. (1982) Exaptation – a missing term in the science of form. *Paleobiology* 8:4–15.

Grant, B. R., and Grant. P. R. (1982) Niche shifts and competition in Darwin's finches: *Geospiza conirostris* and congeners. *Evolution* 36:637–657.

Greegor, D. H. (1975) Renal capabilities of an Argentine desert armadillo. *J. Mammal.* 56:626–632.

Hämäläinen, R. P. (1978) Optimization concepts in models of physiological systems. In *Progress in Cybernetics and Systems Research*, ed. R. Trapl, G. J. Klir, and L. Ricciadi, pp. 539–553. New York: Wiley.

Harlow, H. J. (1984) The influence of harderian gland removal and fur lipid removal on heat loss and water flux to and from the skin of muskrats, *Ondotra zibethicus. Physiol. Zool.* 57:349–356.

Hartl, D. L. (1985) Limits of adaptation. *Genetics* 111:655–675.

Hoppeler, H., and Lindstedt, S. L. (1985) Malleability of skeletal muscle tissue in overcoming limitations: structural elements. *J. Exp. Biol.* 115:355–364.

Hoppeler, H., Lindstedt, S. L., Claasen, H., Taylor, C. R., Mathieu, O., and Weibel, E. R. (1984a) Scaling mitochondrial volume in heart to body mass. *Respir. Physiol.* 55:131–137.

Hoppeler, H., Lindstedt, S. L., Uhlmann, E., Niesel, A., Cruz-Orive, L. M., and Weibel, E. R. (1984b) Oxygen consumption and the compositon of skeletal muscle tissue after training and inactivation in the European woodmouse (*Apodemus sylvaticus*). *J. Comp. Physiol.* 155:51–61.

Hoppeler, H., Jones, J., Lindstedt, S. L., Taylor, C. R., Weibel, E. R., and Lindholm, A. (1986a) Relating $\dot{V}_{O_2max}$ to skeletal muscle mitochondria in horses. (Abstract). Paper presented to the 2nd International Conference on Equine Exercise Physiology, August 7–10, San Diego, Cal.

Hoppeler, H., Roesler, K., and Conley, K. (1986b) The structural basis for energy supply to muscle mitochondria. *Proc. Int. Union Physiol. Sci.*, p. 524 (Abstract).

Jones, J. H., Lindstedt, S. L., Longworth, K. E., Karas, R. H., and Taylor, C. R. (1985) Muscle respiration limits aerobic capacity in goats. *Physiologist* 28:342 (Abstract).

Karas, R. H., Taylor, C. R., Jones, J. H., Lindstedt, S. L., Reeves, R. B., and Weibel, E. R. (1987) Adaptive variation in the mammalian respiratory system: VII. Flow of oxygen across the pulmonary gas exchanger. *Respir. Physiol.* 69:101–115.

Lewontin, R. C. (1977) Adaptation. In *The Encyclopedia Einaudi*, pp. 198–214. Torino: Giuilo Einaudi ed.

Lindstedt, S. L. (1980) Energetics and water economy of the smallest desert mammal. *Physiol. Zool.* 53:82–97.

Lindstedt, S. L., and Swain, S. D. (in press) Body size as a constraint of design and function. In *Population Biology and Life History Evolution of Mammals*, ed. M. S. Boyce. New Haven, Conn.: Yale University Press.

Lindstedt, S. L., Hoppeler, H., Bard, K. M., and Thronson, H. A., Jr. (1985) Estimate of muscle shortening rate during locomotion. *Am. J. Physiol.* 249:R699–R703.

Lipsey, R. G., and Steiner, P. O. (1975) *Economics.* New York: Harper and Row.

Maynard Smith, J. (1978) Optimization theory in evolution. *Annu. Rev. Ecol. Syst.* 9:31–56.

Mayr, E. (1976) *Evolution and the Diversity of Life.* Cambridge, Mass.: Belknap Press.

Mayr, E. (1982) *The Growth of Biological Thought: Diversity, Evolution, Inheritance.* Cambridge, Mass.: Harvard University Press.

Mayr, E. (1983) How to carry out the adaptationist program? *Am. Natur.* 121:324–334.

McMahon, T. M. (1984) *Muscles, Reflexes, and Locomotion.* Princeton, N.J.: Princeton University Press.

Rand, G. M. (1985) Behavior. In *Fundamentals of Aquatic Toxicology*, ed. G. M. Rand and S. R. Petrocelli, pp. 221–263. Washington, D.C.: Hemisphere.

Rosen, R. (1967) *Optimality Principles in Biology.* London: Butterworth.

Schmid, W. D. (1976) Temperature gradients in the nasal passage of some small mammals. *Comp. Biochem. Physiol.* 54:305–308.

Schmidt-Nielsen, K., Hainsworth, F. R., and Murrish, D. E. (1970) Counter-current heat exchange in the respiratory passages: effect on water and heat balance. *Respir. Physiol.* 9:263–276.

Schoemaker, P. J. H. (1984) Optimality principles in science: some epistemological issues. In *Issues in Interdisciplinary Studies*, vol. 1. ed. J. H. P. Paelink and P. H. Vossen, pp. 221–263. Aldershot, U.K.: Grower

Schoener, T. W. (1971) Theory of feeding strategies. *Annu. Rev. Ecol. Syst.* 2:369–404.

Seeherman, H. J., Taylor, C. R., Maloiy, G. M. O., and Armstrong, R. B. (1981) Design of the mammalian respiratory system. II. Measuring maximum aerobic capacity. *Respir. Physiol.* 44:11–23.

Taylor, C. R., and Weibel, E. R. (1981) Design of the mammalian respiratory system. I. Problem and strategy. *Respir. Physiol.* 44:1–10.

Thompson, D'A. (1917) *On Growth and Form.* Cambridge University Press.

Trinh-Trang-Tan, M. M., Diaz, M., Grunfeld, J. P., and Bankir, L. (1981) ADH-dependent nephron heterogeneity in rats with hereditary diabetes insipidus. *Am. J. Physiol.* 240:372–380.

Van Valen, L. (1973) A new evolutionary law. *Evol. Theory* 1:1–30.

Weibel, E. R. (1985) Lung cell biology. In *Handbook of Physiology*, sec. 3, vol. 1, *Respiration*, ed. A. P. Fishman and A. B. Fisher, pp. 47–91, Bethesda, Md.: American Physiological Society.

West, J. B. (1985) *Respiratory Physiology–the Essentials*, 3rd ed. Baltimore: Williams and Wilkins.
Wright, S. (1982) The shifting balance theory and macroevolution. *Annu. Rev. Genet.* 16:1–19.

Discussion

BURGGREN: Any given organ system can serve several different functions. If you analyze the respiratory system, for example, for symmorphosis only in the context of gas exchange, how much of the difficulty that Lindstedt had in matching design to performance could be explained by the fact that this system is optimally designed also for a subservient function other than respiration?

LINDSTEDT: That is obviously a question that neither I nor the inventors of the term have answered. It is a central and important question. We make the assumption that the system was designed for the purpose that we are investigating. Intuitively, the most obvious function of the respiratory system is the delivery of gases to and from the tissues; at least it is a critical function at maximal levels of gas transfer. We can at least start by analyzing those structures relevant to that one function, with the caveat that other subservient functions must be important. Carbon dioxide release, for instance, is never mentioned in Weibel and Taylor's considerations, yet one could argue that CO_2 is probably much more critical than oxygen, perhaps, in respiratory system design. It is certainly very important in control of ventilation.

BENNETT: That was exactly my point. How do you know that the system is not designed for maximal CO_2 release and consequent pH regulation?

LINDSTEDT: You are absolutely right. Much more CO_2 is in the system, and it seems as though the system is protecting this much more than oxygen. In exercise, O_2 and CO_2 are moving in approximately equal magnitudes in opposite directions. By and large, the arguments made for one gas are not necessarily contradictory to the arguments made for the other.

BENNETT: That is true in steady state and at levels of sustainable activity; but during non-steady-state activity, changes in carbon dioxide release are large. Respiratory exchange ratios may exceed 2.0. It may be very important to release CO_2 rapidly to avoid pH disruption. So, peak CO_2 release may be even more important than oxygen consumption in determining the design of the respiratory system. It would be very interesting, I think, to analyze the system for oxygen consumption, and then reanalyze it totally in terms of maximal CO_2 release and see if the predictions are congruent.

ARNOLD: I would like to make a general but minor semantic point concerning the use of the terms "plasticity" and "genetics" in opposition. These are not alternatives. A problem in using "genetically fixed" or "genetic" as opposed to "plastic" as alternatives is that the meaning physiologists imply is not the one that most geneticists would use. It would be better to say "plastic" and "nonplastic," and then look at inheritance as another dimension, so that we could have heritable differences in a plastic trait, just as we could have heritable differences in a nonplastic trait. There is no reason for using the term "genetics" in this context.

LINDSTEDT: One could argue that plasticity is a very important component of the genome such that the heritable trait is plasticity. It makes sense to build a system that way.

RANDALL: In discussions of optimal design, we are really dealing with problems of rate-limiting systems. Under different conditions and in different animals, different steps will be rate-limiting. So, the design of the experiment becomes very critical. Maybe the animal is designed to do everything cheaply, rather than maximizing oxygen uptake. Thus, your selection of goats as experimental animals seems critical if your objective is to investigate limitations to maximal oxygen uptake. I probably would not have chosen goats; I would have chosen an animal that went like a bat out of hell, one that had evolved to maximize oxygen uptake.

LINDSTEDT: We are doing just that. We are rearing pronghorn antelope, which do go like a bat out of hell. In fact, the selection of goats was intentional for the very reasons you mentioned. Goats clearly would not survive in the wild; they are not the animal of choice for any experiment, except for one calling for animals extreme in their aerobic performance. The objective is to see what it is overbuilt or just adequate in this animal and compare that to the obvious choice, such as pronghorn antelope. We certainly now have predictions with regard to the way that pronghorn should behave under similar experimental conditions.

FLORANT: The pronghorn and the goats undergo transitory stresses. A question is whether they are in "extreme environments" for a brief moment in time. One reason why physiological ecologists study animals in extreme environments is to increase the signal-to-noise ratio. Under a stressful situation, solutions to a particular physiological problem may be fewer. You have provided us with an interesting idea regarding fixed and variable resistors. If one looks at fixed and variable components for a particular system that is maximally stressed in an extreme environment, do you see a difference between those animals and the same population not under the same selective pressure?

LINDSTEDT: Let me use another system as an example. General principles of renal physiology might be easier to uncover by looking at desert-adapted animals because of their exaggerated kidneys. If you raise litter mates, some with access to water and others in a more arid situation, one finds very significant differences in the adult animals in their urine-concentrating ability and relative medullary thickness. There is enough plasticity so that raising animals in dissimilar environments may result in two different adult animals with similar genetic makeup.

FLORANT: Is it more plastic, though, than necessary?

LINDSTEDT: I do not know the answer to that question.

METCALFE: I had a humiliating experience with goats. We exercised goats during pregnancy to study their thermodynamics and to see whether they change from the nonpregnant state. The man who taught these animals to exercise on the treadmill used aversive conditioning. He shocked their tail if they did not run hard enough. The data were adequate and we published them. However, the kids born to those animals were small. I wanted to repeat the experiment because I really did not believe it. We had a graduate of an animal-care department with us who refused to teach animals to run on a treadmill by aversive conditioning. She taught them by positive reinforcement. She got them to run much harder and faster, and their babies were not smaller. I had to publish another paper. My point concerns commitment and behavioral background against which an animal performs a given task. The animals were not shocked while they were performing our experiment. They had been shocked only during training, so that from the outside I could not have told the two groups apart.

BENNETT: If one is interested in animal design and function and testing limiting factors, to what extent do we want to use or shun domesticated species? These have been subject to selection for many generations for specific traits – for example, goats for docility, racehorses for speed. Are these the systems we should investigate because of that selection? Or might physiological interactions have been so distorted by directional selection on individual traits that we really should be investigating only wild populations.

ANONYMOUS: The broad-breasted turkey, which has been selected for meat yield, is no longer able to copulate so they must be artificially inseminated. They also have a high degree of aneurysm. As a result, they must be maintained on tranquilizers. That is an extreme case, but it suggests that your concern could be well founded.

ARNOLD: If you are looking at a system that has been put in a state of real maladaptation as a consequence of deliberate selection, then you could get very misleading results. If you are doing ecological genetics, it can be very important to have a natural reference population. In other words, you cannot explore the ecological issues if the history of selection is completely artificial.

FUTUYMA: I can imagine for certain types of questions, a highly aberrant, highly selected strain of turkeys that are incapable of copulating might be just what you need, in the sense that you may have a perturbation of sufficient magnitude that it reveals something about the normal functioning of the system in a way that might otherwise be obscured, except with extraordinary kind of invasive manipulation. Similarly, you use mutants in genetics to learn about developmental pathways.

LINDSTEDT: If ultimately we are interested in principles, it is not a very good test of the principle to uphold it in one really aberrant animal. Nevertheless, if the principle still holds despite all of the domestication, selection, etc., this argues strongly for the importance of that principle.

HUEY: On the other hand, if the principle does not hold in a test with a domesticated animal, it would not be clear whether this invalidates the principle or is merely an artifact of domestication.

POWERS: The example of the turkey is one of hundreds that could be made. Even animals kept in the laboratory for a generation or two, or sometimes just a few months, can show dramatic changes in their physiological performance. Agnathans were thought not to have immunoglobulins. It turned out that these animals were kept in a bucket and were in a poor state of health. When kept properly, they did express immunoglobulins. The immune response of many organisms will go downhill in the laboratory. Many experiments look great in the laboratory; but in the field, only sixty percent of the "laboratory" variables are significant.

BENNETT: One of the original tests of symmorphosis was proposed by Weibel and Taylor and involved the matching of allometric exponents of different variables. They proposed that if the allometric exponents are all congruent with each other, then the system appears to be designed symmorphologically. You [Lindstedt] gave an example of similar exponents of maximal oxygen consumption and ATPase activity. I worry about that kind of argument because of the possibility of false correlative associations. It is very difficult for me to see how ATPase activity, which is a measure of a maximal inherent contractile frequency of muscle, could be associated with maximal oxygen consumption. During burst sprinting in lower tetrapods, limb-cycling frequency may be twenty times as fast as those associated with maximal aerobic speed and maximal oxygen consumption. My guess would be that the congruence in their physiological allometric exponents is due to a false positive correlation with another factor.

LINDSTEDT: That is entirely possible. Those kinds of data are never sufficient evidence, but they are certainly a very good first step. One would hope that we would be able to follow them up with other experimental data.

FUTUYMA: If you have a complex network where many of the components are not rate-limiting, like tracheal diameter, for instance, the interesting implication for evolutionary biologists is that the intensity of stabilizing selection on a character like that is presumably very weak, except for rather broad deviations from the mean. This has very interesting implications for people who are interested in the study of selection in natural populations, individual variation, and patterns of morphological evolution.

BENNETT: That gets back to the problem of more than one function for a particular structure. Human tracheas, for instance, may have certain properties because they are associated with speech or making necks attractive to the opposite sex, or whatever. Any of those factors might provide a constraint, rather than oxygen delivery capacity.

FUTUYMA: If someone interested in evolution wants to find out whether a character departs from some postulated optimum or is not subject to strong stabilizing selection, where does this person get information of that kind, other than from people who are engaged in the detailed study of the function of these features? Are you saying that there is always such an infinity of possible functions for a feature that you can never hope to show that the feature is only weakly constrained by selection?

15 Assigning priorities among interacting physiological systems

DONALD C. JACKSON

Introduction

Physiologists have traditionally studied systems or structures in the body in isolation; that is, either separated from the rest of the organism or under circumstances in which other bodily processes, insofar as possible, are controlled and do not complicate the study's prime objectives. This is clearly essential to understand the responses and mechanisms of a single component; however, it is not how an organism normally functions. An organism is an integrated whole, whose parts interact in complex and poorly understood ways as the animal responds to its environment and to its genetic imperatives.

A major challenge of current physiology, and probably of physiology for the foreseeable future, is to fathom these interactions, particularly as an animal approaches its limits in terms of performance or environmental adaptation. Interactions under these conditions may be viewed in terms of priority considerations, much as a social unit must determine priorities for allocating limited resources. The background for understanding these interactions has been extensively developed. We have considerable, although of course not complete, understanding of physiological functioning of particular systems in resting conditions. In addition, comparative and environmental physiologists have extensively documented, in a descriptive fashion, the adaptations of selected organisms to environmental and performance extremes. The next step, already being vigorously taken, is to sort out the physiological priorities and interactions under these conditions. This is a complex undertaking that will ultimately require collaboration between physiological ecologists and mechanism-oriented physiologists studying all levels of organization down to the molecular.

My objective in this chapter is necessarily more modest than the ambitious ultimate objective just stated. It is to provide a framework or outline for thinking about physiological priorities into which new approaches or problems may be incorporated. This framework is based largely on existing information and ideas, but should be viewed as little more than a tentative start to addressing these problems. My emphasis is at the level of systems physi-

TABLE 15.1 Physiological systems

Organ systems	Control systems
Cardiovascular	Acid-base
Endocrine	Blood gases
Gastrointestinal	Blood pressure
Nervous	Fluid volume
Renal	Osmolality
Respiratory	Specific solutes
Skeletomuscular	Temperature

ology, although priority decisions must occur at both higher and lower levels of organization.

Physiological systems

Two general categories or systems may be defined: organ systems and control systems. Examples of these categories are listed in Table 15.1.

Normal organ system function can be assessed by examining the balance between supply and demand for important blood-borne substances such as metabolic substrates and oxygen. When supply equals demand, a so-called steady-state condition exists. The composition of metabolites and other solutes in the tissue fluids of the organ are constant and the condition of normal homeostasis prevails. In a conflict situation, an organ's demand may exceed its supply, leading to disturbances of fluid composition and tissue function.

Optimal control system function is operationally identified by the maintenance of the regulated variable in the blood or other body fluid compartment at its nominal or normal value although, as will be discussed below, this value may change. The controlled value may be within a very narrow range or within a rather broad range of acceptable values. A disturbance to the control system may result in an inability to maintain the variable within the prescribed range. The well-being of a living organism depends upon stable organ system and control system function.

However, circumstances that strain the capacity to maintain all systems in homeostatic balance are common occurrences for many animals. Indeed, the literature is rich in examples. One thinks of the occasional exploratory dives of Weddell seals that may last for more than an hour (Kooyman, 1981), an extraordinarily long time for a mammal to be deprived of ambient oxygen. Similarly, the male emperor penguin, also from the Antarctic, does not feed during the four months of polar winter when it incubates an egg (LeMayo,

Delclitte, and Chatonnet, 1976; Pinshow, Fedak, Battles, and Schmidt-Nielsen, 1976). How do these animals order their physiological priorities when the uptake of crucial substances is eliminated for long periods? These examples are extreme, but are documented occurrences for these well-adapted animals. Learning how such animals resolve the physiological conflicts that must result from these conditions is a major goal of comparative and adaptational physiologists.

To understand physiological conflicts, both the spectacular and obvious as just discussed and the less obvious as well, it is useful to categorize the general conditions that lead to conflicts. Once these are identified, the physiological solutions to specific cases can be addressed more readily. The categories listed below are somewhat arbitrary, they overlap with one another, and they are not all-inclusive. The intention is to organize generally familiar physiological ideas in a way that will stimulate new experimental approaches. Again, it is not the intention to suggest that these issues are resolved, but rather to provide a context for considering problems in physiological integration.

Categories of conflict-producing conditions

Environmental extremes

For each animal, a range of environmental conditions are compatable with normal homeostasis. If the environment is aqueous, then a range of tolerable temperatures, salinities, hydrostatic pressures, gas tensions, etc. can be defined for an aquatic organism. If aerial, then a range of temperatures, vapor pressures, gas partial pressures, etc. are appropriate for a terrestrial or air-breathing organism. Within these limits, the regulatory systems of an animal can maintain the constancy of its internal environment, although at the extremes the control mechanisms may be hard pressed. In the laboratory, we can isolate environmental conditions and define a limit, for example, to temperature, but in a natural environment this limit may be influenced by uncontrollable changes in other environmental factors that interact to produce a very different response pattern.

Exposure to environmental conditions outside the normal range for any of the ambient characteristics can disturb normal homeostasis. For example, high temperature can exceed the capacity of the heat loss mechanisms of an endotherm and lead to hyperthermia, whereas excessive salinity can disturb osmotic and ionic balance in an aquatic species, etc. Other control systems can be indirectly affected as well. High-temperature exposure increases cardiac output and water loss. If the heat exposure is combined with physical activity, then the stress is compounded for both water balance and cardiovascular function. Evaporative cooling must cope with both exogenous and endogenous heat, and the cardiovascular system is faced with elevated

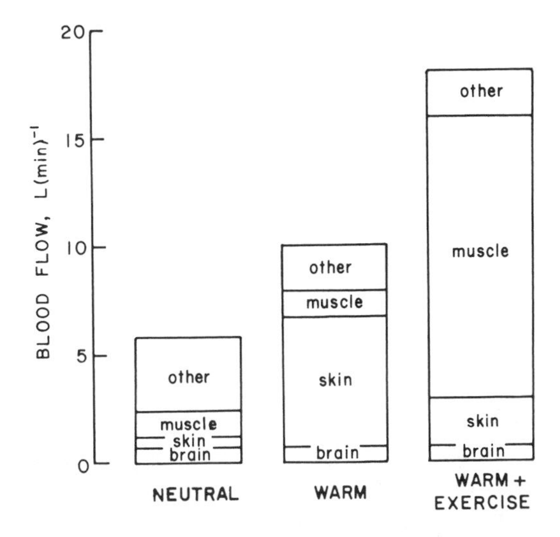

FIGURE 15.1 Effect of heat exposure and combined heat exposure and exercise on total blood flow and on flow to selected tissues of a human subject. Heat exposure alone increases skin flow while flow to "other" tissues falls below normal level. Exercise leads to a large increase in muscle flow, in part at the expense of skin flow. Note that brain flow is unchanged throughout. (Adapted from Houdas and Ring, 1982.)

demands for blood flow from both the active muscles, where O_2 is required, and from the skin, where the heat must be lost. These effects on blood flow distribution are illustrated in Figure 15.1.

Limited resource

Normal physiological function relies upon an adequate supply of various substances, including metabolic substrates, water, and oxygen. This category has obvious overlap with the preceding one, but the distinction in perspective makes it worth separate consideration. An inadequate supply of a resource can result from either a decrease in the availability of the substance or an increase in its loss or consumption. An organism can employ three general strategies to cope with such a shortage. The first is to step up the activity of mechanisms responsible for acquiring or exchanging the substance; for example, by increasing foraging or feeding in the case of inadequate substrate or by increasing ventilatory or circulatory flow in the case of inadequate oxygen. The second strategy, in the event of more severe deprivation, is to conserve the limited supply by selective allocation based on priority considerations. An example of this strategy is the distribution of blood flow during prolonged forced diving. Flow is maintained at near normal level to the ner-

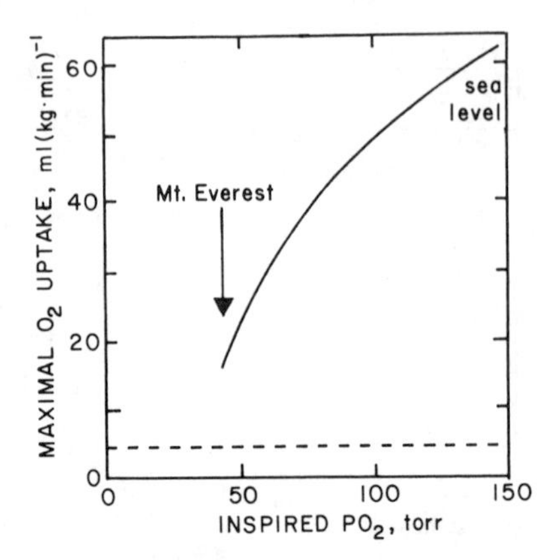

FIGURE 15.2 Maximal O_2 uptake of acclimated human climbers at various altitudes up to the altitude of Mt. Everest. The measurement at the summit elevation (8848 m) was simulated by giving subjects hypoxic gas to breathe at 6300 m. Note the progressive decrement in peak performance above sea level. (Adapted from West et al., 1983.)

vous system, but the skin, the muscles, and other hypoxia-tolerant tissues are deprived of flow. By this means the limited available oxygen is used primarily by those tissues that need it most, and the organism is able to endure the lack of oxygen intake for a much longer period. We know much about these responses during forced laboratory dives, but are only beginning to discover the interactions during voluntary dives in nature.

A mismatch between supply and demand, despite the mobilization of these physiological strategies, usually results in a loss of performance capacity. Oxygen is again a good example. Under conditions of maximal exercise, the O_2 delivery systems are probably operating at their peaks. If ambient oxygen availability falls, the maximal oxygen consumption ($\dot{V}_{O_2max}$) will be expected to fall because no further compensatory increase in the delivery systems is possible. This has been demonstrated clearly by measuring $\dot{V}_{O_2max}$ of human climbers at various elevations (Figure 15.2). Even at the lowest elevation studied, performance fell significantly below the sea level value. A very active area of study at the present time concerns the interactions of physiological systems when an animal is performing at or above its aerobic limit (see Chapter 14). What limits performance and how systems interact and adapt during peak performance are poorly understood, even for extensively studied species, such as humans. For other species, where peak performance occurs naturally in complex behavioral and environmental settings, much work remains to be done.

A third strategy is to abandon normal regulation temporarily. This special approach, utilized by some animals when thermal, hypoxic, or dehydration stress is particularly severe, will be considered later in the chapter.

Shared mechanism

Not uncommonly, one structure or mechanism performs several essentially independent functions, sometimes simultaneously. If one of these functional demands increases and this function has high priority, then other duties performed by this mechanism may suffer. A clear example of this principle is the pulmonary exchange of many mammals and birds, which accomplishes oxygen uptake, carbon dioxide loss, and evaporative heat loss via panting. If ambient oxygen is low, or if the heat load on the animal is high, increased ventilation that disturbs CO_2 homeostasis and acid-base balance will result. Clearly, the maintenance of constant P_{CO_2} and pH has a lower priority than the acquisition of ample oxygen or the elimination of excess heat. In this example the various functions are performed simultaneously and conflict arises if the demands on one function are excessive. There are also shared mechanisms in which only one function can occur at a time. An example is the skeletomuscular apparatus used by the marine turtle, *Chelonia mydas*, for both lung ventilation and locomotion. When an adult female is on the nesting beach, it alternates between breathing and walking (Jackson and Prange, 1979). A high priority of this animal is to complete the nesting process as quickly as possible, but periodically this activity is interrupted, probably in response to threshold stimulation of respiratory chemoreceptors, so that breathing can occur. The two events are related functionally; increased activity leads to increased breathing: increased breathing is required for the activity. They were never observed to occur simultaneously, which indicates (but does not prove) that they cannot do so. In some animals, the distinction between respiratory and locomotory muscles is blurred still further. Fish that employ ram-jet ventilation swim with their mouth open. The same muscles that propel the fish through the water also force water to pass over the gas exchange surfaces of the gills. In shared processes such as this, it may often be difficult to decide which is the primary function and which secondary, and how performance of one function may be compromised by the demands of the other.

Optimal set point

As already discussed, physiological variables are regulated within defined limits to preserve normal function and homeostasis. As a consequence, when the values deviate above or below this range, function may be impaired. An appropriate question to ask is why a particular value is optimal. This question can be approached by considering the consequences of values beyond the normal range. When this is done for a number of systems, the finding is that

TABLE 15.2 Hematocrit values of selected vertebrate species

Species	Hematocrit (%)
Bladdernose seal *(Cystophora cristata)*	63
Human	45
Rabbit *(Oryctolagus cuniculus)*	35
Camel *(Camelus dromedarius)*	25
Turtle *(Pseudemys scripta)*	25
Skate *(Raja punctulata)*	7
Icefish *(Chaenocephalus aceratus)*	0

the normal range or value represents a compromise between conflicting effects. Somero (1986) has denoted the normal range as midrange values. On some scale of cost effectiveness, the normal state represents the optimal state for the whole spectrum of possible values. Some examples can illustrate this.

The fractional (or percent) concentration of red cells in blood, the hematocrit, has a characteristic value for various animals (Table 15.2). Considering that the primary function of red cells (and their contained hemoglobin) is to transport O_2, one might predict that if 45% red cells is good, then 60% cells would be even better. The problem is that increased viscosity of the blood resulting from high hematocrit costs more in cardiac work than the benefit derived from increased blood O_2 carrying capacity. Conversely, blood with low hematocrit is easy for the heart to pump, but the disturbance to O_2 delivery outweighs the advantage. Experimental data confirm this reasoning by showing that the work of delivering O_2 to the cells is minimized at about the normal hematocrit of an animal (Stone, Thompson, and Schmidt-Nielsen, 1968).

Body temperature is an extensively studied variable that is regulated more or less precisely in most vertebrates. In mammals and birds, in which the thermal range is narrow, cellular function is adapted for optimal performance in this temperature range. Temperatures far outside this range can disturb function, especially in the central nervous system. Ectothermic vertebrates regulate temperature primarily by behavioral means, and their precision of control, depending on the characteristics of the environment, is often less than in the endotherms. Regulation in these animals has been associated with upper and lower temperature limits (Heath, 1970; Barber and Crawford, 1977), between which the animals normally maintain their body temperatures primarily by behavioral means. Because of the pervasive effects of temperature on physiological processes, the identification of the particular effects that are optimized is difficult, but the assumption is that the cost/

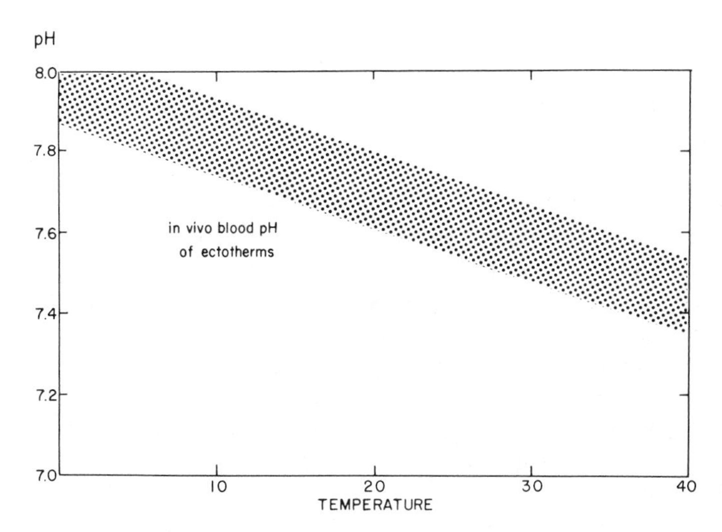

FIGURE 15.3 Blood pH of ectothermic animals as a function of body temperature. The values from a variety of species fall within the shaded band. (From Jackson, 1982.)

benefit equation is most favorable in the regulated range (Huey and Slatkin, 1976). At high temperature, an ectotherm enjoys the benefit of an active metabolism, but the cost to supply active tissues with substrate is high; at low temperature, in contrast, the effort needed to support energy metabolism is greatly diminished, but at the cost of reduced capacity to forage and to evade predators. The interactions are complex and the underlying factors leading to temperature selection may not always be obvious. For example, certain species of Australian skinks have preferred temperatures below the temperature at which peak running speeds are attained (Bennett, Huey, and John-Alder, 1986).

A final example in this category is blood pH, a variable that influences, as does temperature, many physiological functions and to which these functions are adapted. The pH of blood and other body fluids appears to be carefully regulated in all vertebrates, although the higher vertebrates have a narrower range of control and tolerance. Values either above normal (alkalosis) or below normal (acidosis) disturb normal physiology, particularly in the gross functioning of the nervous system and in the various cellular activities involving protein molecules.

The regulated value of blood pH in mammals and birds is close to 7.4, and deviations from this are designated as acid-base disturbance. In ectothermic vertebrates, on the other hand, blood pH varies significantly and predictably as a function of body temperature (Figure 15.3). Hypotheses for the functional significance of this change are (1) that the blood's relative alkalinity is maintained (Howell, Baumgardner, Bondi, and Rahn, 1970) and (2) that net charge state of proteins is held constant (Reeves, 1972). There is evidence that enzymatic activity is optimal at each temperature when pH changes with temperature in this fashion (Park and Hong, 1976; Somero, 1986). A further discussion of these concepts is beyond the scope of this paper, but it should be added that departures from ideal relative alkalinity or alphastat blood acid-base behavior are widespread among vertebrate species that have been studied. Leading figures in comparative acid-base physiology continue to challenge the general concept (Cameron, 1984; Heisler, 1986). Therefore, as with probably every topic discussed here, much work remains to be done.

A more general point to be made in this context is that set points or regulated values are not fixed, but may, like blood pH, change depending on ambient conditions, such as temperature, or because of changing physiological conditions or demands. Additional examples can illustrate this principle. As already discussed, blood hematocrit in humans is close to 45%, and is regulated at that level by a typical feedback mechanism involving the hormone erythropoietin. Chronic hypoxia or high-altitude sojourn, however, leads to a controlled increase in hematocrit that is considered to be an adaptive response, except when it is excessive. Presumably, chronic hypoxia has changed the balance between the conflicting effects of viscosity and O_2 capacity, and the optimum has moved to a higher hematocrit. The behavorial selection of temperature range by ectotherms is also variable and can be adjusted, or that selection can even be abandoned if cost relationships change (Huey and Slatkin, 1976). Hypoxia again has recently been shown to be an effective environmental influence in this regard. Wood, Dupré, and Hicks (1985) have found that a variety of ectotherms select lower temperatures in a thermal gradient when ventilated with an hypoxic gas mixture. This suggests that the optimal point or range for body temperature has been modified by this condition. Finally, an example can be given from human physiology and medicine that involves the size of the heart ventricle at the end of diastole. According to the Frank-Starling relationship, the force of contraction increases, within limits, as end-diastolic volume (EDV) increases. Normally, however, EDV is set at an intermediate level. At lower volumes, contractile force is too low for adequate stroke volume, although cardiac work is reduced; at higher volumes, contractile force is high but ventricular work is excessive (Laplace's law) and upstream pressure in the pulmonary circuit is also high, leading to congestion and edema. Once again, though, changing circumstances can alter the cost/benefit balance. If the heart is weakened and normal cardiac output cannot be maintained at normal EDV, the heart will enlarge to exploit the greater available force. The price that is paid is pulmonary edema and ventilatory difficulties, but the perfusion of the tissues has the highest priority.

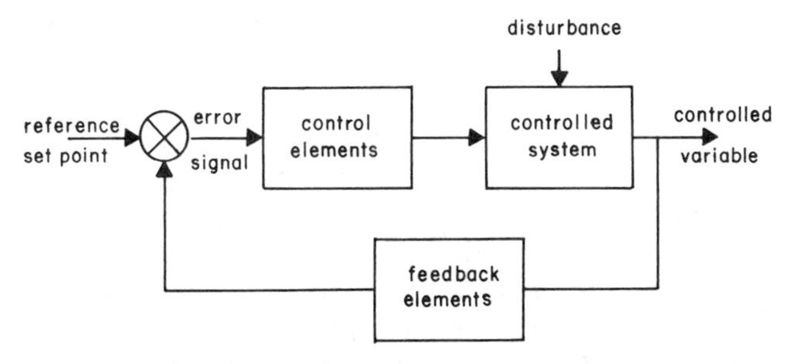

FIGURE 15.4 Generalized control system including a feedback loop to a control center. The loop provides information on the state of the regulated variable that can be compared to the optimal, or set point, level. A discrepancy, or error signal, causes correcting responses in the control elements.

Malfunction

A final category of conflict is one seldom addressed in comparative and physiological ecology, but whose significance nonetheless is considerable. This is the disturbance to physiological function caused by injury, disease, or aging. The compensatory responses and physiological conflicts associated with these natural occurrences have of course been extensively considered in human and veterinary medicine, but sick or injured animals have generally been excluded from physiological studies. Fitness and performance are adversely affected by these conditions, and sick, injured, or aged organisms are forced to prioritize among physiological needs when additional stress is imposed. For example, Porter et al. (1984) found that viral infection and reduced immune response can interact with other environmental challenges, such as moderate limitation of food and water, to reduce growth or survival of young mice. If adequate food and water were available, however, compensatory physiological mechanisms initiated by the sick animals were able to maintain normal performance.

Mechanisms for setting priorities

Control system characteristics

Physiological variables are typically regulated by a negative feedback mechanism involving receptors, central nervous command center, and effectors, with communication links transmitted by nerves or hormones. A generalized diagram is shown in Figure 15.4. Besides identifiable structural features of a control system, operational features can also be described including set point, gain, and time constant. The characteristics and values of these operational features may be important determinants for setting priorities among physiological control systems.

Set point for most systems is not a discrete, immutable entity, but rather a probabilistic, floating target for the controlling mechanisms. An early formulation of this was put forward by Hammel et al. (1963) to explain thermoregulatory behavior in dogs. They proposed a variable temperature set point within the hypothalamus that changed in response to inputs from peripheral thermoreceptors, and receptors responding to exercise, the reticular formation, and other brain regions. As already discussed, various controlled variables show shifts in the regulated state related to conflict-producing circumstances that presumably affect the control system set point. The details of these effects are poorly understood, but have been studied with respect to thermoregulatory control (Heller, 1979).

The gain of a control system describes the magnitude of the effector response as a function of the displacement of the controlled variable from its set point (the error signal). In a proportional control system, the response – for example, lung ventilation in respiratory control – is approximately a linear function of the lung or arterial P_{CO_2} at P_{CO_2} values above the normal level. The slope of this relationship (dV/dP_{CO_2}) is the gain of the controller. If two control systems compete for the same mechanism, such as temperature control and P_{CO_2} control in the case of the respiratory apparatus, the controller with the higher gain will presumably dominate the response. This principle also accounts for the recruitment pattern of control mechanisms when a regulated variable deviates farther and farther from its normal value. An example is blood pressure control in mammals during progressive hemorrhage (Berne and Levy, 1986). Mild hypotension stimulates carotid and aortic baroreceptors, which mediate compensatory mechanisms to restore pressure. With more severe hemorrhage, down to 60-torr arterial pressure, the baroreceptors no longer respond and other mechanisms such as the peripheral chemoreceptors and the cerebral ischemic response come into play to continue compensation. The low gain of these latter reflexes at higher pressures makes them ineffective until hypotension is profound.

The ordering of priorities among regulated variables also relates to their time constant of change. A rapidly changing variable will have priority over a slowly changing one. For example, high-temperature exposure requires the expenditure of water for evaporative cooling. The loss of water can eventually lead to severe dehydration and death if fluid replacement does not occur. The threat of hyperthermia, however, is even more acute because the time constant for temperature change is generally shorter than that for dehydration. This can be illustrated by estimating the time it would take to reach lethal limits if the exchange of selected, regulated variables was interrupted. The variables are oxygen, heat, water, and food. The calculations are based on resting human values and assume that uptake of O_2, water, or food and

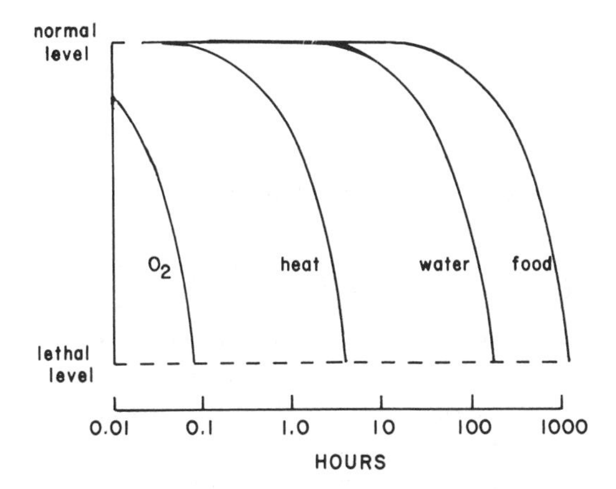

FIGURE 15.5 Idealized representation of rates of change to lethal levels in selected physiological variables under conditions of interrupted exchange. There is O_2 consumption but no breathing, heat production but no heat loss, water excretion and evaporation but no fluid intake, energy metabolism but no food intake. The times shown are realistic for resting, thermoneutral human subjects, but would differ for different conditions or different species. Note that for diagrammatic purposes, heat is paradoxically shown as falling to a lethal level.

loss of heat are discontinued, while consumption of O_2 or food, loss of water, and production of heat continue at normal resting rates. As shown in Figure 15.5 and as is doubtless no surprise to any reader, apnea is the most acute life-threatening event, and physiological responses dealing with this stress would have highest priority, and so on with the other variables. The quantitative details of these relationships would of course differ under different ambient or activity conditions or in different taxa, but the general concept would be similar. This is admittedly a simplistic model that fails to consider how these variables and their priority relationships may interact in real dynamic circumstances.

Central versus local control

An organism can be considered in one sense as a large aggregate of essentially independent cells and organs that carry on their private functions in a normal fashion so long as their local environment remains stable. In another sense, of course, the organism itself is an entity with organization, communication, and division of labor. This is a notion that could be developed further, but for the purposes of this discussion let us just consider the implications for organ system function under conflict-producing circumstances.

Organ function depends critically on blood flow, and mechanisms intrinsic to tissues autoregulate the flow to match the metabolic needs. This corresponds to the independent existence of the tissue. Tissue flow is also controlled via centrally mediated mechanisms subserving organismally relevant variables such as blood pressure. Under stressful situations, these two mechanisms, the local and the central, may conflict, and the actual tissue blood flow will depend on the balance between them. Familiar examples of this are observed during exercise, during breath-hold diving, and during hemorrhage. In hypoxia-tolerant tissues, such as inactive skeletal muscle and skin, central control dominates the local effects and flow will fall in the interest of overall body function. In hypoxia-sensitive tissues, such as brain and heart, and in active skeletal muscle, on the other hand, local regulation rules and flow is sustained to support the needs of these local tissues.

Special adaptations

In some conflict-producing circumstances, the organism essentially abandons the homeostatic struggle because the cost of sustaining it is too high or even impossible to meet. An example of this response is the phenomenon of metabolic arrest (Hochachka, 1986). Many organisms, under severe conditions of cold, anoxia, or dehydration, can profoundly reduce their metabolic rate and the exchange of material with their environment. A specific, familiar example is mammalian hibernation, in which body temperature falls to a level near the cold surroundings so that energy expenditure is greatly reduced. Technically, this is not an abandonment of control, but is rather a drastic and reversible downward shift in the regulated temperature. Nonetheless, it is a strategy that permits survival of the animal under conditions in which the maintenance of normal euthermic homeostasis is not feasible.

Many organisms, both vertebrates and invertebrates, are facultative anaerobes, which means that they can survive periods of total oxygen lack (anoxia). Unlike the tissues of highly O_2-dependent organisms, all of their tissues, including their nervous system, must be able to function temporarily without O_2. Because of the problem of substrate depletion and end-product accumulation, however, an additional key adaptation of these animals is metabolic depression. My colleagues and I have studied these events in the freshwater turtle, *Chrysemys picta*, probably the most anoxia-tolerant vertebrate examined to date. This animal can survive periods of submergence anoxia at 3 °C for up to five months, but at rates of energy utilization that are profoundly reduced (Ultsch and Jackson, 1982). We estimated that the anaerobic metabolic rate at 3 °C was 0.5% of the aerobic rate at 20 °C (Herbert and Jackson, 1985). The underlying mechanisms accounting for this and other examples of metabolic arrest are uncertain and pose important questions for current research.

Privileged sites

In the communal society of *Animal Farm*, Orwell (1946) told us that "some animals are more equal than others." So it is with the community of systems within a living organism. Some, as we have seen, have higher priority than others and are favored when resources are in short supply or when the body is stressed. In terms of organ systems, the central nervous system has the highest priority for blood flow and delivery of O_2 and substrates. Maintenance of brain function is obviously crucial to an animal, and this is ensured in several ways:

1. By whole-body control system function. For many vertebrate control systems, including temperature, fluid balance, food intake, and osmoregulation, the command center and the primary receptor mechanisms are located within the brain. For others, such as blood pressure and P_{O_2}, the receptors are situated on the arterial pathway to the brain. These regulatory processes can therefore be viewed as relating directly to the homeostasis of the local environment of the brain; overall homeostasis of the whole organism follows as a consequence. To paraphrase an old expression: "What's good for the brain is good for the rest of the body."
2. By powerful local control of blood flow. As discussed earlier, local vascular control is dominant in the cerebral circulation over centrally mediated autonomic control. Even during conditions such as diving and strenuous exercise, when cardiac output is selectively distributed, the brain flow is determined chiefly in accordance with the local environment of the brain microcirculation.
3. By local regulation of the fluid environment of the brain. The composition of the cerebrospinal fluid and the brain interstitial fluid is distinctly different from the circulating plasma. Two factors account for this: first, the blood-brain barrier, a selective permeability shield located between the blood and the brain fluid at the capillary endothelium and other interfaces; and second, active transport processes occurring in the choroid plexuses and capillaries that help define the distinctive composition of the cerebrospinal fluid. These mechanisms help protect the brain environment even in the event of disturbance to the whole-body extracellular fluid. Thus, the constancy of the internal environment of the body as defined by the great nineteenth-century physiologist Claude Bernard is really only the first line of defense for the brain; this is backed up by a second line of defense consisting of the blood-brain barrier and the choroid plexus. The organization of the various central and local control systems provide the bases for the high priority of the brain in the economy of the organism.

The second privileged site is the intracellular compartment of the body. This is not a discrete entity, but consists of the contents of each of the multitude of cells in the body. Because of practical difficulties, physiologists have only recently begun to focus their attention on the intracellular space. Traditionally, blood has been studied as an indicator of body homeostasis, because of its ready accessibility and because it has provided a convenient indicator of the internal environment. The cellular environment, though, is clearly the crucial one, because this is where most living processes occur. Technical advances, such as microelectrodes, dyes, and nuclear magnetic resonance, have opened windows into the cellular space, although the variety of cells and the complexity of individual cells still make their study a formidable task.

Like the brain, the cellular compartment possesses local mechanisms that help regulate its state even when the interstitial fluid bathing the cells is altered. The cell membrane is a selective permeability barrier, and together with a host of membrane transport processes, it ensures and preserves a distinctive cellular composition. Such characteristics as pH, ion concentrations, organic substrate concentration, and volume are under active control. Again the analogy to the brain situation is that the cell membrane and its properties represent a second line of defense against environmental perturbations.

Conclusions

The concept of an ordering of priorities among interacting physiological systems is not new. In this chapter, a variety of categories of conflict with examples, as well as some possible mechanisms whereby priorities are set, have been assembled from the existing literature. Doubtless, other writers would have devised different categories and chosen different examples. But each would agree with me, I am sure, that this is a subject that has only begun to be explored.

It is a truism of sorts that every part of the body is linked in some fashion to every other part. But this trite statement is very much to the point being made here. A disturbance to one part reverberates throughout the organism, and produces responses, compromises, and adaptations of various functions. For the ecological physiologist, complexity of internal physiological interaction is compounded by the complexity of social, behavioral, and environmental factors that produce them. An ultimate goal is to understand the interacting physiological responses of animals to their natural dynamic environment. Giant strides have been taken in recent years as a result of technical advances in noninvasive physiological monitoring and in microprocessor-based data acquisition systems that can be attached to free-ranging animals. The application of these and other new techniques by imaginative investigators will further advance the productive collaboration between ecology and physiology.

References

Barber, B. J., and Crawford, E. C., Jr. (1977) A stochastic dual-limit hypothesis for behavioral thermoregulation in lizards. *Physiol. Zool.* 50:53–60.

Bennett, A. F., Huey, R. B., and John-Alder, H. B. (1986) Body temperature, sprint speed, and muscle contraction kinetics in lizards. *Physiologist* 29:179.

Berne, R. M., and Levy, M. N. (1986) *Cardiovascular Physiology.* St. Louis, Mo.: C. V. Mosby.

Cameron, J. N. (1984) Acid-base status of fish at different temperatures. *Am. J. Physiol.* 246:R452–R459.

Hammel, H. T., Jackson, D. C., Stolwijk, J. A. J., Hardy, J. D., and Stromme, S. B. (1963) Temperature regulation by hypothalamic proportional control with an adjustable set point. *J. Appl. Physiol.* 18:1146–1165.

Heath, J. E. (1970) Behavioral regulation of body temperature in poikilotherms. *Physiologist* 13:399–410.

Heisler, N. (1986) Comparative aspects of acid-base regulation. In *Acid-Base Regulation in Animals*, ed. N. Heisler, pp. 397–450. Amsterdam: Elsevier Science.

Heller, H. C. (1979) Hibernation: neural aspects. *Annu. Rev. Physiol.* 41:305–321.

Herbert, C. V., and Jackson, D. C. (1985) Temperature effects on the responses to prolonged submergence in the turtle *Chrysemys picta bellii.* II. Metabolic rate, blood acid-base and ionic changes, and cardiovascular function in aerated and anoxic water. *Physiol. Zool.* 58:670–681.

Hochachka, P. W. (1986) Defense strategies against hypoxia and hypothermia. *Science* 231:234–241.

Houdas, Y., and Ring, E. F. J. (1982) *Human Body Temperature: Its Measurement and Regulation.* New York: Plenum Press.

Howell, B. J., Baumgardner, F. W., Bondi, K., and Rahn, H. (1970) Acid-base balance in cold-blooded vertebrates as a function of body temperature. *Am. J. Physiol.* 218:600–606.

Huey, R. B., and Slatkin, M. (1976) Costs and benefits of lizard thermoregulation. *Q. Rev. Biol.* 51:363–384.

Jackson, D. C. (1982) Strategies of blood acid-base control in ectothermic vertebrates. In *A Companion to Animal Physiology*, ed. C. R. Taylor, K. Johansen, and L. Bolis, pp. 73–90. Cambridge University Press.

Jackson, D. C., and Prange, H. D. (1979) Ventilation and gas exchange during rest and exercise in adult green sea turtles. *J. Comp. Physiol.* 134:315–319.

Kooyman, G. L. (1981) *Weddell Seal: Consummate Diver.* Cambridge University Press.

LeMayo, Y., Delclitte, P., and Chatonnet, J. (1976) Thermoregulation in fasting emperor penguins under natural conditions. *Am. J. Physiol.* 231:913–922.

Orwell, G. (1946) *Animal Farm.* New York: Harcourt Brace Jovanovich.

Park, Y. S., and Hong, S. K. (1976) Properties of toad skin Na-K-ATPase with special reference to effect of temperature. *Am. J. Physiol.* 231:1356–1363.

Pinshow, B., Fedak, M. A., Battles, D. R., and Schmidt-Nielsen, K. (1976) Energy expenditure for thermoregulation and locomotion in emperor penguins. *Am. J. Physiol.* 231:903 -912.

Porter, W. P., Hindill, R., Fairbrother, A., Olson, L. J., Jaeger, J., Yuill, T., Bisgaard, S., Hunter, W. G., and Nolan, K. (1984) Toxicant-disease-environment interactions associated with suppression of immune system, growth, and reproduction. *Science* 224:1014–1017.

Reeves, R .B. (1972) An imidazole alphastat hypothesis for vertebrate acid-base regulation: tissue carbon dioxide content and body temperature in bullfrogs. *Respir. Physiol.* 14:219–236.

Somero, G. N. (1986) Protons, osmolytes, and fitness of internal milieu for protein function. *Am. J. Physiol.* 251:R197–R213.

Stone, H. O., Thompson, H. K., Jr., and Schmidt-Nielsen, K. (1968) Influence of erythrocytes on blood viscosity. *Am. J. Physiol.* 214:913–918.

Ultsch, G. R., and Jackson, D. C. (1982) Long-term submergence at 3 °C of the turtle, *Chrysemys picta bellii*, in normoxic and severely hypoxic water. I. Survival, gas exchange and acid-base status. *J. Exp. Biol.* 96:11–28.

West, J. B., Boyer, S. J., Graber, D. J., Hackett, P. H., Maret, K. H., Milledge, J. S., Peters, R. M., Jr., Pizzo, C. J., Samaja, M., Sarquist, F. H., Schoene, R. B., and Winslow, R. M. (1983) Maximal exercise at extreme altitudes on Mount Everest. *J. Appl. Physiol.* 55:688–698.

Wood, S. C., Dupré, R. K., and Hicks, J. W. (1985) Voluntary hypothermia in hypoxic animals. *Acta Physiol. Scand.* 124:46.

Discussion

FLORANT: Peter Hochachka suggested that certain regulatory mechanisms may be suspended or arrested during periods of dormancy or periods of low temperature. I do not really feel that it is a suspension or an arrest. Many regulatory functions occur even at low temperatures. They are just turned down, as Craig Heller has described.

JACKSON: Certainly in hibernation, the thermoregulatory system is still functioning, although at a lower level, and certainly sensory systems are still operating. So, I do not think that you can say that it is turned off. It is turned down, certainly, but that is part of what Hochachka is talking about.

BURGGREN: Mammalian physiologists often regard any change from what they view as the homeostatic set point as a "deterioration" or "lack of regulation." Many such physiologists fail to realize that ectotherms, or animals that generally experience a very broad range of environmental factors still regulate, but sharply adjust their set point.

JACKSON: Many variables, including variables in mammals, are not regulated at a fixed value. They tend to change with the circumstances.

BENNETT: Claude Bernard must be spinning in his grave because you did not include the extracellular fluid compartment as one of your defended areas. I think that this exclusion was conscious, in the sense that you now feel that the central nervous system and intracellular compartments are more strongly defended. Is this the result of the appreciation of new information that was unavailable to Bernard? When you give introductory physiology classes, do you still talk about the internal environment or is it passé?

JACKSON: The regulation of the internal environment, as Bernard described, is a given for all of us. That is what we have assumed to be the object of the regulatory systems. Beyond that regulation is further regulation of these other sites. The internal environment is bathed in the external environment and is stabilized from changes in the ambient environment by these physiological functions. The cells and the central nervous system are bathed in the internal environment, and those environments are in turn regulated more by their own special processes.

BENNETT: In many animals other than mammals, almost nothing is regulated at exactly one set point. Now we have information that the classical view of the constancy of the internal environment does not apply to most animals.

JACKSON: This has certainly been an interest of mine, in seeing how regulation proceeds at different temperatures in an ectotherm. Some things do seem to change with temperature; for example, pH and P_{CO_2} are both regulated at these new set points. But other factors, for example, ion concentrations, are not that changeable with temperature.

BARTHOLOMEW: One positive thing to emerge from the study of animals in the natural environment is that the physiology of interest at the present time is the physiology of the nonsteady state. It is the steady state that the physiologists have assumed and then imposed upon their preparations. Right now the exciting thing is the physiology of regulation in an unsteady world with unsteady processes.

RANDALL: The shift from Claude Bernard's view is that there are changes in states in all animals, including mammals. There are different optima for each parameter at each state. In comparative physiology, we have very obvious examples, for instance open and closed bivalve mussels, where you have enormous changes in almost everything as the bivalve closes. Then it opens up and shifts to another set of steady states. Studies of the shift between one state and another is an area of considerable interest.

DAWSON: Another area of interest is the response of animals to oscillatory types of phenomena, and these responses may be different from those to step changes. Generally studies are carried out using step changes rather than oscillations, even though the latter are common in the natural environment.

FUTUYMA: The normal state of most animals is to be parasitized, and yet little is known of the effects of these parasites on physiological function of the host. This is an area that has not been explored extensively.

16 Physiological changes during ontogeny

JAMES METCALFE AND MICHAEL K. STOCK

Introduction

Ontogeny, by which we mean the development of an individual organism, offers many fascinating questions to the physiological ecologist. Successful reproduction is, of course, essential for species survival. For successful ontogeny, a series of complex processes must be integrated among at least three individuals: the mother, the father, and the embryo or fetus.

Traditionally, ontogeny refers to that extremely complex series of carefully orchestrated processes that result in an independent individual of the same species as its parents. Although the genetic components involved in reproduction have recently attracted special attention, the environment of the developing ovum, embryo, and fetus is of extreme importance for its survival and health. In many instances, indeed in most, the environment provided to its unborn offspring by the female is dramatically different from that occupied by the adult organism. One example of this difference is the relatively poor oxygen supply available to the developing organism. Recent studies have stressed the importance of oxygen availability for the development of the embryo, and we will use this work to illustrate current investigations concerning the influence of the prenatal environment.

The prenatal environment: oxygen as a regulator of growth

The most highly oxygenated blood of the sheep fetus in utero (the blood returning from the placenta in the umbilical veins) contains less than half as much oxygen as arterial blood does in a healthy adult. We cannot conclude from this, however, that oxygen availability is unimportant for normal fetal growth and development. Relatively small decrements in oxygen availability result in fetal growth retardation, and larger decrements cause congenital deformities.

The mammalian fetus is poorly suited for studies of the importance of oxygen supply. The experimental difficulties imposed by the physical protection of the mother's body have been surmounted with ingenious catheterization techniques. However, there is a more insidious handicap: the maternal organism responds to experimental changes in environmental oxygen concentration by physiological adaptations that limit the impact of those changes on the fetus. For example, evidence derived from pregnant sheep in the Peruvian Andes, where oxygen availability is lowered by the decreased atmospheric pressure (Barron et al., 1963), indicates that adaptations in maternal breathing, maternal blood composition, and maternal blood flow maintain fetal oxygen supply at a value close to that characteristic of sea level. The fetus, in fact, appears physiologically unaware that its mother is at high altitude. More recently, studies performed on pregnant women at high altitude suggest that many women can accomplish the same protective functions for their fetus (Moore et al., 1982, 1984).

To evaluate the effects of changes in oxygen availability on the fetus, the hen's egg (*Gallus domesticus*) has been used as an experimental model. This model has the considerable advantage that no maternal organism intervenes to thwart experimental manipulations of oxygenation in the developing embryo. On the other hand, oxygen availability is apparently quite different in the hen's egg than in the mammalian uterus. The partial pressure of oxygen (oxygen tension) in chorioallantoic venous blood (analogous to umbilical venous blood in the mammal) exceeds 80 torr during much of the twenty-one–day incubation period (Metcalfe and Bissonnette, 1981), a value close to that in adult arterial blood; however, as the time for hatching approaches, the oxygen tension falls to levels similar to those found throughout gestation in the mammalian umbilical vein (Metcalfe, Stock, and Ingermann, 1984). In synchrony with the fall in oxygen tension, embryonic growth slows and oxygen consumption, which has previously been rising in parallel with embryonic growth, now plateaus contemporaneously with the decline in growth rate (Metcalfe et al., 1984). On the nineteenth day of incubation, when the chick embryo begins to breathe air, the oxygen tensions in chorioallantoic arterial and venous blood rise, oxygen consumption increases, and growth rate (expressed as grams per day of gain in wet mass) spurts to levels above any seen earlier in incubation (Metcalfe et al., 1984).

On the basis of the synchronous changes in these three indices of embryonic metabolism (oxygen tensions in chorioallantoic blood, growth rate, and oxygen consumption), we postulated that the changes in growth rate and oxygen consumption occurred secondarily to the changes in oxygen availability. To test this hypothesis, we exposed hen's eggs beginning on the sixteenth day of incubation to an atmosphere containing 60% oxygen (hyperoxia, compared to 21% oxygen in the air we breathe), thus increasing oxygen availability during the several days prior to the onset of pulmonary respiration. The effects on embryonic growth and embryonic oxygen consumption, summarized in Figure 16.1, are consistent with our hypothesis. Compared with controls maintained in 21% oxygen, the hyperoxic embryos grew faster and maintained a higher level of oxygen consumption per gram of body

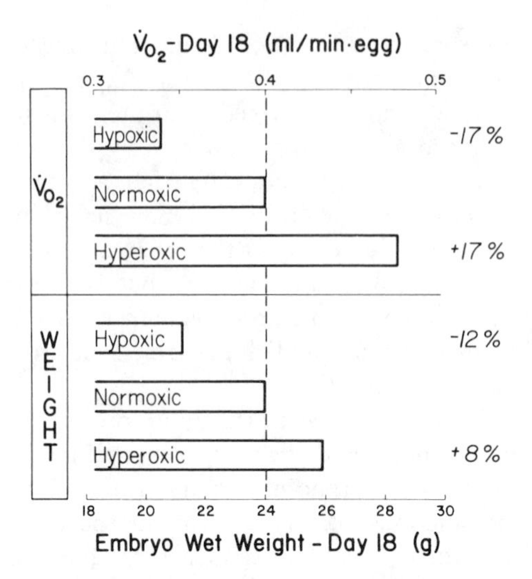

FIGURE 16.1 The effects of a three-day alteration of oxygen availability on oxygen consumption and growth of the chick embryo. Hyperoxic eggs were exposed to 60% oxygen beginning on the sixteenth day of incubation. Hypoxic eggs were exposed to 15% oxygen beginning on day 16. Measurements were made on day 18.

mass. These data appear to confirm the dependence of prenatal growth on oxygen availability.

Figure 16.1 also contains data for animals exposed to 15% oxygen (hypoxia) beginning on the sixteenth day of incubation. In contrast to the hyperoxic embryos, both growth and oxygen consumption declined relative to controls in the hypoxic embryos. Parenthetically, it is interesting and important to recognize that when faced with a restriction of oxygen availability, the avian embryo (and the mammalian fetus) has one mechanism for conserving oxygen that is not available to the adult organism: it can stop growing, maintaining life in the hope of improved oxygen availability at some future time.

The mechanism by which oxygen availability regulates growth of the avian embryo in the days prior to hatching is currently the subject of investigation in our laboratory. Three general possibilities are apparent. First, oxygen might act upon some specific group of cells in the embryo to increase the production of a growth-regulating substance like insulin, growth hormone, or an insulin-like growth factor. Second, oxygen might increase the sensitivity of cells to substances that regulate growth. Third, oxygen might act to increase the rate of adenosine triphosphate (ATP) generation in all the body cells; higher concentrations of ATP in the cells of a growing organism might be used, at least in part, for accelerated growth. Preliminary evidence derived from the nucleated red blood cells of the chick is consistent with the hypothesis that hyperoxia stimulates ATP generation (Ingermann, Stock, Metcalfe, and Shih, 1983). In this regard, two further characteristics of the metabolic response to hyperoxia are worthy of mention. First, although oxygen consumption of the near-hatch embryo begins to rise within three hours of initial exposure to hyperoxia, the magnitude of the response continues to increase for at least seventy-two hours, even when corrected for the gain in mass during that time (Stock and Metcalfe, 1987). Second, the stimulation of oxygen consumption induced by hyperoxic exposure does not disappear immediately when the egg is returned to a normoxic environment. After three hours of exposure to 60% oxygen, the elevated rate of oxygen consumption does not return to normal levels for at least three hours after return to a 21% oxygen environment, a much longer period of time than we would expect intracellular oxygen tension to remain elevated (Stock, Asson-Batres, and Metcalfe, 1985). Taken together, these two findings suggest that the activity of some regulator of oxygen consumption, perhaps one of the mitochondrial respiratory enzymes, is stimulated and that after stimulation, the increased enzyme activity cannot be "turned off" promptly.

Control of growth by oxygen availability may occur in vertebrates other than the chicken. The growth rates of the embryos of several species of fish can be increased by increasing the concentrations of dissolved oxygen in the water they breathe (Hamor and Garside, 1976). Experimental methods of inducing chronic hyperoxia in mammalian fetuses have not yet been developed; however, the growth rate of the human fetus slows perceptibly in the last two weeks before parturition (Babson, Behrman, and Lessel, 1970), then rises within the first month after the onset of air breathing (Babson and Benda, 1976). It is certainly possible that oxygen availability normally regulates the growth of the mammalian fetus at the end of pregnancy.

The maternal role: physiological adaptations during pregnancy

The maternal organism plays an important role in determining the environment of her developing offspring. In egg-laying species, for example, the selection of an appropriate nesting site and the performance of carefully regulated nesting behavior are of supreme importance. In placental mammals, the maternal organism must provide for the needs of her offspring in the uterus during the most formative period of its life. To provide a suitable milieu for embryonic development, important alterations in maternal physiology occur. These deviations from nonpregnant physiology may be so profound that they jeopardize the ability of the female to react to stress.

To illustrate the challenge of this aspect of ontogeny, we will deal with some of the changes that occur in the maternal cardiovascular and respiratory systems during human pregnancy. These changes are critically impor-

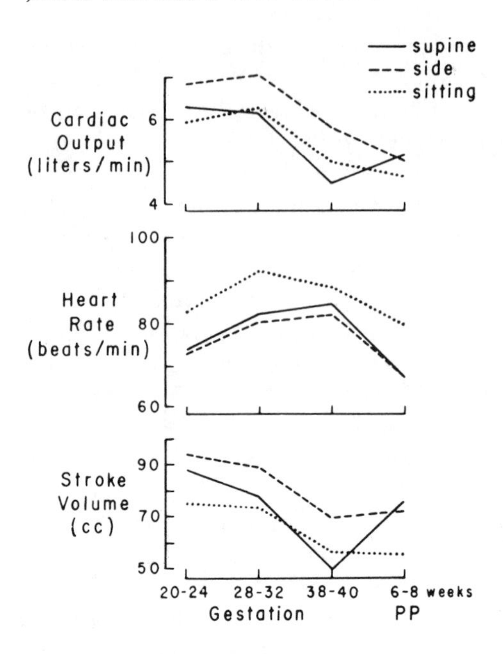

FIGURE 16.2 Measurements of heart rate and cardiac output (by dye dilution) were made in the same women during pregnancy and six to eight weeks post partum. Each subject was studied at rest in three body positions. Knowledge of cardiac output and heart rate allowed calculation of stroke volume. (From Metcalfe, McAnulty, and Ueland, 1981.)

tant in providing the fetus with an acceptable environment for its development.

Maternal cardiac output

Cardiac output is the product of heart rate multiplied by the amount of blood ejected with each systolic contraction ("stroke volume"). Resting cardiac output increases by an average of 40% above nonpregnant values during pregnancy in normal human subjects (Ueland, Novy, Peterson, and Metcalfe, 1969). The change in cardiac output has several unexpected features, as illustrated in Figure 16.2. First, most of the increase occurs relatively early in gestation. The highest values in all body positions are observed by mid pregnancy. Second, as term approaches, maintenance of the supine position causes a striking fall in cardiac output, to levels below those found post partum. This characteristic is probably unique to humans and is attributed to compression of the inferior vena cava by the pregnant uterus, resulting in a decrease in venous return to the heart. In the sitting position as well, cardiac output tends to be lower than in lateral recumbency, probably as a result of (1) venous pooling in the veins of the lower extremities secondary to gravitational effects, and also (2) the increase in venous distensibility that occurs during pregnancy (Wood and Goodrich, 1964; McCalden, 1975; Fawer et al., 1978).

Early in pregnancy the increase in cardiac output depends upon increases in both heart rate and stroke volume. Figure 16.2 shows that, as pregnancy progresses toward term, stroke volume declines whether subjects are studied in the supine, the sitting, or the lateral recumbent position; heart rate rises progressively, so that the decline in cardiac output is less dramatic than is the simultaneous fall in stroke volume (Ueland et al., 1969).

The mechanism by which stroke volume increases early in pregnancy illustrates the importance of placental hormones in regulating maternal physiology. The volume of blood contained in the left ventricle at the end of diastole (the end-diastolic volume) increases during pregnancy (Katz, Karliner, and Resnik, 1978). Although this change is attributed by some workers to an increase in the filling pressure of the left ventricle secondary to the demonstrated increase in maternal blood volume that accompanies pregnancy (Longo, 1983), no direct evidence for this hypothesis has been presented. An alternative explanation is based upon the premise that the left ventricle enlarges during pregnancy. When left ventricular pressure-volume curves from pregnant and nonpregnant guinea pigs are compared, ventricular volume is larger at each pressure level for the pregnant animals (Morton et al., 1984). These data are presented graphically in Figure 16.3. A similar increase in ventricular capacity can be evoked by the administration of estrogenic substances to nonpregnant guinea pigs (Hart, Hosenpud, Hohimer, and Morton, 1985). Apparently the steroid hormones of pregnancy cause the mother's heart to grow, so that her cardiac output is increased without calling upon any of the emergency mechanisms that all of us need for "fight or flight."

Distribution of the increased cardiac output

For the average woman, the maximum increase in resting cardiac output during gestation amounts to approximately two liters per minute. Early in pregnancy the majority of this increment supplies maternal tissues other than the pregnant uterus. As pregnancy advances, however, maternal uterine blood flow increases progressively; current evidence indicates that in the final weeks of gestation most of the increased cardiac output flows to the pregnant uterus (Metcalfe, McAnulty, and Ueland, 1986).

Maternal blood volume

Maternal blood volume increases by an average of 40% above nonpregnant values, mostly as a result of an increase in plasma volume (Pritchard, 1965). Plasma volume increases earlier and to a proportionately greater degree than

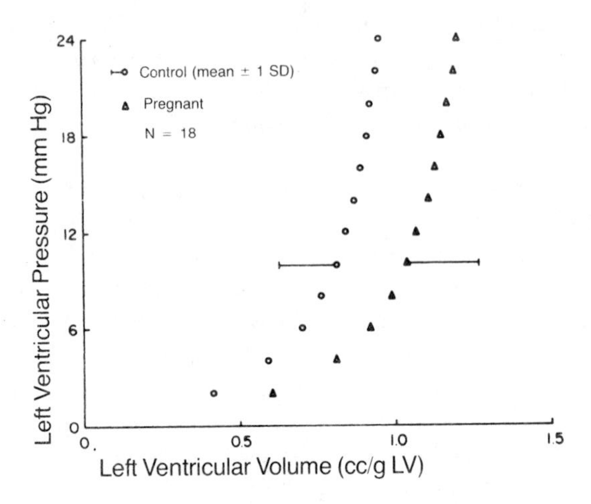

FIGURE 16.3 Left ventricular pressure is plotted against left ventricular volume corrected for left ventricular mass. At 10mm Hg distending pressure, left ventricular volume, referred to left ventricular mass, was 28% greater in pregnant guinea pigs than in nonpregnant controls. (From Morton et al., 1984.)

total red cell volume (Scott, 1972). The retention of sodium and water that necessarily accompanies the increase in plasma volume can be attributed to the effects of both estrogens and progesterone; however, the responses of pregnant women to changes in posture or blood volume are similar to those of nonpregnant individuals, suggesting that the increase in blood volume only compensates for the increase in vascular capacity (due to growth of uterine veins and to increased venous compliance) that accompanies pregnancy (Metcalfe et al., 1986).

Vascular resistance and arterial blood pressure

Despite the increase in cardiac output, there is a decline in resting mean arterial blood pressure during pregnancy, due largely to a fall in the diastolic value (Gallery, Hunyor, Ross, and György, 1977). From a hemodynamic standpoint, the fall in mean arterial pressure in the face of increased cardiac output means that peripheral vascular resistance declines proportionately more than cardiac output increases. Because peripheral vascular resistance depends mainly upon arteriolar resistance, this combination of findings is interpreted as showing arteriolar dilation early in pregnancy, a finding that, like the left ventricular enlargement and increase in venous distensibility already mentioned, is attributed to estrogenic compounds (Ueland and Parer, 1966; Resnik, Battaglia, Makowski, and Meschia, 1974).

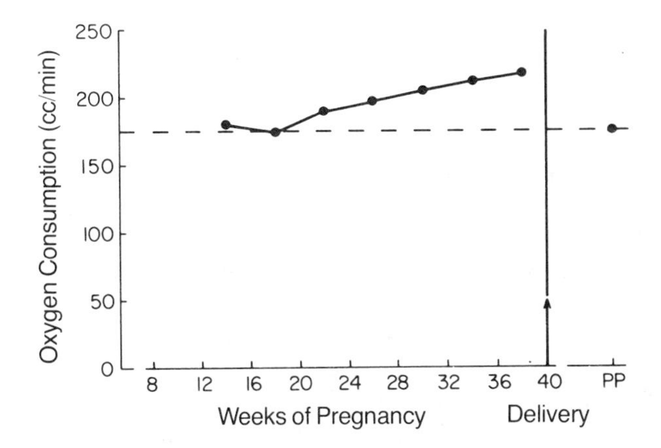

FIGURE 16.4 Maternal oxygen consumption increases during pregnancy. The average postpartum value is taken as the normal nonpregnant value and is represented by the horizontal broken line. Each point on the curve represents the average of a number of determinations in six individuals, each of whom was studied repeatedly during pregnancy and the puerperium under basal conditions. (From Burwell and Metcalfe, 1958.)

Taken together, these findings indicate that the remarkable increase in resting cardiac output that accompanies human pregnancy is achieved without calling upon "emergency" mechanisms such as increased sympathetic stimulation. Cardiac reserve is apparently maintained during pregnancy, at least in the resting state. Through this constellation of carefully integrated hemodynamic adjustments, the mother's cardiovascular system provides for the needs of the fetus without jeopardizing blood flow to maternal tissues. As we have indicated, most of the cardiovascular changes are evoked by steroid hormones originating from the placenta, a tissue peculiar to pregnancy and discarded at birth.

Changes in maternal respiration

As shown in Figure 16.4, maternal oxygen consumption increases progressively as pregnancy advances. At term, the average oxygen consumption is 15% greater than the postpartum value when measurements are made in basal subjects. This increase (Burwell et al., 1938) is accompanied by a proportionately greater increase in alveolar ventilation (the amount of air that reaches the gas exchange surfaces each minute). This increased breathing is called hyperventilation. As a result, arterial carbon dioxide tension falls (Lucius et al., 1970; Pernoll et al., 1975). Similar hyperventilation can be induced in male human subjects by the administration of progesterone (Döring, Loeschcke, and Ochwadt, 1950), and the degree of hyperventilation is closely related to the progesterone level in arterial blood (Machida, 1981).

Importance of maternal adjustments

Considerable circumstantial evidence suggests that the changes in maternal physiology that have been summarized are important for the normal development of the fetus. When maternal cardiac output is limited by heart disease, fetal growth and development are demonstrably jeopardized (Whittemore, Hobbins, and Engle, 1982; Whittemore, 1983).

The same handicap occurs among women living at high altitude who do not increase their rate or depth of breathing adequately during pregnancy (Moore et al., 1982). Similarly, the depression of breathing caused by methadone addiction may explain the lower birth mass and the high incidence of sudden infant death syndrome observed in the offspring of methadone-dependent women (Chavez, Ostrea, Stryker, and Smialek, 1979).

The magnitude of the increment in plasma volume during normal pregnancy correlates better with the birth mass of the infant (Hytten and Paintin, 1963), with placental mass (Rovinsky and Jaffin, 1965), and with the combined mass of the fetus and placenta (Ueland, 1976) than it does with maternal size.

The dependence of fetal growth and development upon the magnitude of changes in maternal cardiac output and ventilation suggests that an adequate supply of well-oxygenated blood to the pregnant uterus is essential for optimum fetal health and development. This suggestion is strengthened by the well-established growth-retarding effect of maternal cigarette smoking (Abel, 1980; Nieburg, Marks, McLaren, and Remington, 1985) or arterial hypoxemia (Whittemore et al., 1982; Whittemore, 1983). However, because maternal blood flow to the uterus is an essential link in the chain by which all substances are supplied to the developing fetus, an assessment of the importance of oxygen for fetal growth has required specific investigation.

New directions in ontogenetic ecophysiology

The effects of altering the prenatal environment (oxygen concentration) on growth and development of the chick embryo illustrate several possible new directions for research. First, a careful definition of the environment during ontogeny, especially in its early stages, needs to be continued, together with descriptions of the normal variations that occur in environmental temperature, humidity (in the case of the avian egg), oxygen supply, and the availability of nutrients. Second, the effects of deviations from those normal limits, like the deviations in availability of oxygen that we have discussed at length, are worthy of exploration. In this regard, the long-term consequences as well as the immediate effects on embryonic development should be assessed. Data obtained from humans handicapped prenatally show sustained effects that persist into adulthood (Babson and Phillips, 1973; James, 1982).

In general, the interactions of heredity and environment are of great importance and can be assessed only in the whole organism. The attention of physiological ecologists is properly directed to the embryo and fetus because the organism appears to be most sensitive to environmental change during its early development.

We have dealt in detail with the human cardiovascular and respiratory adaptations to pregnancy in order to illustrate the complex strategy which the maternal organism of one species uses to accomplish successful reproduction. Changes of equal magnitude and importance occur in other maternal systems, including digestion and neurological function. Behavioral changes are also apparent. For many of these alterations the mechanism remains to be established.

It is important to stress the diversity of strategies employed by different species for successful reproduction. Contrast, for example, the immaturity of the opossum when it emerges from its mother's uterus with the maturational robustness of the guinea pig. This diversity presents exciting challenges for the physiological ecologist. For example, the energetics of reproduction needs to be considered in the light of the ecological niche occupied by each species. Quantitative comparisons of the energy needs of egg-laying, as compared to viviparity, must be made.

In a broader biological context, the requisites for successful reproduction often extend beyond the fetus and its mother to involve the paternal organism. In some species, for example the pipefish and the seahorse, the paternal organism takes an active role in ensuring the successful development of the young (Hartman, 1956). Clearly, the mechanisms of these dramatic physiological changes offer an exciting opportunity for systematic investigation.

References

Abel, E. L. (1980) Smoking during pregnancy: a review of effects on growth and development of offspring. *Human Biol.* 52:593–625.

Babson, S. G., and Benda, G. I. (1976) Growth graphs for the clinical assessment of infants of varying gestational age. *J. Pediatr.* 89:814–820.

Babson, S. G., and Phillips, D. S. (1973) Growth and development of twins dissimilar in size at birth. *N. Engl. J. Med.* 289:937–940.

Babson, S. G., Behrman, R. E., and Lessel, R. (1970) Fetal growth: liveborn birth weights for gestational age of white middle class infants. *Pediatrics* 45:937–944.

Barron, D. H., Metcalfe, J., Meschia, G., Huckabee, W., Hellegers, A., and Prystowsky, H. (1963) Adaptation of pregnant ewes and their fetuses to high altitude. In *Physiological Effects of High Altitude*, ed. W. H. Weihe, pp. 115–129. New York: Pergamon.

Burwell, C. S., and Metcalfe, J. (1958) *Heart Disease and Pregnancy: Physiology and Management*, 1st ed. Boston: Little, Brown.

Burwell, C. S., Strayhorn, W. D., Flickinger, D., Corlette, M. B., Bowerman, E. P., and Kennedy, J. A. (1938) Circulation during pregnancy. *Arch. Intern. Med.* 62:979–1003.

Chavez, C. J., Ostrea, E. M., Jr., Stryker, J. C., and Smialek, Z. (1979) Sudden infant death syndrome among infants of drug-dependent mothers. *J. Pediatr.* 95:407–409.

Döring, G. K., Loeschcke, H. H., and Ochwadt, B. (1950) Weitere Untersuchungen über die Wirkung der Sexualhormone auf die Atmung. *Pflügers Arch.* 252:216–230.

Fawer, R., Dettling, A., Weihs, D., Welti, H., and Schelling, J. L. (1978) Effect of the menstrual cycle, oral contraception and pregnancy on forearm blood flow, venous distensibility and clotting factors. *Eur. J. Clin. Pharmacol.* 13:251–257.

Gallery, E. D. M., Hunyor, S. N., Ross, M., and Györy, A. Z. (1977) Predicting the development of pregnancy-associated hypertension: the place of standardized blood-pressure measurement. *Lancet* 1:1273–1275.

Hamor, T., and Garside, E. T. (1976) Developmental rates of embryos of Atlantic salmon, *Salmo salar* L., in response to various levels of temperature, dissolved oxygen, and water exchange. *Can. J. Zool.* 54:1912–1917.

Hart, M. V., Hosenpud, J. D., Hohimer, A. R., and Morton, M. J. (1985) Hemodynamics during pregnancy and sex steroid administration in guinea pigs. *Am. J. Physiol.* 249:R179–R185.

Hartman, C. G. (1956) The after-dinner address. *Trans. Am. Gynecol. Soc.* 79:98–105.

Hytten, F. E., and Paintin, D. B. (1963) Increase in plasma volume during normal pregnancy. *J. Obstet. Gynaecol. Br. Commonw.* 70:402–407.

Ingermann, R. L., Stock, M. K., Metcalfe, J., and Shih, T.-B. (1983) Effect of ambient oxygen on organic phosphate concentrations in erythrocytes of the chick embryo. *Respir. Physiol.* 51:141–152.

James, W. H. (1982) The IQ advantage of the heavier twin. *Br. J. Psychol.* 73:513–517.

Katz, R., Karliner, J. S., and Resnik, R. (1978) Effects of a natural volume overload state (pregnancy) on left ventricular performance in normal human subjects. *Circulation* 58:434–441.

Longo, L. D. (1983) Maternal blood volume and cardiac output during pregnancy: a hypothesis of endocrinologic control. *Am. J. Physiol.* 245:R720–R729.

Lucius, H., Gahlenbeck, H., Kleine, H.-O., Fabel, H., and Bartels, H. (1970) Respiratory functions, buffer system, and electrolyte concentrations of blood during human pregnancy. *Respir. Physiol.* 9:311–317.

Machida, H. (1981) Influence of progesterone on arterial blood and CSF acid-base balance in women. *J. Appl. Physiol.* 51:1433–1436.

McCalden, T. A. (1975) The inhibitory action of oestradiol-17-β and progesterone on venous smooth muscle. *Br. J. Pharmacol.* 53:183–192.

Metcalfe, J., and Bissonnette, J. M. (1981) A comparison of chorioallantoic and placental respiration. In *Advances in Physiological Sciences*, vol. 10: *Respiration*, ed. I. Hutas and L. A. Debreczeni, pp. 127–134. Budapest: Pergamon.

Metcalfe, J., McAnulty, J. H., and Ueland, K. (1981) Cardiovascular physiology. *Clin. Obstet. Gynecol.* 24:693–710.

Metcalfe, J., Stock, M. K., and Ingermann, R. L. (1984) The effects of oxygen on growth and development in the chick embryo. In *Respiration and Metabolism of Embryonic Vertebrates,* ed. R. S. Seymour, pp. 204–230. Dordrecht, The Netherlands: Dr W. Junk.

Metcalfe, J., McAnulty, J. H., and Ueland, K. (1986) *Burwell and Metcalfe's Heart Disease and Pregnancy: Physiology and Management*, 2nd ed., chapter 2. Boston: Little, Brown.

Moore, L. G., Rounds, S. S., Jahnigen, D., Grover, R. F., and Reeves, J. T. (1982) Infant birth weight is related to maternal arterial oxygenation at high altitude. *J. Appl. Physiol.* 52:695–699.

Moore, L. G., Brodeur, P., Chumbe, O., D'Brot, J., Hofmeister, S., and Monge, C. (1984) Maternal ventilation, hypoxic ventilatory response and infant birth weight during high altitude pregnancy. *Fed. Proc.* 43:434.

Morton, M., Tsang, H., Hohimer, R., Ross, D., Thornburg, K., Faber, J., and Metcalfe, J. (1984) Left ventricular size, output, and structure during guinea pig pregnancy. *Am. J. Physiol.* 246:R40–R48.

Nieburg, P., Marks, J. S., McLaren, N. M., and Remington, P. L. (1985) The fetal tobacco syndrome. *JAMA* 253:2998–2999.

Pernoll, M. L., Metcalfe, J., Kovach, P. A., Wachtel, R., and Dunham, M. J. (1975) Ventilation during rest and exercise in pregnancy and postpartum. *Respir. Physiol.* 25:295–310.

Pritchard, J. A. (1965) Changes in the blood volume during pregnancy and delivery. *Anesthesiology* 26:393–399.

Resnik, R., Battaglia, F. C., Makowski, E. L., and Meschia, G. (1974) The effect of actinomycin-D on estrogen-induced uterine blood flow. *Gynecol. Obstet. Invest.* 5:24.

Rovinsky, J. J., and Jaffin, H. (1965) Cardiovascular hemodynamics in pregnancy. I. Blood and plasma volumes in multiple pregnancy. *Am. J. Obstet. Gynecol.* 93:1–15.

Scott, D. C. (1972) Anemia in pregnancy. In *Obstretics and Gynecology Annual: 1972*, ed. R. M. Wynn, pp. 219–244. New York: Appleton-Century-Crofts.

Stock, M. K., and Metcalfe, J. (1987) Modulation of growth and metabolism of the chick embryo by a brief (72 h) change in oxygen availability. *J. Exp. Zool.* [*Suppl.*] 1:351–356.

Stock, M. K., Asson-Batres, M. A., and Metcalfe, J. (1985) Stimulatory and persistent effect of acute hyperoxia on respiratory gas exchange of the chick embryo. *Respir. Physiol.* 62:217–230.

Ueland, K. (1976) Maternal cardiovascular dynamics. VII. Intrapartum blood volume changes. *Am. J. Obstet. Gynecol.* 126:671–677.

Ueland, K., and Parer, J. T. (1966) Effects of estrogen on the cardiovascular system of the ewe. *Am. J. Obstet. Gynecol.* 96:400–406.

Ueland, K., Novy, M. J., Peterson, E. N., and Metcalfe, J. (1969) Maternal cardiovascular dynamics. IV. The influence of gestational age on the maternal cardiovascular response to posture and exercise. *Am. J. Obstet. Gynecol.* 104:856–864.

Whittemore, R. (1983) Congenital heart disease: its impact on pregnancy. *Hosp. Pract.* 18:65–80.

Whittemore, R., Hobbins, J. C., and Engle, M. A. (1982) Pregnancy and its outcome in women with and without surgical treatment of congenital heart disease. *Am. J. Cardiol.* 50:641–651.

Wood, J. E., III, and Goodrich, S. M. (1964) Dilation of the veins with pregnancy or with oral contraceptive therapy. *Trans. Am. Clin. Climatol. Assoc.* 76:174–180.

Discussion

BURGGREN: The importance of ontogeny to some of the characters we are studying in lower vertebrates deserves emphasis. Some of the ontogenetic differences within a species can be far greater than those between species, between genera, and in some instances, even between families. One of the major differences that we see when comparing lower vertebrates with birds and mammals is that many of the organs are simultaneously called on both to develop and to function from the very start, whereas in mammals, at least, the lungs and the liver to some extent, may go through "dress rehearsals" before birth but are spared the need to function by the action of the placenta. The combination of organs having to develop and function simultaneously, together with the precocial release into the environment, results in a very immature animal subject to the vagaries of the environment. Perhaps as a consequence, development is very plastic during the free-living larval stages. Morphology or physiology can be permanently altered in the ensuing adult forms by environmental perturbation during the free-living larval stages. Thus, if we are rearing populations and taking animals through many different generations, we have to be particularly careful about establishing an environment that is appropriate for the normal development. This is particularly true for fishes and amphibia.

METCALFE: Whether the variability of surviving adults is greater as you get to lower vertebrate forms is an hypothesis that should be tested. I have no data that support or refute this hypothesis, but intuitively it might very well be correct.

POWERS: In lower vertebrates other components contribute to this condition. Lower vertebrates lay eggs, and the characteristics of those eggs are controlled by the maternal genome. Immediately after fertilization, as soon as you can make a measurement, there is already a difference between genotypes that is totally dependent upon the maternal contribution. In *Fundulus*, those enzymes that control early development are essentially 100% maternal for the first couple of days. In addition, in many lower vertebrates oxygen accelerates development but a drop in oxygen promotes hatching. Will chicken eggs hatch if they are maintained in a very high partial pressure of oxygen? At least some fish eggs will not hatch unless the partial pressure of oxygen drops.

METCALFE: Yes, they hatch. Their hatching rate is lower and they hatch sooner.

ARNOLD: A study by Karn and Penrose [*Ann. Eugen.* 16:147–164, 1951] and perhaps more recent ones show an inverted "U"-shaped relationship between postnatal mortality and birth mass. Survivorship peaks at about the mean birth mass and then falls off in either direction from that peak. When a mother smokes and her offspring consequently have a lower birth mass, is the survivorship of the child predicted by this graph or is there an additional effect of maternal smoking on infant mortality? Do the offspring of smoking mothers have a completely different curve?

METCALFE: I do not know if that survivorship curve has been drawn, but the mortality is related to many things other than birth weight. For example, placenta previa is more common in smoking women, and so is toxemia of pregnancy; thus there are other contributing factors.

BURGGREN: What do you view to be the major detrimental effects of hyperoxia on embryos? You suggested that one response might be to increase growth to protect against excess oxygen.

METCALFE: If hyperoxia occurs at the end of incubation, the hatchability of the eggs is greatly reduced. I do not understand the mechanism by which they fail to hatch. It seems to be associated with a lack of retraction of the yolk in the body cavity. The animal sticks to the shell and just can not get loose. There is some fragmentary evidence that they are handicapped in other ways. Those that survive, that manage to hatch, have a slower acquisition of learning the combination of a heat reward and a color, so that there may be a slowing of development in the central nervous system.

COLLINS: The conditional state of the organism, whether it is pregnant or not, affects physiological traits. This is not a new idea for physiologists. As we investigate variation among individuals, however, there is an increased burden on the investigator to control for such background variables, which may otherwise confound analysis of interindividual variation.

METCALFE: Yes, I hope I emphasized that enough. Pig farmers, for example, know that a runt pig is an entirely different animal from the rest of the litter, so much so that the runt is almost always destroyed. The smallest animal in the litter of five guinea pigs can be shown to be different from the largest animal in the litter. Certainly singlet offspring of guinea pigs are different from offspring in a litter of five.

17 Interacting physiological systems: a discussion

DAVID J. RANDALL

RANDALL: I would like to start this general discussion by presenting my views on the characteristics of physiologists, physiological ecologists, and population biologists. Physiology is studied by people interested in proximate factors; physiologists tend to be mechanistic and are interested in describing systems or the interactions between systems. They are also interested in, but do not pay much attention to, so-called ultimate factors, which dominate studies of population biology and evolution. Physiologists have teamed with ecologists; and the area of physiological ecology has proved fruitful, especially from a mechanistic point of view. Ecologists have emphasized the importance of studying physiological systems within an environmental context. Physiologists have helped to explain some aspects of animal dispersion, why animals are found where they are, and the limitations physiological systems place on animal distribution. This has been a successful marriage of convenience, one that has existed for many years and will continue to exist.

This meeting has been discussing a change in direction for physiologists. This new direction is to apply a physiological approach to problems in population biology and evolution. Moreover, if you want to change fields, you have to change techniques. This seems to me to be self-evident. One must always use the most appropriate techniques to solve the problem at hand. New problems usually require the application of new techniques, and they must be used no matter how expensive or complicated. The question we should address is whether or not the physiological approach is applicable to studies in population biology and evolution. As I see it, evolution is an area with theories for which it is difficult to develop paradigms. The previous marriage between physiologists and ecologists resulted in the study of an area with many paradigms and few theories. What I do not know is whether a detailed description of systems characteristic of the physiological approach is really going to help in answering some of the questions in population biology and evolution. How much do you need to know? Will this marriage of physiologists and population biologists work? How can physiologists approach some of the questions? Can physiologists add to studies of evolution?

POWERS: The answer is yes. What are the big questions in evolutionary and population biology? There are a whole series of them, but one that comes and goes is whether or not all of this genetic variation means anything. Is it genetic noise, or is it a function of the life history of the organism? Are there functional differences between all these genetic variants, and, if so, do these differences mean anything in terms of evolution, or are they there just because of fine tuning? Perhaps there are functional differences, but are they swamped out by migration and other factors? As someone who straddles biochemistry, physiology, and population biology, I see great dangers in trying to deal with these fields without proper expertise in all of them. It is easy to be naive about the importance of migration and life histories and so forth, and then to make much of physiological differences of no real significance. It is important to consult with people who can be critical of such studies. There is a synergistic potential for learning and helping each other to generate new questions and new approaches. You might start with genetic differences at a series of loci. Are these functionally equivalent biochemically? Are they physiologically equivalent and are there fitness differences? Then you can build in the dimension of multilocus systems, such as Koehn discussed. Does it mean anything physiologically, or is it just an artifact of the genetic structure of the organism? This really calls for an interdisciplinary approach. I would strongly encourage people to move in that direction.

BENNETT: Physiologists have a tremendous part to play here. For me, the organizational paradigm for this goes back to Steve Arnold's paper in *American Zoologist* in 1983 [see Chapter 7 for reference]. That paper is an extremely important contribution and a road map for us to follow as to how physiologists might form effective bridges to both evolutionary biology and ecology. He proposes examining the influence of physiological and morphological characters on performance; this he calls the "performance gradient." The influence of performance on fitness is in turn called the "fitness gradient." This construct puts physiology in a unique part of this continuum, because one can do investigations in both directions, down to the genetic or morphological level or up to the ecological and evolutionary. It is a great leap from the genetic or morphological levels into purported fitness effects. You must make many assumptions, which sometimes are very difficult to test. Because of the past success of physiological ecology, we understand many of the potential ecological consequences of physiological factors and understand that the physiological characters may map fairly well onto performance and fitness in natural environments. Opportunities are unparalleled for physiologists to look at the consequences for fitness and for population structure of many physiological variables. What we study is closer than the traditional morphological characters or genetic characters to actual fitness. We have a unique role to play in this analysis.

FUTUYMA: I think that there will be a bidirectional influence between evolutionary studies and physiological studies. On the one hand, evolutionary biology can serve as one framework, not the only possible framework, but one possible framework of organizing physiological information for suggesting hypotheses about physiological phenomena. In the same way, for example, the study of behavior has been reorganized by taking an evolutionary approach, which is rather different from the more classical ethological approach. From the other direction, what does physiology have to offer to people who want to know about evolution? Evolutionary biology is trying to understand why there is such a great diversity of organisms, which, in large part, amounts to trying to understand what the function is of the characteristics that differentiate various taxa from one another. A detailed understanding of function is necessary to understand the morphological diversity among taxa. It is precisely the physiologist who possesses that detailed knowledge of function. So, physiology seems to be a really pivotal field, and it is very unfortunate that many evolutionary biologists have only rudimentary understanding of physiology.

BISHOP: In that regard, then, you are really advocating the reductionist approach to get at specifics, in the way that Powers does, and then to come back to the general importance of the character.

FUTUYMA: No argument there from me.

ARNOLD: One of the things that has struck me in the course of my collaborations with Bennett and Huey is the realization that studies of selection on many key variables in physiology and in physiological ecology have never been conducted. We do not know how strong selection acts on sprint speed, or whether there is a correlational selection between pairs of variables. We are not talking about random variables; we are talking about the pivotal measurements that figure in virtually every discussion in physiology and in environmental physiology. For some reason, focus on these variables has never been combined with demographic studies to find out how these variables map onto fitness. Moreover, these variables have been ignored by demographers. Physiologists have not had the demographic situations to ask the questions. I think that we are going to learn some very surprising things as we try to combine measurements of these important factors with demographic studies of fitness. The same can be said for a genetic focus on these same variables. Few geneticists have tried to do genetic studies on these complex organismal traits. Animal breeding has a history of looking at traits like efficiency of food utilization and growth rate, but is still some distance away from the variables at the central focus of environmental physiology, in particular, respiration and exercise physiology. We have historically neglected an important relationship: how selection acts on these variables. It is not going to be easy to measure selection on these traits. The point is that we can pose the problem and try and make some progress.

PACKARD: I have an unanswered question that stems from environmental control of the process of sexual differentiation in certain reptiles. Does some physiological variation induced by environmental variation have ecological but not evolutionary significance?

RANDALL: It must depend upon the time scale. Short-term changes might be of no evolutionary significance, but over a longer time, they seem almost bound to be coupled.

POWERS: I agree. I would say, though, that even plasticity might be considered an evolutionary process. So, it may be important to be able to respond to the environment in a "nonevolutionary" way in the short term, whereas in fact in the long term, it is an important evolutionary adaptation.

FUTUYMA: Along that line, one of the most interesting things that physiologists have studied is the capacity for homeostasis versus conformity to changes in salinity. Here is a whole rich area: under what conditions do you expect regulation versus conformity to evolve? There are some very interesting evolutionary questions here. I know of only one or two attempts to begin even to theorize, much less to perform tests of those theories.

FEDER: When we talk about the future of ecological physiology, or new directions in the field, we must recognize that ecological physiology is one case of a general kind of science, organismal biology. There are two important points to be made with reference to organismal biology. First, a key characteristic of organismal biology is that it is, and should be, unbounded. That is, it admits information at both higher levels and lower levels. This distinguishes it from other fields and other approaches, such as molecular biology, which in the hands of most practitioners, is often bounded at the molecular level of organization. In specifying directions, such as population genetics or evolutionary biology or reductionistic approaches, we are constraining ourselves too much. We should be without constraint. If we want to move more in the direction of medical technology, for example, that should be permitted. Second, just as there are certain emergent properties of each level of the biological hierarchy, there are emergent properties of organisms. There are emergent questions in organismal biology that are different from those at lower and higher levels. Three major questions are: First, can we explain the evolution of physiological traits through selection? "Can" is a key word here. Is natural selection a necessary explanation, a sufficient explanation, or both? Second, what is the explanation of the discontinuity in physiological variation? We do not see all possible combinations of traits at random. We see coadapted character complexes, suites of adaptation. Why is it that we see this limitation? If it reflects a constraint, where does the constraint come from? Third, what is the explanation of physiological diversity, or more properly, to what extent do various kinds of explanatory factors (phylogeny, ontogeny, history, chance) contribute to overall variation and physiological diversity? Again, these are emergent and unique questions to the field of organismal biology that could be addressed in a physiological context, a morphological context, a behavioral context, and an endocrinological context, and probably should be addressed in all of these contexts to do a good job.

DAWSON: I want to commend the group that organized this meeting. It is an exceedingly important effort. I had my doubts initially, but the extent of dialogue and outcrossings of priesthoods that seems to be going on is exceedingly healthy. Some genealogy is involved in this, but it has been the interaction of some of the principals here with other people who came from other schools and other traditions that seems to be the valuable thing. I want to conclude my remarks by endorsing Feder's advocacy of coherence and a central theme in organismal biology, namely the obligation to incorporate data generated at any level into an organismal framework. Moreover, organismal biology should not be just a service organization for other fields. Organismal biology can yield insights not to be found at other levels of study. Although these insights clearly have important implications for a much larger arena of disciplines, our major reason for searching for them should be that they will enhance the scope and vigor of organismal biology per se.

BENNETT: As things appear to have wound down, I will quote Dawson from a conference he organized many years ago on avian energetics. His main comment was that he felt like a kid in a candy store, at having been able to invite so many people that he respected and wanted to hear and got to sit down and listen to their thoughts for awhile. I really feel very much the same way, and I know that the other organizers of this conference concur. This has been a very exciting intellectual event for me, and I think for all of us.

18 New directions in ecological physiology: conclusion

MARTIN E. FEDER

Ecological physiology has grown to maturity largely around one central concept: adaptation to the environment. The major achievements of the field have been empirical: the identification and examination of adaptation in diverse physiological systems, species, and environments. "Adaptation to the environment" ought to and undoubtedly will remain the mainstay of ecological physiology for the forseeable future. Nonetheless, if the field is to continue to prosper, it must incorporate and cultivate conceptual approaches that are novel, at least to ecological physiology. Such new approaches need not be exclusive of the study of adaptation, and in all likelihood will complement it.

The directions outlined in this book are but a small sample of possible directions. [For example, Sibly and Calow (1986) have urged the adoption of optimality theory as an analytical framework for ecological physiology. See also Chapter 14.] It would be presumptuous to assert that these directions are the only suitable ones. As George Bartholomew suggests in Chapter 2, imposition of a definition or direction upon a field can result in sterility rather than growth. Our conviction is firm, however, that a more widespread and critical debate about conceptual issues and directions of growth would be a healthy development for the field of ecological physiology.

As outlined in the Introduction and in Chapter 1, a major factor that led the editors to organize this volume was a widespread (but, in the editors' opinion, inaccurate) perception of a lack of intellectual and conceptual momentum in animal ecological physiology (e.g., Ross, 1981), a feeling of no new physiological worlds left to conquer. Bennett's cautionary tale of comparative anatomy (Chapter 1) is instructive in this respect. Another concern was that ecological physiology might become a "service industry" for other disciplines, relegated to demonstrating the functional significance of physiological features that emanate from molecular biologists' and evolutionary biologists' pursuit of larger questions (cf. Wake, 1982; Prosser, 1986). Will a critical debate of new directions lead to an invigoration of ecological physiology? Perhaps it already has, albeit in an way that portends the transformation of ecological physiology and related fields (e.g., functional morphology, comparative endocrinology, and neuroethology) from free-standing disciplines to parts of a larger synthetic endeavor, organismal biology.

For me, the central question of organismal biology, and one that expresses the essence of the potential new directions of ecological physiology, is, How can we explain the evolution of diverse complex organisms? ("Complex" is used here to represent multiple interacting organismal systems and components, each in turn comprised of multiple interacting components, and not the distinction between unicellular and multicellular organisms.) This question incorporates a host of subordinate questions concerning multiple interrelated processes. How are the individual systems and adaptations within a complex organism prioritized and reconciled with one another when their functions conflict (cf. Chapter 16)? How is this prioritization and reconciliation maintained and adjusted in the face of functional and ontogenetic change (see Chapters 15 and 17)? Is our current understanding of natural selection and other evolutionary processes adequate to explain the evolution of organismal complexity (e.g., Chapters 3 and 9)? How satisfactory is the evidence for each of the components in this process: individual variation in complex traits (Chapter 7), genetic variation in complex traits (Chapter 8), population divergence and other modes of speciation (Chapter 5), and the accumulation of diversity in various taxa (Chapter 4)? These questions, although hardly novel, are both important and largely unanswered.

This volume highlights an existing trend toward emphasis on such questions, but largely with respect to physiological issues. In the past, ecological physiology usually emphasized understanding the function and significance of individual adaptations (themselves often complex) that comprise organisms, as opposed to understanding the evolution of organismal complexity. As previous chapters suggest, many ecological physiologists have grown increasingly dissatisfied with the wholesale enumeration of physiological adaptations as an end in itself, and for some time have been exploring the evolution of physiological complexity. Importantly, ecological physiologists have not restricted themselves to any particular level of organization but have pursued the explanation of physiological complexity wherever it has taken them, from biological molecules to communities and ecosystems (Figure 18.1; see also Bartholomew, 1986).

Even so, given the nature of these questions, ecological physiologists may be able to address these questions most effectively if they begin to think of themselves as organismal biologists first and ecological physiologists second. Why?

First, like most scientists, ecological physiologists have limits. The "mindset" of ecological physiologists is to focus on the phenomena of physiology, albeit in an ecological or comparative context. Accordingly, ecological physiologists too often either overlook or only vaguely perceive the commonalities with other subdisciplines of organismal biology and the larger questions

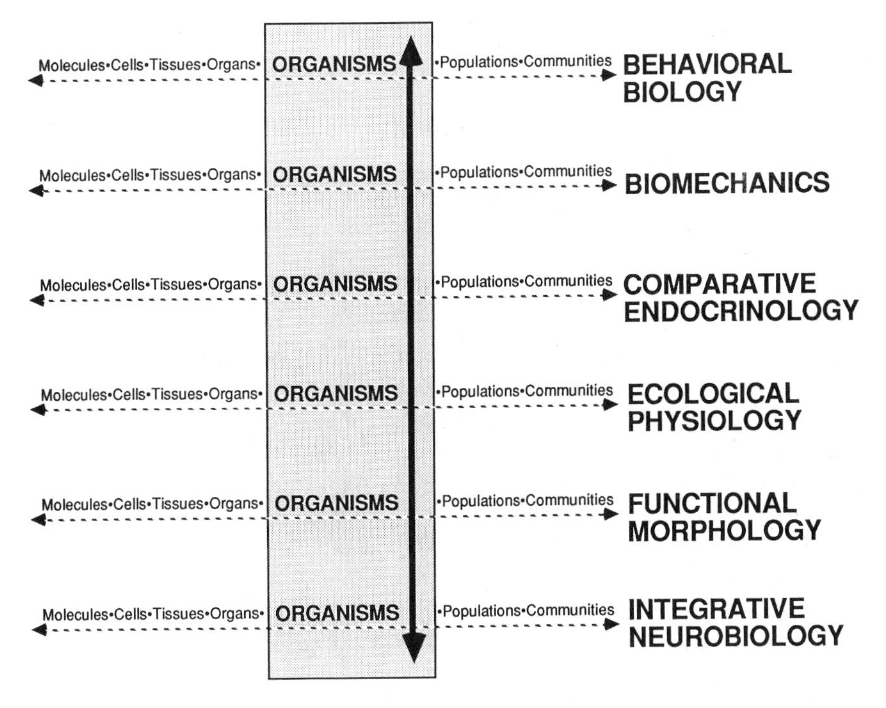

FIGURE 18.1 A prospective community of interest for organismal biology. To the right is a *partial* (alphabetical) listing of subdisciplines that compose organismal biology. Each prides itself on its interest in and ability to integrate diverse levels of organization; thus, the broken lines represent the predominant axes of intellectual interaction for each subdiscipline. However, in so doing, each subdiscipline restricts itself to particular phenomena of greatest interest (e.g., physiological systems for ecological and comparative physiology; hormonal regulation for comparative endocrinology), operates in relative isolation, and thereby diffuses focus away from common central problems of organismal biology (e.g., the evolution of organismal complexity). For these reasons, this chapter advocates a synthesis of the subdisciplines into a community of organismal biologists, whose focus (solid arrow) would be on the emergent questions at the organismal level of organization. This focus should be complementary to (and not exclusive of) the foci of the component subdisciplines. Ecological physiologists can and ought to play an important role in this community.

inherent in the organismal level of organization (Figure 18.1). As a group, ecological physiologists too seldom confront the question, How can we explain the evolution of diverse complex organisms? Many ecological physiologists are inadequately informed of the strides made by other subdisciplines in addressing central questions, and in turn fail to communicate their own achievements to a wider audience. These limitations are understandable consequences of trends towards increasing specialization and the "information overload" that each subdiscipline (including ecological physiology) produces (Bartholomew, 1982).

Second, physiology has intrinsic boundaries as a means of investigating organisms. Organisms have attributes other than physiological ones, which necessitates a broad approach to organismal questions. Ecophysiological attributes are sometimes not the most suitable subjects for an examination of the evolution of organismal complexity. For example, physiological attributes seldom undergo fossilization and, therefore, necessitate the indirect (and, as Huey concludes, problematic) analyses outlined in Chapter 4 for addressing historical issues. Physiological attributes have their own unusual properties. For example, many physiological attributes are relatively plastic, may undergo both acute responses and rapid acclimatory modifications in response to environmental challenges, and may therefore present a very different portrait of organismal complexity than would a study of more slowly responding morphological attributes.

Ecological physiologists clearly are not unique with respect to these problems (Figure 18.1). Many other subdisciplines are interested in organisms but study selected portions of their biology. As outlined above for ecological physiology, this restricted focus may inhibit the solution of central questions in organismal biology. As with the fabled blind men inspecting an elephant, each subdiscipline characterizes its corresponding portion of the whole thoroughly but neglects the whole as an object of organized study.

This characterization of subdisciplines within organismal biology is not intended as a disparagement, but rather as an argument for a synthetic field of organismal biology. There are both a real need for and real advantages to the coalescence of the subdisciplines within organismal biology, with the concomitant recognition of common goals and directions for organismal biology. The evolution of organismal complexity is an important question and an extremely difficult one. Its solution is beyond the capabilities of any single specialized field such as ecological physiology. By contrast, the *potential* community of interest in the evolution of organismal complexity is clearly larger and probably better able to address this issue than any single subdiscipline. Curiously, molecular biology, cell biology, and population biology each has an obvious community of interest with evident central themes, dogmas, and directions; whereas organismal biology lacks the same.

A science of organismal biology and a community of organismal biologists is needed to understand the evolution of organismal complexity. Ecological physiologists can and ought to play an important role in this community. The editors hope that this book will stimulate such undertakings.

Acknowledgments

I thank Albert Bennett, Warren Burggren, Raymond Huey, and Philip Ulinski for their insightful comments on the manuscript, and Kirk Miller for a particularly thoughtful and provocative review of Chapter 3, which laid the groundwork for this chapter. The parable of the elephant and the blind men as a metaphor for biological investigation is borrowed from Carl Gans. Writing was supported by NSF Grant DCB 84–16121.

References

Bartholomew, G. A. (1982) Scientific innovation and creativity: a zoologist's point of view. *Am. Zool.* 22:227–235

Bartholomew, G. A. (1986) The role of natural history in contemporary biology. *Bioscience* 36:324–329.

Prosser, C. L. (1986) The challenge of adaptational biology. *Physiologist* 29:2–4.

Ross, D. M. (1981) Illusion and reality in comparative physiology. *Can. J. Zool.* 59:2151– 2158.

Sibly, R. M., and Calow, P. (1986) *Physiological Ecology of Animals: An Evolutionary Approach.* Oxford: Blackwell Scientific.

Wake, D. B. (1982) Functional and evolutionary morphology. *Perspect. Biol. Med.* 25:603–620.

Author index

Abel, E. L., 336
Ahlquist, J., 81
Aiello, A., 56
Alberch, P., 50, 58, 60, 206
Albon, S. D., 52
Albrecht, G. H., 234
Alexander, R. McN., 289, 291, 301
Allison, R., 261
Anderson, D. E., 221, 233
Anderson, D. G., 175
Anderson, P. R., 175
Andrews, R. M., 153
Armstrong, R. B., 296
Arnheim, N., 177
Arnold, S. J., 50, 53, 85, 149, 152, 153, 158, 159, 161, 197, 202–3
Aschoff, J., 264
Asson-Batres, M. A., 331
Atchley, W. R., 197, 199–200, 218, 221, 232
Atz, J. W., 76, 93
Axenrod, T., 267

Babson, S. G., 331, 337
Baker, C. M. A., 179
Bankir, L., 301
Barbancho, M., 105
Barber, B. J., 316
Bard, K. M., 296, 299
Barkley, M. S., 196
Barnes, P. T., 160, 175
Barron, D. H., 329
Barrowclough, G. F., 88, 93
Bartels, H., 335
Bartholomew, G. A., 15, 21, 22, 23, 25, 27, 28, 29, 33, 42, 47, 59, 62, 63, 65, 198, 259, 348, 350
Battaglia, F. C., 334
Battles, D. R., 312
Baulov, M., 179
Baumgardner, F. W., 318
Bayne, B. L., 177, 179
Beatley, J. D., 24
Becker, W. A., 196, 209
Behrman, R. E., 331
Bellman, R., 290
Benda, G. I., 331
Benedict, F. G., 4
Benesch, R., 115
Benesch, R. E., 115
Bennett, A. F., 44, 61, 64, 78, 79, 81, 84, 85, 88, 90, 93, 94, 149, 152, 153, 154, 155, 161, 317
Berger, E., 175
Berman, S. L., 153
Berne, R. M., 320
Berven, K. A., 56
Best, R. C., 261
Bewley, G. C., 153, 175
Bijlsma, R., 105
Bisgaard, S., 319
Bishop, S. H., 176
Bissonette, J. M., 329
Blem, C. R., 218, 232–3
Blower, S. M., 105, 174
Bock, W. J., 60
Bogert, C. M., 17, 92
Bondi, K., 318
Bonefeld, B. A., 179
Bouchard, C., 161
Bowerman, E. P., 335
Boyer, S. J., 314
Bradford, G. E., 196
Braun, E. J., 299, 301
Brocklehurst, K., 126
Brodeur, P., 329
Brown, D. C., 105–6, 108–9, 111
Brown, W. W., 92
Bucher, T. L., 33, 198
Bulmer, M. G., 194
Burggren, W. W., 64
Burghardt, G. M., 77, 197
Burkhoff, A. M., 109
Burri, P. H., 294
Burton, R. S., 176
Burwell, C. S., 335
Bush, G. L., 80

Butler, P. J., 261, 269
Buth, D. G., 92
Butterworth, F. M., 105

Cadle, J. E., 91
Calder, W. A., III, 4, 155, 216, 291, 300
Calow, P., 2, 347
Cameron, J. N., 318
Carey, C., 180–2, 200
Carlson, S., 92
Carter, P. A., 51, 105, 174, 190
Case, S. M., 80
Casey, T. M., 22, 27
Cashon, R. E., 105–6, 108–9, 111, 113, 115, 118
Cassin, R. C., 105, 174
Cavalli-Sforza, L. L., 88
Cavener, D. R., 105
Ceccarelli, G., 267
Chambers, G. K., 175
Chanutin, A., 115
Chapman, F., 218, 232
Chappell, M. A., 160
Charlesworth, B., 206
Chatonnet, J., 311–12
Chavez, C. J., 336
Cheeseman, C. L., 269
Cheverud, J. M., 82, 197, 199, 203, 205, 206
Christian, K. A., 85, 153
Chumbe, O., 329
Cibischino, M., 153
Claasen, H., 295
Clarke, B., 104–6, 111, 120
Clausen, J., 56
Clegg, M. T., 105
Cleland, W. W., 123–5
Close, R., 155
Clutton-Brock, T. H., 52, 61, 63, 81
Cochran, W. G., 218, 220, 223, 229, 233, 234
Cockerham, C. C., 196
Conley, K., 296
Conrad, M., 302
Corces, V., 109
Corlette, M. B., 335
Cornish-Bowden, A., 126
Cowles, R. B., 17
Cox, D. R., 218, 220, 234
Cracraft, C., 77
Crawford, D. L., 113, 115
Crawford, E. C., Jr., 316
Crisci, J. V., 86, 87
Crook, H. H., 81
Crow, J. F., 194
Crowder, E., 261
Crowley, P. H., 126
Crowley, S. R., 152
Cruz-Orive, L. M., 295
Curnish, R. R., 115
Curtsinger, J. W., 105
Cuvier, G., 290

Dalessio, P. M., 116
Daly, K., 105
Dann, S., 264
Darwin, C., 77, 81, 189, 198, 291
Davies, D. G., 280
Davis, M., 179
Dawson, W. R., 43
D'Brot, J., 329
Deaton, L. E., 177
Dejours, P., 294
Delclitte, P., 311–12
Dellaripa, P., 153
Dettling, A., 333
deVlaming, V., 218, 232
Diaz, M., 301
Dickerson, G. E., 161, 194, 199, 202
Diehl, W. J., 179, 180–2
DiMichele, J. A., 105–6, 108–9, 111, 116, 117
DiMichele, L., 105, 106, 113, 115–20, 160
Dingle, H., 198
Dinkel, C. A., 218, 232
Dobson, F. S., 76–7
Dobzhansky, T., 12, 42
Donhoffer, S., 216
Donoghue, M. J., 88
Donohue, K., 51, 190
Dorado, G., 105
Döring, G. K., 336
Dow, M. M., 82
Dumont, J. P. C., 76–7, 83, 85, 90
Dunham, A. E., 81–3
Dunham, M. J., 335
Dupré, R. K., 318
Dutton, R. E., 280

Eckert, C. G., 91–2
Edwards, A. W. F., 88
Eisen, E. J., 201
Eldridge, N., 42, 48, 53, 77
Else, P. L., 155
Elsner, R., 260–1
Emsley, A., 161
Endler, J. A., 51, 56, 80, 107–8, 158–9, 189
Engle, M. A., 336
Epple, A., 76, 93
Everse J., 112

Fabel, H., 335
Faber, J.. 333–4
Fairbrother, A., 319
Falconer, D. S., 160, 194–6, 198–201, 208–9
Farhi, L. E., 281
Farris, J. S., 88

Fawer, R., 333
Fedak, M. A., 312
Feder, M. E., 64, 65
Feldman, M. W., 176
Felsenstein, J., 77, 80–2, 87, 94, 154
Ferguson, G. W., 160
Fersht, A., 122, 125, 126
Finney, D. J., 218
Fisher, R. A., 191, 218
Fleagle, J. G., 218
Flickenger, D., 335
Fox, M. A., 200
Fox, S. F., 160
Fox, W., 200
Friedman, A. H., 262
Fujio, Y., 179
Futuyma, D. J., 199

Gadian, D. G., 267
Gaffney P. M., 179–82
Gahlenbeck, H., 335
Galileo, G., 290
Gallery, E. D. M., 334
Gallivan, G. J., 261
Gans, C. 42, 47, 65, 80, 83
Garland, T., Jr., 152, 157–8
Garside, E. T., 331
Garton, D. W., 179–80
Gaskins, C. T., 221
Geohegan, W. H., 116
Gibson, J. B., 175
Gilbert, D. G., 105
Gilbert, N., 161
Gillen, R. G., 115
Gillespie, J. H., 92
Gonor, J. J., 218, 232
Gonzalez-Villasenor, L. I., 105–6, 108–9, 111
Gooden, B., 260–1
Goodrich, S. M., 333
Gordon, C., 200
Gorman, G. C., 92
Gould, F., 199
Gould, S. J., 39, 42, 48, 49, 52, 53, 59, 60, 62, 63, 64–6, 154, 206, 216, 218, 293–5, 302
Graber, D. J., 314
Grant, B. R., 302
Grant, P. R., 302
Greaney, G. S., 115, 118
Greegor, D. H., 300
Green, R. H., 179, 182
Greenberg, M. J., 251, 254, 269
Greene, H. W., 65, 77, 90
Greer, A. E., 85, 88, 89
Gromko, M. H., 105
Groos, G. A., 264
Grossman, G., 218, 232
Grover, R. F., 329, 336
Grunfeld, J. P., 300
Guinness, F. E., 52
Guralnik, D. B., 275
Guries, R. P., 179
Gurtner, G. H., 280
Gyles, N. R., 202
Györy, A. Z., 334

Hackett, P. H., 314
Hainsworth, F. R., 300
Haldane, J. B. S., 92, 127
Hall, J. G., 174–6
Hämälänen, R. P., 289
Hammel, H. T., 320
Hamor, T., 331
Hanken, J., 153
Hardy, J. D., 320
Harlow, H. J., 292
Harris, H., 104
Hart, M. V., 333
Hartl, D. L., 302
Hartman, C. G., 337
Harvey, P. H., 61, 63, 81
Hasler, A. D., 254
Heath, J. E., 259, 316
Heinrich, B., 27, 28
Heisler, N., 318
Hellegers, A., 329
Heller, H. C., 320
Hennig, W., 77, 86
Herbert, C. V., 322
Hertz, P. E., 152
Hicks, B., 179, 182
Hicks, J. W., 318
Hiesey, W. W., 56
Highton, R., 92
Hilbish, T. J., 177–8
Hill, W. G., 196
Hillman, S. S., 44
Hindill, R., 319
Hobbins, J. C., 336
Hobish, M. K., 118
Hochachka, P. W., 260, 322
Hofmeister, S., 329
Hohimer, A. R., 333
Hohimer, R., 333–4
Hong, S. K., 318
Hoppeler, H., 295–6
Hosenpud, J. D., 333
Houdas, Y., 313
Howell, B. J., 318
Huckabee, W., 329
Huey, R. B., 42, 56, 61, 78, 79, 80, 81, 84, 85, 88, 90, 92, 93, 94, 152, 155, 317–18
Hughes, M. B., 105
Hunter, W. G., 319
Hunyor, S. N., 334

Huxley, J. S., 12, 107, 217
Hytten, F. E., 336

Immerman, F. W., 105, 177
Ingermann, R. L., 329, 331
Innes, D. J., 176

Jablonski, D., 53, 61
Jackson, D. C., 40, 315, 317, 320, 322
Jaeger, J., 319
Jaffin, H., 336
Jahnigen, D., 329, 336
James, W. H., 337
Janzen, D. H., 77
John-Alder, H. B., 84, 152, 317
Johnson, F. M., 175
Johnston, I. A., 43
Jones, J., 295–6

Kanwisher, J. W., 261
Kaplan, N. O., 112
Karas, R. H., 296
Karliner, J. S., 333
Kashyap, T. S., 161
Katz, R., 333
Keck, D. D., 56
Kempster, H. L., 202
Kenagy, G. J., 25
Kennedy, J.A., 335
Kimura, M., 194
Kinder, G. B., 202
King, J. W. B., 161
King, M.-C., 80–1
Kingsolver, J. G., 60
Kleiber, M., 4
Kleine, H.-O., 335
Kluge, A. G., 88
Knibb, W. R., 175
Knowles, T., 179
Kocher, T. D., 180–2
Koehl, M. A. R., 60
Koehn, R. K., 51, 105, 111, 174–82
Kohane, M., 175
Kooyman, G. L., 311
Kovach, P. A., 335
Kramer, D. L., 64, 257
Krebs, H. A., 78, 147
Krogh, A., 1, 147

LaBarbera, M. C., 61
Lacy, F. C., 161
Lande, R., 53, 159, 194, 197–8, 203–6
Langlois, B., 161
Larson, A., 88, 92
Lauder, G. V., 57, 58, 62, 63, 65, 77, 79, 83, 85, 90, 153
Laurie-Ahlberg, C. C., 105, 153, 160
Leach, G. J., 116
Leamy, L., 197
Ledig, F. T., 179, 233
Lee, E., 116
Legates, L. E., 201
LeMayo, Y., 311–12
Lessel, R. 331
Leutenegger, W., 82, 205
Levin, D. A.. 179
Levy, M. N., 320
Lewontin, R. C., 42, 48, 59, 66, 103, 104, 154, 233, 293, 298–9, 302
Licht, P., 43
Liem, K. F., 62, 76
Lighton, J. R. B., 22
Lindholm, A., 295–6
Lindstedt, S. L., 291, 295–6, 299–300
Linhart, Y. B., 179
Lipsey, R. G., 289
Loeschcke, H. H., 336
Lofdahl, K. L., 199
Longo, L. D., 333
Longworth, K. E., 296
Lotrich, V. A., 117
Lucchesi, J. C., 105, 153
Lucius, H., 335
Lydic, R.. 233
Lynch, C. B., 161, 190

McAnulty, J. H., 332–4
McCalden, T. A., 333
McCuaig, J., 179, 182
McDonald, J. H., 180
Mace, G. M., 81
Machida, H., 336
MacIntyre, R. J., 105
McKechnie, S. W., 175
McLaren, N. M., 336
McMahon, T. M., 292
McNab, B. K., 56
Maddison, D. R., 88
Maddison, W. P., 88
Makaveev, T., 179
Makowski, E. L., 334
Malina, R. M., 161
Maloiy, G. M. O., 296
Manwell, C., 179
Maret, K. H., 314
Marks, J. S., 336
Marlowe, T. J., 161
Maroni, G., 153
Marsh, R. L., 198
Martin, P. S., 77
Maslin, T. P., 86
Mathieu, O., 296
Maxson, L. R., 92
Maxson, R. D., 92

Maynard Smith, J., 293, 302
Mayr, E., 12, 33, 62, 77, 148, 291, 302
Medawar, P. B., 47
Meredith, W. H., 117
Meschia, G., 329, 334
Metcalfe, J., 329, 331–4, 335
Meyer, M., 282
Meyers, R. S., 40
Mied, P. A., 118
Miles, D. B., 81, 82, 83
Miles, H. E., 179
Milkman, R. D., 176
Milledge, J. S., 314
Miller, R., 197
Miller, S., 105, 175
Minyard, J. A., 218, 232
Mitson, R. B., 269
Mitton, J. B., 111, 176, 179–82
Moalli, R., 40
Mohren, W., 23
Mommsen, T. P., 28
Monge, C, 329
Montren, L., 153
Moore, L. G., 329, 336
Moore, M. N., 177
Morgan, K. R., 27, 28
Morris, W., 275
Morton, M., 333–4
Moynihan, M., 91
Muma, K. E., 92
Murrish, D. E., 300

Nagy, K. A., 268
Nei, M., 110
Newell, R. I. E., 177, 180
Nieburg, P., 336
Niesel, A., 295
Niesel, D. W., 175
Nolan, K., 319
Norris, K. S., 91
Notter, D. R., 161
Novy, M. J., 332

Oakeshott, J. G., 175
O'Brien, S. J., 105
Ochwadt, B., 336
Olson, E. C., 91, 197
Olson, L. J., 319
Orwell, G., 323
Osgood, D. W., 200
Ostrea, E. M., Jr., 336

Paintin, D. B., 336
Palmer, J. O., 198
Pang, P. K. T., 76, 93
Parer, J. T., 333–4
Park, Y. S., 318
Patton, J. L., 77, 80, 82
Pearcy, R. W., 105, 175
Pernoll, M. L., 335
Peters, R. H., 4
Peters, R. M., Jr., 216, 314
Peterson, C. R., 85
Peterson, E. N., 332
Peterson, R. S., 23
Peterson, S., 199
Phillips, D. S., 337
Phillips, S. C., 175
Pierce, B. A., 179
Pietruszka, R. D., 152
Piiper, J., 278–82
Pinshow, B., 312
Pizzo, C. J., 314
Place, A. R., 105, 108, 113, 115, 118, 125
Platt, J. R., 47, 64–6
Polkuhowich, J. J., 116
Porter, W. P., 153, 319
Pough, F. H., 62, 153
Powers, D. A., 105–6, 108–9, 111, 113, 115–20, 125, 160
Prange, H. D., 315
Pritchard, J. A., 333
Prosser, C. L., 38, 39, 42, 251, 256, 266, 269, 347
Prystowsky, H., 329
Putnam, R. W., 44, 152

Rahn, H., 41, 318
Rand, G. M., 291
Randall, D. J., 251, 266
Rausher, M. D., 199
Reeves, J. T., 318, 329, 336
Reeves, R. B., 40–1
Reichman, O. J., 24
Reid, M. L., 91, 92
Reist, J. D., 218, 232
Remington, P. L., 336
Resnik, R., 333–4
Richmond, R. C., 105
Ridley, M., 59, 61–3, 66, 76, 77, 80, 81, 86, 90
Riggs, A., 115
Ring, E. J. F., 313
Ringler, N. H., 153
Riska, B., 200
Robertson, A., 194, 196, 198
Robertson, R. M., 76, 77, 83, 85, 90
Robinson, O. W., 201
Rodhouse, P. G., 180
Roe, A., 233
Roesler, K., 296
Rohlf, F. J., 218, 234
Ropson, I., 105–6, 108–9
Rosen, R., 289

Rosenthal, E., 105
Ross, D. 333–4
Ross, D. M., 46, 66, 251, 254, 269–70, 347
Ross, M., 334
Roth, G., 61
Rounds, S. S., 329, 336
Rovinsky, J. J., 336
Ruben, J. A., 64
Rutledge, J. J., 197, 199–201
Ryan, M. J., 198

Sacktor, B., 105
Samaja, M., 314
Sarquist, F. H., 314
Schaal, B. A., 179
Schaffer, H. E., 175
Scheid, P., 278–82
Schelling, J. L., 333
Schmalhausen, I. F., 206
Schmid, W. D., 300
Schmidt-Nielsen, K., 1, 4, 155, 216, 280, 300, 312, 316
Schoemaker, P. J. H., 289, 302
Schoene, R. B., 314
Schoener, T. W., 289–90
Scholander, P. F., 261
Scholz, A. T., 264
Schopf, T. J. M., 41, 62, 65, 66
Scott, D. C., 334
Scott, T. M., 179–80
Seeherman, H. J., 296
Segal, I. H., 125
Shaffer, H. B., 153
Sheehan, K. B., 105
Shih, T.-B., 331
Shoemaker, V. H., 43
Shumway, S. E., 179–80
Sibley, C. G., 81
Sibly, R. M., 347
Siebenaller, J. F., 105, 177
Silberglied, R. E., 56
Simpson, G. G., 12, 47, 57, 59, 62, 80, 89, 92, 159, 233
Singh, S. M., 179, 182
Slatkin, M., 79, 206, 317–18
Smialek, Z., 336
Smith, D., 54, 161
Smith, E. N., 261
Smith, R. J., 216, 217
Snedecor, G. W., 218, 234
Snyder, G. K., 43
Snyder, L. R. G., 160
Sokal, R. R., 77, 87, 218, 234
Somero, G. N., 316, 318
Song, S. H., 281
Soulé, M., 92
Spearow, J. L., 196
Stanley, S. M., 53, 80
Stearns, S. C., 77, 81, 82
Stebbins, G. L., 104
Steel, R. G. D., 218, 234
Steiner, P. O., 289
Stevenson, R. D., 42, 80, 85
Stock, M. K., 329, 331
Stolwijk, J. A. J., 320
Stone, H. O., 316
Straney, D. O., 77, 82
Strauss, R. E., 218, 221, 232, 234
Strayhorn, W. D., 335
Stromme, S. B., 320
Stryker, J. C., 336
Stuessy, T. F., 86, 87
Sturgeon, K. B., 179
Sugita, Y., 115
Sullivan, B. K., 153, 159
Sulzbach, D. S., 190
Swain, S. D., 292, 300
Swan, M. S., 105, 174

Taigen, T. L., 44, 51, 149, 153, 159, 198
Tanner, J. M., 218, 221, 232
Taylor, C. R., 4, 291, 293–7
Taylor, M. A., 91
Taylor, M. H., 106, 116, 117
Thomas, R. D. K., 91
Thompson, D'A., 290, 294
Thompson, D. W., 217
Thompson, H. K., Jr., 316
Thompson, R. J., 179
Thornburg, K., 333–4
Thorpe, R. S., 233
Throckmorton, L. H., 80
Thronson, H. A., Jr., 296
Tolley, E. A., 161
Toolson, E. C., 153
Torrie, J. H., 218, 234
Townsend, C. R., 2
Tracy, C. R., 85, 153
Trillmich, F., 23
Trinh-Trang-Tan, M. M., 301
Tsang, H., 333–4
Tsuji, J. S., 85, 161
Tucker, V. A., 29, 259
Tuma, H. J., 218, 232
Turelli, M., 206

Ueland, K., 332–6
Uhlmann, E., 295
Ultsch, G. R., 40, 323

van Berkum, F. H., 80, 81, 88, 161
Van Beneden, R. J., 105–6, 108–9, 111, 118
van de Graaff, K. M., 24
Van Valen, L., 80, 295

Vawter, L. T., 92
Venev, I., 179
Via, S., 198–9, 205
Vigue, C. L., 105
Vrba, E. S., 53, 60, 292

Wachtel, R., 335
Waddington, C. H., 206
Wake, D. B., 59, 61, 66, 76, 92, 347
Walesby, N. J., 43
Walsberg, G. E., 153, 159
Wanntorp, H.-E., 77
Waterman, T. H., 76, 93, 251, 269
Watrous, L. E., 87
Watt, W. B., 51, 105, 160, 174, 175, 190
Weathers, W. W., 43, 259
Webb, P. W., 85
Webster, T. P., 92
Weibel, E. R., 4, 293–7
Weihs, D., 333
Weir, B. S., 153
Weisgram, P. A., 105
Wells, K. D., 44, 51, 149, 153, 159, 198
Welti, H., 333
West, J. B., 294, 314
Wheeler, Q. D., 87
White, F. N., 259
White, T. J., 92
Whittemore, R., 336
Widdows, J., 179
Wiley, E. O., 77, 87
Wilke, F., 23
Wilkins, J. R., 175
Willham, R. L., 199
Wilson, A. C., 80–1, 92
Wilson, E. O., 59
Wilson, L. L., 218, 232
Winberg, G. C., 179
Windsor, D. M., 56
Winslow, R. M., 314
Woakes, A. J., 269
Wood, J. E., III, 333
Wood, S. C., 318
Worth, H., 281
Wright, S., 60, 194, 208, 294
Wyles, J. S., 92

Yang, S. Y., 92
Young, J. P. W., 177
Yuill, T., 319

Zera, A. J., 174–6
Zeuthen, E., 279
Zouros, E., 179

Subject index

acidosis, 317
adaptation, 4–7, 9–10, 13–4, 16–8, 20–1, 23–4, 26–7, 29–30, 32, 35–6, 38, 45–50, 53–6, 61, 64, 71, 73–4, 78–9, 82–3, 85–6, 88–91, 107, 120, 135, 138, 140, 143, 145, 155, 160–1, 171–2, 174, 176, 178, 189, 196, 198, 200–6, 208, 213–14, 225, 240–6, 253–4, 256, 259–60, 262–3, 270, 291, 293, 300, 302, 307, 310, 312, 322, 324, 329, 331, 337, 345, 347, 348
Agassiz, Alexander, 12, 74
alkalosis, 317
allele, 50, 104, 107, 112, 113, 175–8, 194–5
allometry, 4, 22, 61, 82, 155, 216–20, 224–5, 228, 230, 233, 237–40, 291, 294, 296, 302, 308
allozyme, 105, 175
American Journal of Physiology, 2
American Physiological Society, 2, 143
American Society of Zoologists, 2, 143
amino acid, 171, 175–8, 269
analysis, statistical, *see* statistics
analytical precision, 260
anesthesia, 167, 256, 258, 261–2, 268
animal distribution, 265, 342, 344; *see also* population
anoxia, 322
assimilation, 179, 252; *see also* food
ATP (adenosine triphosphate), 112, 115–16, 118–20, 267, 295, 330–1

behavior, 1–5, 13–15, 17–19, 23–4, 28, 36–9, 42, 55, 59, 63, 76, 81, 98, 102, 104, 130, 135–6, 142–3, 147, 149, 152, 155, 161, 168, 180, 197, 199, 200, 206, 214, 236, 240, 241, 247, 253–7, 259–61, 264, 267, 273, 276, 290, 293, 307, 314, 316, 318, 320, 324, 331, 337, 344, 346; *see also* ethology
Bernard, Claude, 15, 323, 326–7
Bigelow, Henry, 12
biochemistry, 2, 28, 102, 104, 141, 172–3, 175, 179, 242, 244, 253–4, 256, 269, 343
biological noise, 263–4
black box, 255, 259–60, 265, 267, 276, 288
blood
  chemistry, 13, 40, 59, 178, 184, 261
  circulation and pressure, 29, 167, 258–9, 262, 268–9, 274, 313, 320, 322–3, 329, 332, 335–6
  hemoglobin and hematocrit, 157, 317
  gases, 260, 280, 312
  pH, 40–1, 119, 317–18
  respiratory properties, 73, 116, 118–20, 225, 270, 280, 328
  vessels, 267, 280
  volume, 333–4
body size, 4, 22, 25, 60–1, 64, 146, 155, 190–6, 198, 200, 202, 204–5, 216–34, 236, 238, 297–300, 302
brain, 91, 313, 320, 322–4

Camp, Joseph, 12
captivity, 32, 169
cardiac output, 296, 312, 318, 323, 332–6
catheter implantation, 268
CAT (computerized axial tomography) scanning, 267
cell volume regulation, 176–7
central tendency, 149
cladistics, 53, 71–3, 77, 86, 96, 133, 135
clines, 12, 20, 106–7, 176, 185, 245
coadaptation, 48, 83, 84–6, 88–91, 100, 345
coefficient of variation, 220
collaboration, 93, 132, 137, 139–40, 144, 249, 264–5, 273–4, 310, 324, 344
Comparative Biochemistry and Physiology, 2
Condor, 2
control systems, 249, 311–12, 318–20, 323; *see also* set point
convergence, *see* evolution
Copeia, 2

Darwin, Charles, 42, 45, 47, 74, 81, 102, 189, 242, 291, 301
Darwinian fitness, *see* fitness
data collection, *see* methodology, techniques
desert, 1, 4, 6, 16–17, 24–6, 38, 88, 100, 140, 160, 257, 299–300, 306

development, 9, 18, 48, 50, 58, 60–1, 71, 105, 111, 116, 118–19, 121, 135, 168, 198–9, 206, 213, 298, 308, 328, 331–2, 336, 340–1; *see also* ontogeny
diversity, 7, 9, 14–15, 19–20, 29, 38–9, 41–2, 46–50, 59, 61–2, 64, 72, 76, 80, 83, 106–7, 109–11, 135, 147, 178–9, 241, 302, 337, 344, 346, 348
diving, 259–61

Ecological Society of America, 2
ecologist, 13, 17, 31, 72, 120, 132, 134–40, 142–3, 158, 160, 170, 172–3, 186, 240, 242–3, 245, 247, 251, 257, 261, 264–7, 270–2, 306, 310, 328, 337, 342
Ecology, 2
ectothermy, 3, 27–8, 41, 79, 316–18, 326–7
education, 264–5
egg, 27, 55, 57, 62, 116–17, 131–2, 200–2, 208, 311, 329–31, 336–7, 340–1
electromyography, 256, 269
electrophoresis, 49, 92, 103, 107, 176, 179–80, 190, 266
embryo, 116, 328–31, 336–7
endothermy, 3, 27–8, 59, 64, 78, 312, 316
energetics, 2, 21, 178, 252, 254, 337, 346
environment
  external, 149, 251, 327
  internal, 152, 256, 299, 312, 323–4, 327
  micro, 18, 200
  natural, 1, 3, 5–6, 16, 147, 159–60, 173, 214, 252–3, 256, 267, 312, 327, 343
  simulated, 252
  variables of, 5, 20, 42–3, 45, 56, 74, 88, 169, 245, 252, 255–7, 260–1, 273
enzyme, 175–80, 182, 186–7, 198, 214, 241, 247, 269, 294, 331, 340
  activity, 132, 157–8, 177, 318, 331
  K_m (Michaelis-Menten constant), 112, 121–3, 125–6, 174–5
  kinetics, 111, 121, 129, 266
  polymorphism, 49, 172, 179
  respiratory, 332
V_{max}, 111–12, 114, 122–6, 174–5
equilibrium, 42–5, 48–9, 51, 56, 72, 83–6, 88, 90–1, 100, 120, 122, 126–7, 138, 282
ethology, 148, 266, 344, 348; *see also* behavior
evaporative cooling, 136, 312, 320
evolution
  convergent, 20–2, 45–6, 85, 87–8, 93, 137
  divergent, 20, 21, 85, 92, 137, 348
  mosaic, 89
  parallel, 20, 27, 57–8, 71, 87, 89, 245
  pathways of, 254
  physiological, 56, 59, 62, 77, 83, 91–4, 189–90, 270
  saltatory, 89
  study of, 12, 45, 66, 102, 138, 148, 159, 240, 244, 253–4, 286, 288–9, 343–5
extinction, 7, 14, 53–4, 61, 63, 65, 293

fecundity, 140, 180, 188
feedback, 256, 275, 318–19
fetus, 328–32, 335–7
field ecology, 255, 266
fitness, 14, 20, 42, 50–3, 55–6, 59–60, 63, 72–4, 98–9, 105, 119, 131, 137, 140, 143, 159–61, 172–4, 186–8, 190, 202, 204, 206, 213, 241, 244–6, 292, 299, 302, 319, 343–4
food
  availability of, 31, 177
  utilization of, 344
  see also assimilation
foraging, 2, 19, 44, 51, 56, 61, 66, 153, 247, 252, 259, 290, 313, 317
fossil record, 42, 53, 58, 71, 76, 91–2

Galapagos Islands, 23–4
gene
  heritability, 49–50, 160–1, 189, 191–2, 197, 199, 207–9, 212, 242, 244, 306
  heterozygosity, 116–17, 136, 178–82, 186, 188
  genotype, 109, 111, 113, 115–20, 131, 132, 172–7, 182, 186, 191, 195, 214, 241, 245, 256–7, 291–2, 298, 340
  polygenes, 171, 191
  see also allele
genetic correlation, 60, 145, 189–209, 212, 240
genetic polymorphism, 174–5
genetic variation, 14, 50–1, 60, 72, 103–7, 117–18, 131–2, 145, 170–2, 174–6, 178, 186–7, 200, 205, 240–1, 244, 246, 343
genetics
  population, 33, 105, 172, 241–2, 286–7, 345
  quantitative, 82, 100, 102, 160, 173, 186, 190–1, 193–4, 196, 207, 209, 214
  study of, 13, 15, 73, 136, 139, 148, 170–1, 174, 176, 178, 189, 244, 269, 285, 307–8
geographic distribution, 106
Goldschmidt, Richard, 12
Grinnell, Joseph, 12
growth, 3–4, 18, 42, 48, 150, 159, 178, 182, 199–200, 216, 319, 328–30, 334, 336, 341
  plant, 24–6
  rate of, 64, 161, 179, 186, 196, 247, 330–1, 344
  scope-for-, 179–80

heat, *see* temperature

heart, 105, 178, 262, 316, 318, 322
  aerobic capacity, 295–6
  cardiac output, 296, 312, 318, 323, 332–6
  disease, 336
  mass, 158
  rate, 4, 29, 157, 214, 261, 269, 332–3
hibernation, 25–6, 257, 260–1, 322, 326
high altitude, 6, 73, 225, 329, 336
homeostasis, 253, 256, 287, 311–12, 315, 322–4, 326, 345; *see also* equilibrium
hormones, 276, 319, 333, 335, 349
  antidiuretic, 220, 300
  erythropoietin, 318
  estrogen, 333–4
  growth, 330
  insulin, 330
  progesterone, 334, 336
  testosterone, 262
humidity, 18, 336
hyperoxia, 329–31, 341
hypothermia, 36
hypoxia, 64, 118, 284, 314–15, 318, 322, 330–1

intracellular pH, 267

jargon, 94, 209, 214, 252
journal, 2, 65, 75, 137, 141–3, 265
  editors, 65, 265
  reviewers, 141, 265
Journal of Comparative Physiology, 2, 148, 232
Journal of Experimental Biology, 2
Journal of Experimental Zoology, 2, 232
Journal of Thermal Biology, 2

Krogh (August) Principle, 9, 78, 147

Lauder, George, 35, 71
Liem, Karel, 35
locomotion, 21, 23, 25, 44, 48, 51, 56, 73, 85, 141, 149, 150–3, 155, 157–9, 161, 213, 247, 252, 294, 315
  sprint speed, 51, 79–81, 85–6, 90, 98–9, 168, 247, 308, 344
  training, 168

Markin, Juan, 254, 271
metabolism
  aerobic, 44, 112, 157–8, 260, 295–8, 308
  anaerobic, 112, 157, 260, 322
  arrest of, 322
  efficiency of, 180, 260
  intermediary, 172, 174
  rate of, 4, 22, 39, 50, 79, 115, 177, 180–2, 198, 211, 237–8, 241, 264, 322
  "scope-for-activity," 181–2
  "scope-for-growth," 179–80
methodology, 1, 72, 105, 160, 179, 186, 251–6, 258, 263, 265–6, 270, 273, 312, 331; *see also* techniques
Michaelis-Menten constant (K_m), 112, 121–3, 125–6, 174–5
molecular biology, 266, 286, 350
morphology, 4, 7, 15, 17, 21–2, 28, 37, 42, 72, 76, 83, 99, 101, 103–4, 141, 145, 147–9, 152–3, 155, 157, 161, 186, 197, 207, 213, 216, 240–1, 243–4, 247, 267, 289–90, 308–9, 340, 343–4, 346–7, 350–1
muscle, 21, 27, 43, 69–70, 112, 118–19, 153, 155, 157–8, 247, 256, 292, 294–5, 296–8, 308, 313–15, 322

National Science Foundation, 2, 267
natural history, 1, 13–14, 19, 32, 51, 65, 135, 288
natural selection, 5, 13–5, 20–1, 42, 45–6, 49, 56, 59–61, 63, 71, 103–4, 106, 109, 111, 119, 149, 159, 189, 201–2, 207, 241, 245–6, 287–8, 291, 293, 299, 345
Nature, 65
nuclear magnetic resonance, (NMR), 121, 133, 267, 274

Oecologia, 2
ontogeny, 48, 71, 216, 243, 249, 328–37, 340, 346; *see also* development
organismal biology, 47, 66, 144–5, 150, 161–2, 273, 345–6, 349–50
outliers, 242, 261
oxygen
  anoxia, 322
  availability of, 312, 314–15, 321, 328–31, 336
  consumption of, 22, 180–1, 216, 229, 232, 237–8, 261, 305, 306, 315, 330–1, 335; *see also* metabolism
  hemoglobin binding, 52, 115–20, 270, 281; *see also* blood
  hyperoxia, 329–31, 341
  hypoxia, 64, 118, 284, 315–16, 318, 322, 330–1
  maximal consumption of, 51, 151, 153, 157, 159, 166, 274, 306, 308
  partial pressure, 16, 262, 341
  stores of, 260
  transport of, 3, 105, 151, 225, 309; *see also* blood
  and ventilatory control, 305–6

Paleobiology, 65
paradigms, 342–3

parsimony, 87, 93, 293; *see also* cladistics
phenotype, 36–7, 58, 60, 72, 111, 118–19, 131, 137, 140, 146, 168, 171–5, 178, 191, 193–5, 197–200, 213–14, 242, 245–6, 252, 253, 256–7, 262, 291–2, 294, 298, 300
photoperiod, 223, 252, 257, 261, 264; *see also* environment
phylogenetic analysis, 240
phylogenetic artifacts, 100–1
phylogenetic constraints, 24–5, 27
phylogenetic history, 5, 154
phylogenetic inertia, 59, 61
phylogenetic perspective, 10, 76, 137; *see also* evolution
phylogenetic relationships, 45, 135, 143, 259
phylogenetic tree, 46, 84, 87, 89–92, 138
phylogeny, 19, 41, 48, 58, 63, 73, 76–94, 100–1, 242, 346
physiological adaptation, 7, 16–18, 20, 27, 38, 46, 48, 54–6, 78, 83, 202, 241, 254, 329, 331
physiological conflicts, 312, 319
physiological diversity, 9, 20, 39, 41, 46–50, 59, 61–2, 64, 83, 346
physiological ecology, 1–7, 10–2, 15, 17, 19–20, 33, 38–9, 42, 45–8, 53–7, 59, 62, 64–6, 71–2, 75–6, 94, 102–3, 106, 121, 132, 135–45, 147, 150, 160, 170, 172–5, 182, 236, 240–7, 249, 251–4, 258–71, 289, 319, 342–5, 347
physiological performance, 99–100, 136, 143, 170, 172–4, 178, 181–2, 242, 249, 308
physiological phenotype, 173–4, 178, 262
physiological systems, 48, 83, 194, 208, 249, 310–11, 314, 324, 342, 347
Physiological Zoology 2, 65, 232
physiologist, 2–6, 9–10, 17–8, 31–3, 36, 39, 45–55, 58–9, 62, 64, 69, 74–6, 80, 82, 90, 93–4, 103, 132, 140, 142, 148, 150, 158, 170–1, 178–9, 189, 193, 201, 207, 233, 242–3, 252, 257, 262, 263, 266–7, 274, 275–6, 306, 310, 312, 323, 326–7, 341, 342, 343–5, 348–50
physiology slash-and-burn, 254; *see also* techniques, invasive and noninvasive
placenta, 328, 331, 333, 335–6, 340–1
plasticity, 37, 117, 242, 303, 305–7, 345
poikilothermy, 109
polymorphism, genetic, 174–5
population, 14, 20, 39
  density, 117, 188, 252
  study of, 9, 23–6, 32–3, 38, 45, 65, 73, 99, 103, 105, 115, 139, 143, 145, 155, 273, 342–3, 350
  variation among and within, 9–10, 41–2, 49–53, 60, 100–1, 103–4, 106–7, 111, 121, 130–1, 135, 138, 140, 146–7, 154–55, 159–61, 168, 170, 189–90, 199, 204–8, 213, 216, 225, 227, 239–41, 244, 254, 259, 263, 306
  see also genetics
positron emission transverse tomography, (PETT), 267
predation, 27, 55–6, 66, 73, 85, 99, 160, 206, 252, 317
pregnancy, 26, 200, 212–13, 307, 329, 331–7, 341

radioactive microspheres, 268
receptors
  baro-, 320
  chemo-, 315, 320
reproduction
  effort of, 26, 180, 252
  potential for, 25–6, 158
  state of, 257
  timing of, 24–6, 180
respiration, 2, 19, 57, 62, 64, 73, 179–80, 182, 252, 269–70, 275–7, 288, 293–301, 305, 311, 315, 320, 329, 331, 335, 337, 344
Respiration Physiology, 2
rhythms
  activity, 13, 16
  circadian, 264
  respiratory, 289
Romer, Alfred, 12

salinity, 16, 36, 54, 62, 106, 160, 175–8, 312, 345, *see also* environment
scaling, 6, 21–2, 79, 82, 216–23, 226, 228–33, 246, 290, 294–300; *see also* body size
Schmidt-Nielsen, Bodil, 17
Schmidt-Nielsen, Knut, 1, 17, 143, 243
Scholander, Per, 1, 17, 261
Science, 65
season, 5, 13, 23–6, 32, 175, 225, 229, 238, 257, 260, 262, 264; *see also* environment
set point, 315, 318–20, 326–7
sexual differentiation, 205, 345; *see also* development
"Slash-and-burn physiology," 254
Society for Experimental Biology, 2
sprint speed, *see* locomotion
statisticians, 266
statistics
  analysis of variance, 61, 81–2, 100, 117, 169, 218–21, 223, 225–31, 233
  averaging-rule algorithm, 88–9
  BMDP (computer program), 233
  covariance, 146, 192–3, 196, 201–4, 218, 221–5, 227, 229, 231–4, 237, 243
  extrapolations, 50, 186, 229

statistics (*cont.*)
  golden mean, 148, 150
  grand mean, 221–2, 227, 229, 231
  intraclass correlation coefficient, 167
  isometric variation, 43, 155, 216–18, 225, 233
  mass-corrected residuals, 155, 157, 162
  median rule algorithm, 87
  multiple regression analysis, 157–8
  multivariate statistics, 157–9, 257
  nested analysis of variance, 81
  partial correlation, 61, 234
  profile analysis of variance, 169
  regression, 22, 78, 81–2, 86, 91, 148, 157–8, 162–3, 167, 192, 208, 218, 222, 226, 233–4, 236–8, 242, 265, 287–8, 297
  relational analyses, 79
  repeatability, 145, 149, 166–9, 242
  sample size, 51, 167, 181, 186, 219, 221, 224, 228, 230, 242, 263
  SAS (computer program), 233
  slope, 23, 41, 148, 191–2, 221–5, 227, 229, 231, 233, 236, 297, 320
  SPSS (computer program), 233
  stepwise multiple regression, 157, 167, 242
  transformational analyses, 58, 83–4, 91
surgery, 253, 256–7, 260–2; *see also* techniques
surgical intervention, 254, 257, 267
survivorship, 51, 53–4, 60, 105, 111, 140, 150, 158–9, 178, 190, 199, 245, 341
systematics, 12–13, 19, 46, 65, 77, 93, 102, 135–8, 140, 143, 242–3, 253, 266, 269; *see also* cladistics

tautology, 5, 245, 298
techniques
  catheter implantation, 268
  Computerized axial tomography (CAT) scanning, 267
  electromyography, 256, 269
  electrophoresis, 49, 92, 103, 107, 176, 179–80, 190, 266
  invasive and noninvasive, 121, 249, 251–74, 308, 324
  miniaturization, 261–2, 267
  nuclear magnetic resonance (NMR), 121, 133, 267, 274
  reductionist approach, 10, 48, 102, 132, 134, 144, 269–70, 344–5
  restraint, 253, 256–8, 261, 269
  radioactive microspheres, 268
  telemetry, 7, 261, 268–70
  see also methodology
telemetry, 7, 261, 268–70
temperature
  body, 3, 23, 27–9, 31, 41, 43, 55–6, 64, 79–80, 86, 99, 118, 152–4, 157, 200, 203, 268, 275, 316–18, 322
  ectothermy, 3, 27–8, 41, 79, 317–19, 327–8
  endothermy, 3, 27–8, 59, 64, 78, 312, 316
  evaporative cooling, 136, 312, 320
  heat exposure, 312–13
  heat prostration, 23
  heat retention, 23
  hypothermia, 36
  optimum, 55, 81, 85–90, 98–100, 175
  poikilothermy, 109
  thermal acclimation, 115
  thermal history, 252, 257, 262
  thermal preferenda, 81, 84, 85, 89–92, 99, 214, 265–6, 316
  thermal tolerance, 42, 55, 60, 201
  see also environment
training
  academic and research, 133, 139, 251, 254, 257, 264, 266
  of animals, 168–9, 261, 307
  see also locomotion
transducer
  Doppler, 267–8, 274
  electromagnetic blood flow, 267–8, 274

uterus, 328–29, 332–4, 336–7

variation
  interindividual, 49, 52, 146–52, 154–9, 161, 168, 180, 241–2, 341
  intraindividual, 167
  phenotypic, 60, 72, 131, 138, 178, 257
  see also population, statistics
ventilation 162, 232, 236, 279, 281–2, 285, 293–4, 305, 313, 315, 318, 320, 325, 335–6

water
  availability, 1, 4, 16, 261, 307, 313, 319
  exchange of, 16, 252
  flux, 31, 220, 268
  loss of, 19, 59, 136, 153, 200–1, 291–2, 299–300, 312, 320–1
  temperature of, 109–10, 113
  see also environment